CAMBRIDGE LIBRARY COLLECTION

Books of enduring scholarly value

Physical Sciences

From ancient times, humans have tried to understand the workings of the world around them. The roots of modern physical science go back to the very earliest mechanical devices such as levers and rollers, the mixing of paints and dyes, and the importance of the heavenly bodies in early religious observance and navigation. The physical sciences as we know them today began to emerge as independent academic subjects during the early modern period, in the work of Newton and other 'natural philosophers', and numerous sub-disciplines developed during the centuries that followed. This part of the Cambridge Library Collection is devoted to landmark publications in this area which will be of interest to historians of science concerned with individual scientists, particular discoveries, and advances in scientific method, or with the establishment and development of scientific institutions around the world.

Climate and Time in their Geological Relations

This first book by James Croll (1821–90), published in 1875, includes many of the original geophysical theories that he had formulated throughout the early years of his career. A self-educated amateur, Croll obtained work at the Glasgow Andersonian Museum, which gave him leisure time to pursue his scientific interests. The fluidity of scientific disciplines at the time allowed him to virtually invent the field of geophysics, and his unique insights united ideas previously thought unconnected, such as using physics to explore the causes of the glacial epochs, climatic changes and the circulation and temperature of ocean currents. Croll, whose *Stellar Evolution and Its Relations to Geological Time* is also reissued in this series, later became a Fellow of the Royal Society and of St Andrew's University, but (possibly because of his non-scientific background) he writes in a style which makes his works accessible to a lay readership.

Cambridge University Press has long been a pioneer in the reissuing of out-of-print titles from its own backlist, producing digital reprints of books that are still sought after by scholars and students but could not be reprinted economically using traditional technology. The Cambridge Library Collection extends this activity to a wider range of books which are still of importance to researchers and professionals, either for the source material they contain, or as landmarks in the history of their academic discipline.

Drawing from the world-renowned collections in the Cambridge University Library and other partner libraries, and guided by the advice of experts in each subject area, Cambridge University Press is using state-of-the-art scanning machines in its own Printing House to capture the content of each book selected for inclusion. The files are processed to give a consistently clear, crisp image, and the books finished to the high quality standard for which the Press is recognised around the world. The latest print-on-demand technology ensures that the books will remain available indefinitely, and that orders for single or multiple copies can quickly be supplied.

The Cambridge Library Collection brings back to life books of enduring scholarly value (including out-of-copyright works originally issued by other publishers) across a wide range of disciplines in the humanities and social sciences and in science and technology.

Climate and Time in their Geological Relations

A Theory of Secular Changes of the Earth's Climate

James Croll

CAMBRIDGE UNIVERSITY PRESS

Cambridge, New York, Melbourne, Madrid, Cape Town,
Singapore, São Paolo, Delhi, Mexico City

Published in the United States of America by Cambridge University Press, New York

www.cambridge.org
Information on this title: www.cambridge.org/9781108048378

This edition first published 1875
This digitally printed version 2012

ISBN 978-1-108-04837-8 Paperback

The original edition of this book contains a number of colour plates, which have been reproduced in black and white. Colour versions of these images can be found online at www.cambridge.org/9781108048378

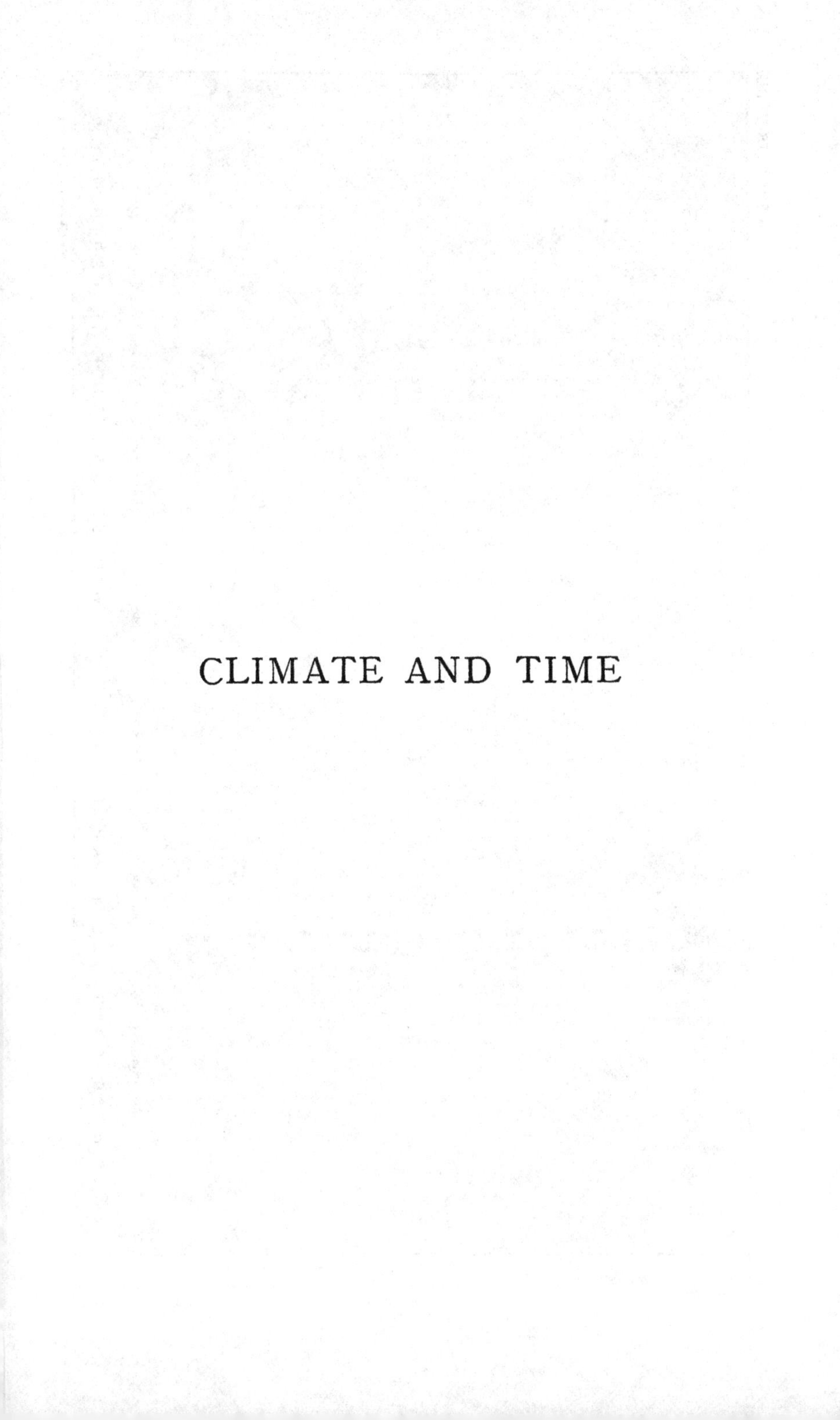

CLIMATE AND TIME

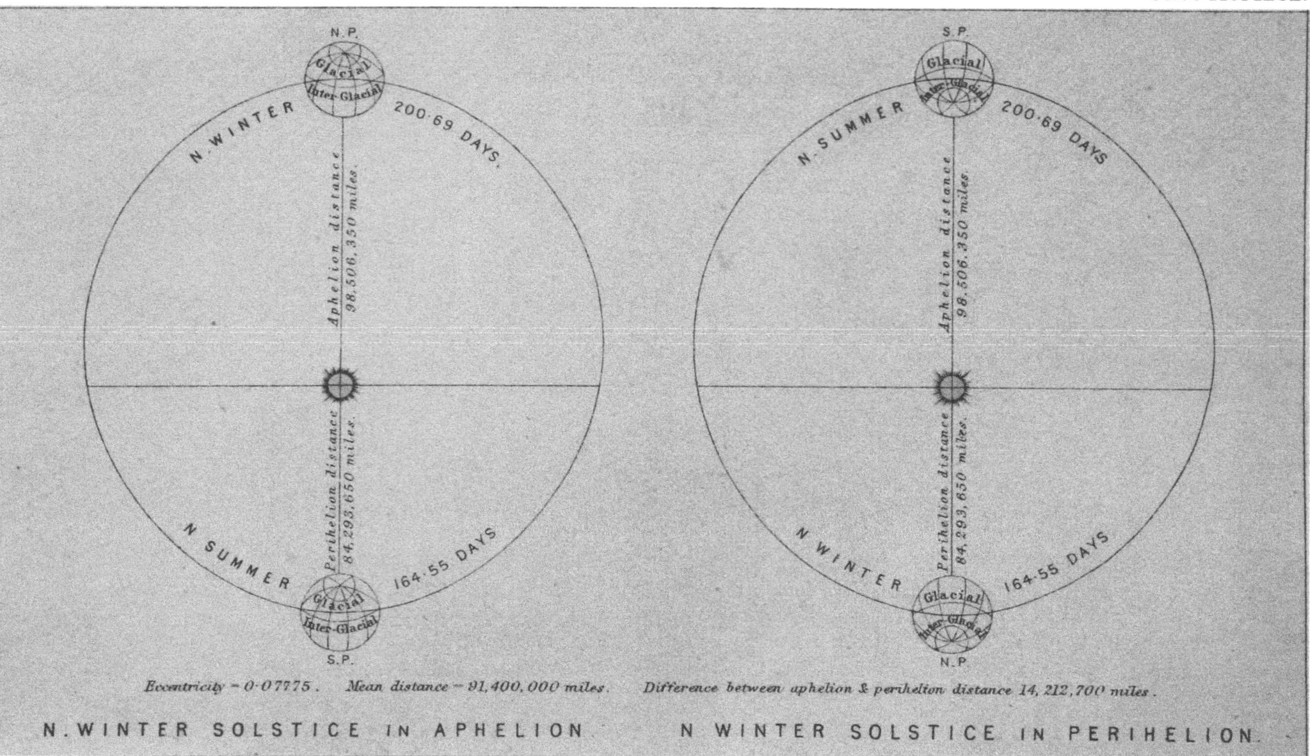
N.P.
Glacial
Inter-Glacial
N.WINTER
200·69 DAYS.
Aphelion distance
98,506,350 miles.
Perihelion distance
84,293,650 miles.
N SUMMER
164·55 DAYS
S.P.
S.P.
N.SUMMER
200·69 DAYS
N WINTER
164·55 DAYS
N.P.
Eccentricity = 0·07775. Mean distance = 91,400,000 miles. Difference between aphelion & perihelion distance 14,212,700 miles.
N.WINTER SOLSTICE IN APHELION
N WINTER SOLSTICE IN PERIHELION.

W & A.K. Johnston, Edinr. and London.

CLIMATE AND TIME

IN THEIR GEOLOGICAL RELATIONS

A THEORY OF

SECULAR CHANGES OF THE EARTH'S CLIMATE

BY JAMES CROLL

OF H.M. GEOLOGICAL SURVEY OF SCOTLAND

LONDON

DALDY, ISBISTER, & CO.

56, LUDGATE HILL

1875

LONDON:
PRINTED BY VIRTUE AND CO.,
CITY ROAD.

PREFACE.

In the following pages I have endeavoured to give a full and concise statement of the facts and arguments adduced in support of the theory of Secular Changes of the Earth's Climate. Considerable portions of the volume have already appeared in substance as separate papers in the Philosophical Magazine and other journals during the past ten or twelve years. The theory, especially in as far as it relates to the cause of the glacial epoch, appears to be gradually gaining acceptance with geologists. This, doubtless, is owing to the greatly increased and constantly increasing knowledge of the drift-phenomena, which has induced the almost general conviction that a climate such as that of the glacial epoch could only have resulted from cosmical causes.

Considerable attention has been devoted to objections, and to the removal of slight misapprehensions, which have naturally arisen in regard to a subject comparatively new and, in many respects, complex, and beset with formidable difficulties.

I have studiously avoided introducing anything of a hypothetical character. All the conclusions are based either on known facts or admitted physical principles. In short, the aim of the work, as will be shown in the introductory chapter, is to prove that secular changes of climate follow, as a necessary effect, from admitted physical agencies, and that these changes, in as far as the past climatic condition of the globe is concerned, fully meet the demand of the geologist.

The volume, though not intended as a popular treatise, will be found, I trust, to be perfectly plain and intelligible even to readers not familiar with physical science.

I avail myself of this opportunity of expressing my obligations to my colleagues, Mr. James Geikie, Mr. Robert L. Jack, Mr. Robert Etheridge, jun., and also to Mr. James Paton, of the Edinburgh Museum of Science and Art, for their value able assistance rendered while these pages were passing through the press. To the kindness of Mr. James Bennie I am indebted for the copious index at the end of the volume, as well as for many of the facts relating to the glacial deposits of the West of Scotland.

JAMES CROLL.

EDINBURGH, *March*, 1875.

CONTENTS.

CHAPTER I.

INTRODUCTION.

CHAPTER II.

OCEAN-CURRENTS IN RELATION TO THE DISTRIBUTION OF HEAT OVER THE GLOBE.

CHAPTER III.

OCEAN-CURRENTS IN RELATION TO THE DISTRIBUTION OF HEAT OVER THE GLOBE.—(*Continued.*)

CHAPTER IV.

OUTLINE OF THE PHYSICAL AGENCIES WHICH LEAD TO SECULAR CHANGES OF CLIMATE.

CHAPTER V.

REASON WHY THE SOUTHERN HEMISPHERE IS COLDER THAN THE NORTHERN.

CHAPTER VI.

EXAMINATION OF THE GRAVITATION THEORY OF OCEANIC CIRCULATION.—LIEUT. MAURY'S THEORY.

CHAPTER VII.

EXAMINATION OF THE GRAVITATION THEORY OF OCEANIC CIRCULATION.—LIEUT. MAURY'S THEORY.—(*Continued.*)

CHAPTER VIII.

EXAMINATION OF THE GRAVITATION THEORY OF OCEANIC CIRCULATION.—DR. CARPENTER'S THEORY.

CHAPTER IX.

EXAMINATION OF THE GRAVITATION THEORY OF OCEANIC CIRCULATION.—THE MECHANICS OF DR. CARPENTER'S THEORY.

CHAPTER X.

EXAMINATION OF THE GRAVITATION THEORY OF OCEANIC CIRCULATION.—DR. CARPENTER'S THEORY.—OBJECTIONS CONSIDERED.

CHAPTER XI.

THE INADEQUACY OF THE GRAVITATION THEORY PROVED BY ANOTHER METHOD.

CHAPTER XII.

MR. A. G. FINDLAY'S OBJECTIONS CONSIDERED.

CHAPTER XIII.

THE WIND THEORY OF OCEANIC CIRCULATION.

CHAPTER XIV.

THE WIND THEORY OF OCEANIC CIRCULATION IN RELATION TO CHANGE OF CLIMATE.

CHAPTER XV.

WARM INTER-GLACIAL PERIODS.

CHAPTER XVI.

WARM INTER-GLACIAL PERIODS IN ARCTIC REGIONS.

CHAPTER XVII.

FORMER GLACIAL EPOCHS.—REASON OF THE IMPERFECTION OF GEOLOGICAL RECORDS IN REFERENCE TO THEM.

CHAPTER XVIII.

FORMER GLACIAL EPOCHS; GEOLOGICAL EVIDENCE OF.

CHAPTER XIX.

GEOLOGICAL TIME.—PROBABLE DATE OF THE GLACIAL EPOCH.

CHAPTER XX.

GEOLOGICAL TIME.—METHOD OF MEASURING THE RATE OF SUBAËRIAL DENUDATION.

CHAPTER XXI.

THE PROBABLE AGE AND ORIGIN OF THE SUN.

CHAPTER XXII.

A METHOD OF DETERMINING THE MEAN THICKNESS OF THE SEDIMENTARY ROCKS OF THE GLOBE.

CHAPTER XXIII.

THE PHYSICAL CAUSE OF THE SUBMERGENCE AND EMERGENCE OF THE LAND DURING THE GLACIAL EPOCH.

CHAPTER XXIV.

THE PHYSICAL CAUSE OF THE SUBMERGENCE AND EMERGENCE OF THE LAND DURING THE GLACIAL EPOCH.—(*Continued.*)

CHAPTER XXV.

THE INFLUENCE OF THE OBLIQUITY OF THE ECLIPTIC ON CLIMATE AND ON THE LEVEL OF THE SEA.

CHAPTER XXVI.

COAL AN INTER-GLACIAL FORMATION.

CHAPTER XXVII.

PATH OF THE ICE-SHEET IN NORTH-WESTERN EUROPE AND ITS RELATIONS TO THE BOULDER CLAY OF CAITHNESS.

CHAPTER XXVIII.

NORTH OF ENGLAND ICE-SHEET, AND TRANSPORT OF WASTDALE CRAG BLOCKS.

CHAPTER XXIX.

EVIDENCE FROM BURIED RIVER CHANNELS OF A CONTINENTAL PERIOD IN BRITAIN.

CHAPTER XXX.

THE PHYSICAL CAUSE OF THE MOTION OF GLACIERS.—THEORIES OF GLACIER-MOTION.

CHAPTER XXXI.

THE PHYSICAL CAUSE OF THE MOTION OF GLACIERS.—THE MOLECULAR THEORY.

APPENDIX.

LIST OF PLATES.

CHAPTER I.

INTRODUCTION.

The Fundamental Problem of Geology.—Geology a Dynamical Science.—The Nature of a Geological Principle.—Theories of Geological Climate.—Geological Climate dependent on Astronomical Causes.—An Important Consideration overlooked.—Abstract of the Line of Argument pursued in the Volume.

The Fundamental Problem of Geology.—The investigation of the successive changes and modifications which the earth's crust has undergone during past ages is the province of geology. It will be at once admitted that an acquaintance with the agencies by means of which those successive changes and modifications were effected, is of paramount importance to the geologist. What, then, are those agencies? Although volcanic and other subterranean eruptions, earthquakes, upheavals, and subsidences of the land have taken place in all ages, yet no truth is now better established than that it is not by these convulsions and cataclysms of nature that those great changes were effected. It was rather by the ordinary agencies that we see every day at work around us, such as rain, rivers, heat and cold, frost and snow. The valleys were not produced by violent dislocations, nor the hills by sudden upheavals, but were actually carved out of the solid rock, silently and gently, by the agencies to which we have referred. "The tools," to quote the words of Professor Geikie, "by which this great work has been done are of the simplest and most every-day order—the air, rain, frosts, springs, brooks, rivers, glaciers, icebergs, and the sea. These tools have been at work from the earliest times of which any geological record has been preserved. Indeed, it is out of the accumulated chips and dust which they have made, afterwards hardened into solid rock and upheaved, that the very framework of our continents has been formed." *

It will be observed—and this is the point requiring particular attention—that the agencies referred to are the ordinary meteorological or climatic agencies. In fact, it is these agencies which constitute climate. The various peculiarities or modifications of climate result from a preponderance of one or more of these agencies over the rest. When heat, for example, predominates, we have a hot or tropical climate. When cold and frost predominate, we have a rigorous or arctic climate. With moisture in excess, we have a damp and rainy climate; and so on. But this is not all. These climatic agencies are not only the factors which carved out the rocky face of the globe into hill and dale, and spread over the whole a mantle of soil; but by them are determined the character of the *flora* and *fauna* which exist on that soil. The flora and fauna of a district are determined mainly by the character of the climate, and not by the nature of the soil, or the conformation of the ground. It is from difference of climate that tropical life differs so much from arctic, and both these from the life of temperate regions. It is climate, and climate alone, that causes the orange and the vine to blossom, and the olive to flourish, in the south, but denies them to the north, of Europe. It is climate, and climate alone, that enables the forest tree to grow on the plain, but not on the mountain top; that causes wheat and barley to flourish on the mainland of Scotland, but not on the steppes of Siberia.

Again, if we compare flat countries with mountainous, highlands with lowlands, or islands with continents, we shall find that difference of climatic conditions is the chief reason why life in the one differs so much from life in the other. And if we turn to the sea we find that organic life is there as much under the domain of climate as on the land, only the conditions are much less complex. For in the case of the sea, difference in the temperature of the water may be said to constitute almost the only

* Trans. of Edin. Geol. Soc., vol. ii. p. 252.

difference of climatic conditions. If there is one fact more clearly brought out than another by the recent deep-sea explorations, it is this, that nothing exercises so much influence on organic life in the ocean as the temperature of the water. In fact, so much is this the case, that warm zones were found to be almost equivalent to zones of life. It was found that even the enormous pressure at the bottom of the ocean does not exercise so much influence on life as the temperature of the water. There are few, I presume, who reflect on the subject that will not readily admit that, whether as regards the great physical changes which are taking place on the surface of our globe, or as regards the growth and distribution of plant and animal life, the ordinary climatic agents are the real agents at work, and that, compared with them, all other agencies sink into insignificance.

It will also be admitted that what holds true of the present holds equally true of the past. Climatic agents are not only now the most important and influential; they have been so during all past geological ages. They were so during the Cainozoic as much as during the present; and there is no reason for supposing they were otherwise during the remoter Mesozoic and Palæozoic epochs. They have been the principal factors concerned in that long succession of events and changes which have taken place since the time of the solidification of the earth's crust. The stratified rocks of the globe contain all the records which now remain of their action, and it is the special duty of the geologist to investigate and read those records. It will be at once admitted that in order to a proper understanding of the events embodied in these records, an acquaintance with the agencies by which they were produced is of the utmost importance. In fact, it is only by this means that we can hope to arrive at their rational explanation. A knowledge of the agents, and of the laws of their operations, is, in all the physical sciences, the means by which we arrive at a rational comprehension of the effects produced. If we have before us some complex and intricate effects which have been produced by heat, or by light, or by electricity, &c., in order to understand them we must make ourselves acquainted with the agents by which they were produced and the laws of their action. If the effects to be considered be, for example, those of heat, then we must make ourselves acquainted with this agent and its laws. If they be of electricity, then a knowledge of electricity and its laws becomes requisite.

This is no mere arbitrary mode of procedure which may be adopted in one science and rejected in another. It is in reality a necessity of thought arising out of the very constitution of our intellect; for the objective law of the agent is the conception by means of which the effects are subjectively united in a rational unity. We may describe, arrange, and classify the effects as we may, but without a knowledge of the laws of the agent we can have no rational unity. We have not got the higher conception by which they can be *comprehended*. It is this relationship between the effects and the laws of the agent, a knowledge of which really constitutes a science. We might examine, arrange, and describe for a thousand years the effects produced by heat, and still we should have no science of heat unless we had a knowledge of the laws of that agent. The effects would never be seen to be necessarily connected with anything known to us; we could not connect them with any rational principle from which they could be deduced *à priori*. The same remarks hold, of course, equally true of all sciences, in which the things to be considered stand in the relationship of cause and effect. Geology is no exception. It is not like systematic botany, a mere science of classification. It has to explain and account for effects produced; and these effects can no more be explained without a knowledge of the laws of the agents which produced them, than can the effects of heat without a knowledge of the laws of heat. The only distinction between geology and heat, light, electricity, &c., is, that in geology the effects to be explained have almost all occurred already, whereas in these other sciences effects actually taking place have to be explained. But this distinction is of no importance to our present purpose, for effects which have already occurred can no more be explained without a knowledge of the laws of the agent which produced them than can effects which are in the act of occurring. It is, moreover, not strictly true that all the effects to be explained by the geologist are already past. It falls within the scope of his science to account for the changes which are at present taking place on the earth's crust.

No amount of description, arrangement, and classification, however perfect or accurate, of the facts which come under the eye of the geologist can ever constitute a science of geology any more than a description and classification of the effects of heat could constitute a science of heat. This will, no doubt, be admitted by every one who reflects upon the subject, and it will be maintained that geology, like every other science, must possess principles applicable to the facts. But here confusion and misconception will arise unless there be distinct and definite ideas as to what ought to constitute a geological principle. It is not every statement or rule that may apply to a great many facts, which will constitute a geological principle. A geological principle must bear the same characteristics as the principles of those sciences to which we have referred. What, then, is the nature of the principles of light, heat, electricity, &c.? The principles of heat are the laws of heat. The principles of electricity are the laws of electricity. And these laws are nothing more nor less than the ways according to which these agents produce their effects. The principles of geology are therefore the laws of geology. But the laws of geology must be simply the laws of the geological agents, or, in other words, the methods by which they produce their effects. Any other so-called principle can be nothing more than an empirical rule, adopted for convenience. Possessing no rationality in itself, it cannot be justly regarded as a principle. In order to rationality the principle must be either resolvable into, or logically deducible from, the laws of the agents. Unless it possess this quality we cannot give the explanation *à priori*.

The reason of all this is perfectly obvious. The things to be explained are effects; and the relationship between cause and effect affords the subjective connection between the principle and the explanation. The explanation follows from the principle simply as the effect results from the laws of the agent or cause.

Theories of Geological Climate.—We have already seen that the geological agents are chiefly the ordinary climatic agents. Consequently, the main principles of geology must be the laws of the climatic agents, or some logical deductions from them. It therefore follows that, in order to a purely scientific geology, the grand problem must be one of geological climate. It is through geological climate that we can hope to arrive ultimately at principles which will afford a rational explanation of the multifarious facts which have been accumulating during the past century. The facts of geology are as essential to the establishment of the principles, as the facts of heat, light, and electricity are essential to the establishment of the principles of these sciences. A theory of geological climate devised without reference to the facts would be about as worthless as a theory of heat or of electricity devised without reference to the facts of these sciences.

It has all along been an admitted opinion among geologists that the climatic condition of our globe has not, during past ages, been uniformly the same as at present. For a long time it was supposed that during the Cambrian, Silurian, and other early geological periods, the climate of our globe was much hotter than now, and that ever since it has been gradually becoming cooler. And this high temperature of Palæozoic ages was generally referred to the influence of the earth's internal heat. It has, however, been proved by Sir William Thomson * that the general climate of our globe could not have been sensibly affected by internal heat at any time more than ten thousand years after the commencement of the solidification of the surface. This physicist has proved that the present influence of internal heat on the temperature amounts to about only 1-75th of a degree. Not only is the theory of internal

* Phil. Mag., January, 1863.

heat now generally abandoned, but it is admitted that we have no good geological evidence that climate was much hotter during Palæozoic ages than now; and much less, that it has been becoming *uniformly* colder.

The great discovery of the glacial epoch, and more lately that of a mild and temperate condition of climate extending during the Miocene and other periods to North Greenland, have introduced a complete revolution of ideas in reference to geological climate. Those discoveries showed that our globe has not only undergone changes of climate, but changes of the most extraordinary character. They showed that at one time not only an arctic condition of climate prevailed in our island, but that the greater part of the temperate region down to comparatively low latitudes was buried under ice, while at other periods Greenland and the Arctic regions, probably up to the North Pole, were not only free from ice, but were covered with a rich and luxuriant vegetation.

To account for these extraordinary changes of climate has generally been regarded as the most difficult and perplexing problem which has fallen to the lot of the geologist. Some have attempted to explain them by assuming a displacement of the earth's axis of rotation in consequence of the uprising of large mountain masses on some part of the earth's surface. But it has been shown by Professor Airy,* Sir William Thomson,† and others, that the earth's equatorial protuberance is such that no geological change on its surface could ever possibly alter the position of the axis of rotation to an extent which could at all sensibly affect climate. Others, again, have tried to explain the change of climate by supposing, with Poisson, that the earth during its past geological history may have passed through hotter and colder parts of space. This is not a very satisfactory hypothesis. There is no doubt a difference in the quantity of force in the form of heat passing through different parts of space; but space itself is not a sub-

* *Athenæum*, September 22, 1860.

† Trans. Glasgow Geol. Soc., vol. iv., p. 313.

stance which can possibly be either cold or hot. If, therefore, we were to adopt this hypothesis, we must assume that the earth during the hot periods must have been in the vicinity of some other great source of heat and light besides the sun. But the proximity of a mass of such magnitude as would be sufficient to affect to any great extent the earth's climate would, by its gravity, seriously disarrange the mechanism of our solar system. Consequently, if our solar system had ever, during any former period of its history, really come into the vicinity of such a mass, the orbits of the planets ought at the present day to afford some evidence of it. But again, in order to account for a cold period, such as the glacial epoch, we have to assume that the earth must have come into the vicinity of a cold body.* But recent discoveries in regard to inter-glacial periods are wholly irreconcilable with this theory.

A change in the obliquity of the ecliptic has frequently been, and still is, appealed to as an explanation of geological climate. This theory appears, however, to be beset by a twofold objection: (1), it can be shown from celestial mechanics, that the variations in the obliquity of the ecliptic must always have been so small that they could not materially affect the climatic condition of the globe; and (2), even admitting that the obliquity could change to an indefinite extent, it can be shown † that no increase or decrease, however great, could possibly account for either the glacial epoch or a warm temperate condition of climate in polar regions.

The theory that the sun is a variable star, and that the glacial epochs of the geologists may correspond to periods of decrease in the sun's heat, has lately been advanced. This theory is also open to two objections: (1), a general diminution of heat ‡ never could produce a glacial epoch; and (2), even if it could, it would not explain interglacial periods.

The only other theory on the subject worthy of notice is that

* See Mr. Hopkin's remarks on this theory, Quart. Journ. Geol. Soc., vol. viii.

† See Chap. xxv.

‡ See Chap. iv.

of Sir Charles Lyell. Those extraordinary changes of climate are, according to his theory, attributed to differences in the distribution of land and water. Sir Charles concludes that, were the land all collected round the poles, while the equatorial zones were occupied by the ocean, the general temperature would be lowered to an extent that would account for the glacial epoch. And, on the other hand, were the land all collected along the equator, while the polar regions were covered with sea, this would raise the temperature of the globe to an enormous extent. It will be shown in subsequent chapters that this theory does not duly take into account the prodigious influence exerted on climate by means of the heat conveyed from equatorial to temperate and polar regions by means of ocean currents. In Chapters II. and III. I have endeavoured to prove (1), that were it not for the heat conveyed from equatorial to temperate and polar regions by this means, the thermal condition of the globe would be totally different from what it is at present; and (2), that the effect of placing all the land along the equator would be diametrically the opposite of that which Sir Charles supposes.

But supposing that difference in the distribution of land and water would produce the effects attributed to it, nevertheless it would not account for those extraordinary changes of climate which have occurred during geological epochs. Take, for example, the glacial epoch. Geologists almost all agree that little or no change has taken place in the relative distribution of sea and land since that *epoch*. All our main continents and islands not only existed then as they do now, but every year is adding to the amount of evidence which goes to show that so recent, geologically considered, is the glacial epoch that the very contour of the surface was pretty much the same then as it is at the present day. But this is not all; for even should we assume (1), that a difference in the distribution of sea and land would produce the effects referred to, and (2), that we had good geological evidence to show that at a very recent period a form of distribution existed which would produce the necessary glacial conditions, still the glacial epoch would not be explained, for the phenomena of warm inter-glacial periods would completely upset the theory.

Geological Climate depending on Astronomical Causes.—For a good many years past an impression has been gradually gaining ground amongst geologists that the glacial epoch, as well as the extraordinary condition of climate which prevailed in arctic regions during the Miocene and other periods, must some way or other have resulted from a cosmical cause; but all seemed at a loss to conjecture what that cause could possibly be. It was apparent that the cosmical cause must be sought for in the relations of our earth to the sun; but a change in the obliquity of the ecliptic and the eccentricity of the earth's orbit are the only changes from which any sensible effect on climate could possibly be expected to result. It was shown, however, by Laplace that the change of obliquity was confined within so narrow limits that it has scarcely ever been appealed to as a cause seriously affecting climate. The only remaining cause to which appeal could be made was the change in the eccentricity of the earth's orbit—precession of the equinoxes without eccentricity producing, of course, no effect whatever on climate. Upwards of forty years ago Sir John Herschel and a few other astronomers directed their attention to the consideration of this cause, but the result arrived at was adverse to the supposition that change of eccentricity could greatly affect the climate of our globe.

As some misapprehension seems to prevail with reference to this, I would take the liberty of briefly adverting to the history of the matter,—referring the reader to the Appendix for fuller details.

About the beginning of the century some writers attributed the lower temperature of the southern hemisphere to the fact that the sun remains about seven days less on that hemisphere than on the northern; their view being that the southern hemisphere on this account receives seven days less heat than the northern. Sir Charles Lyell, in the first edition of his "Principles," published in 1830, refers to this as a cause which might produce some slight effect on climate. Sir Charles's remarks seem to have directed Sir John Herschel's attention to the subject, for in the latter part of the same year he read a paper before the Geological Society on the astronomical causes which may influence geological phenomena, in which, after pointing out the mistake into which Sir Charles had been led in concluding that the southern hemisphere receives less heat than the northern, he considers the question as to whether geological climate could be influenced by changes in the eccentricity of the earth's orbit. He did not appear at the time to have been aware of the conclusions arrived at by Lagrange regarding the superior limit of the eccentricity of the earth's orbit; but he came to the conclusion that possibly the climate of our globe may have been affected by variations in the eccentricity of its orbit. "An amount of variation," he says, "which we need not hesitate to admit (at least provisionally) as a possible one, may be productive of considerable diversity of climate, and may operate during great periods of time either to mitigate or to exaggerate the difference of winter and summer temperatures, so as to produce alternately in the same latitude of either hemisphere a perpetual spring, or the extreme vicissitudes of a burning summer and a rigorous winter."

This opinion, however, was unfortunately to a great extent nullified by the statement which shortly afterwards appeared in his "Treatise on Astronomy," and also in the "Outlines of Astronomy," to the effect that the elliptic form of the earth's orbit has but a very trifling influence in producing variation of temperature corresponding to the sun's distance; the reason being that whatever may be the ellipticity of the orbit, it follows that equal amounts of heat are received from the sun in passing over equal angles round it, in whatever part of the ellipse those angles may be situated. Those angles will of course be described in unequal times, but the greater proximity of the sun exactly compensates for the more rapid description, and thus an equilibrium of heat is maintained. The sun, for example, is much nearer the earth when he is over the southern hemisphere than he is when over the northern; but the southern hemisphere does not on this account receive more heat than the northern; for, owing to the greater velocity of the earth when nearest the sun, the sun does not remain so long on the southern hemisphere as he does on the northern. These two effects so exactly counterbalance each other that, whatever be the extent of the eccentricity, the total amount of heat reaching both hemispheres is the same. And he considered that this beautiful compensating principle would protect the climate of our globe from being seriously affected by an increase in the eccentricity of its orbit, unless the extent of that increase was very great.

"Were it not," he says, "for this, the eccentricity of the orbit would materially influence the transition of seasons. The fluctuation of distance amounts to nearly 1-30th of its mean quantity, and consequently the fluctuation in the sun's direct heating power to double this, or 1-15th of the whole. Now the perihelion of the orbit is situated nearly at the place of the northern winter solstice; so that, were it not for the compensation we have just described, the effect would be to exaggerate the difference of summer and winter in the southern hemisphere, and to moderate it in the northern; thus producing a more violent alternation of climate in the one hemisphere, and an approach to perpetual spring in the other. As it is, however, no such inequality subsists, but an equal and impartial distribution of heat and light is accorded to both." *

Herschel's opinion was shortly afterwards adopted and advocated by Arago † and by Humboldt.‡

Arago, for example, states that so little is the climate of our globe affected by the eccentricity of its orbit, that even were the orbit to become as eccentric as that of the planet Pallas (that is, as great as 0·24), "still this would not alter in

* "Treatise on Astronomy," § 315; "Outlines," § 368.

† *Annuaire* for 1834, p. 199. Edin. New Phil. Journ., April, 1834, p. 224.

‡ "Cosmos," vol. iv. p. 459 (Bohn's Edition). "Physical Description of the Heavens," p. 336.

any appreciable manner the mean thermometrical state of the globe."

This idea, supported by these great authorities, got possession of the public mind; and ever since it has been almost universally regarded as settled that the great changes of climate indicated by geological phenomena could not have resulted from any change in the relation of the earth to the sun.

There is, however, one effect that was not regarded as compensated. The total amount of heat received by the earth is inversely proportional to the minor axis of its orbit; and it follows, therefore, that the greater the eccentricity, the greater is the total amount of heat received by the earth. On this account it was concluded that an increase of eccentricity would tend to a certain extent to produce a warmer climate.

All those conclusions to which I refer, arrived at by astronomers, are perfectly legitimate so far as the direct effects of eccentricity are concerned; and it was quite natural, and, in fact, proper to conclude that there was nothing in the mere increase of eccentricity that could produce a glacial epoch. How unnatural would it have been to have concluded that an increase in the quantity of heat received from the sun should lower the temperature, and cover the country with snow and ice! Neither would excessively cold winters, followed by excessively hot summers, produce a glacial epoch. To assert, therefore, that the purely astronomical causes could produce such an effect would be simply absurd.

Important Consideration overlooked.— The important fact, however, was overlooked that, although the glacial epoch could not result *directly* from an increase of eccentricity, it might nevertheless do so *indirectly*. Although an increase of eccentricity could have no direct tendency to lower the temperature and cover our country with ice, yet it might bring into operation physical agents which would produce this effect.

If, instead of endeavouring to trace a direct connection between a high condition of eccentricity and a glacial condition of climate, we turn our attention to the consideration of what are the physical effects which result from an increase of eccentricity, we shall find that a host of physical agencies are brought into operation, the combined effect of which is to lower to a very great extent the temperature of the hemisphere whose winters occur in aphelion, and to raise to nearly as great an extent the temperature of the opposite hemisphere, whose winters of course occur in perihelion. Until attention was directed to those physical circumstances to which I refer, it was impossible that the true cause of the glacial epoch could have been discovered; and, moreover, many of the indirect and physical effects, which in reality were those that brought about the glacial epoch, could not, in the nature of things, have been known previously to recent discoveries in the science of heat.

The consideration and discussion of those various physical agencies are the chief aim of the following pages.

Abstract of the Line of Argument pursued in this Volume.—I shall now proceed to give a brief abstract of the line of argument pursued in this volume. But as a considerable portion of it is devoted to the consideration of objections and difficulties bearing either directly or indirectly on the theory, it will be necessary to point out what those difficulties are, how they arose, and the methods which have been adopted to overcome them.

Chapter IV. contains an outline of the physical agencies affecting climate which are brought into operation by an increase of eccentricity. By far the most important of all those agencies, and the one which mainly brought about the glacial epoch, is the *Deflection* of Ocean Currents. The consideration of the indirect physical connection between a high state of eccentricity and the deflection of ocean currents, and also the enormous influence on climate which results from this deflection constitute not only the most important part of the subject, but the one beset with the greatest amount of difficulties.

The difficulties besetting this part of the theory arise mainly from the imperfect state of our knowledge, (1st) with reference to the absolute amount of heat transferred from equatorial to temperate and polar regions by means of ocean currents and the influence which the heat thus transferred has on the distribution of temperature on the earth's surface ; and (2nd) in connection with the physical cause of ocean circulation.

In Chapters II. and III. I have entered at considerable length into the consideration of the effects of ocean currents on the distribution of heat over the globe. The only current of which anything like an accurate estimate of volume and temperature has been made is the Gulf-stream. In reference to this stream we have a means of determining in absolute measure the quantity of heat conveyed by it. On the necessary computation being made, it is found that the amount transferred by the Gulf-stream from equatorial regions into the North Atlantic is enormously greater than was ever anticipated, amounting to no less than one-fifth part of the entire heat possessed by the North Atlantic. This striking fact casts a new light on the question of the distribution of heat over the globe. It will be seen that to such an extent is the temperature of the equatorial regions lowered, and that of high temperate, and polar regions raised, by means of ocean currents, that were they to cease, and each latitude to depend solely on the heat received directly from the sun, only a very small portion of the globe would be habitable by the present order of beings. This being the case, it becomes obvious to what an extent the deflection of ocean currents must affect temperature. For example, were the Gulf-stream stopped, and the heat conveyed by it deflected into the Southern Ocean, how enormously would this tend to lower the temperature of the northern hemisphere, and raise the temperature south of the equator.

Chapters VI., VII., VIII., IX., X., and XIII., are devoted to the consideration of the physical cause of oceanic circulation. This has been found to be the most difficult and perplexing part of the whole inquiry. The difficulties mainly arise from the great diversity of opinion and confusion of ideas prevailing in regard to the mechanics of the subject. There are two theories propounded to account for oceanic circulation; the one which may be called the *Wind* theory, and the other the *Gravitation* theory; and this diversity of opinion and confusion of ideas prevail in connection with both theories. As the question of the cause of oceanic circulation has not only a direct and important bearing on the subject of the present volume, but is further one of much general interest, I have entered somewhat fully into the matter.

The Gravitation theories may be divided into two classes. The first of these attributes the Gulf-stream and other sensible currents of the ocean to difference of specific gravity, resulting from difference of temperature between the sea in equatorial and polar regions. The leading advocate of this theory was the late Lieutenant Maury, who brought it so much into prominence in his interesting book on the "Physical Geography of the Sea." The other class does not admit that the sensible currents of the ocean can be produced by difference of specific gravity; but they maintain that difference of temperature between the sea in equatorial and polar regions produces a general movement of the upper portion of the sea from the equator to the poles, and a counter-movement of the under portion from the poles to the equator. This form of the gravitation theory has been ably and zealously advocated by Dr. Carpenter, who may be regarded as its representative. The Wind theories also divide into two classes. According to the one ocean currents are caused and maintained by the impulse of the trade-winds, while according to the other they are due not to the impulse of the trade-winds alone, but to that of the prevailing winds of the globe, regarded as a general system. The former of these is the one generally accepted; the latter is that advocated in the present volume.

The relations which these theories bear to the question of secular change of climate, will be found stated at length in Chapter VI. It will, however, be better to state here in a few words what those relations are. When the eccentricity of the earth's orbit attains a high value, the hemisphere, whose winter solstice occurs in aphelion, has, for reasons which are explained in Chapter IV., its temperature lowered, while that of the opposite hemisphere is raised. Let us suppose the northern hemisphere to be the cold one, and the southern the warm one. The difference of temperature between the equator and the North Pole will then be greater than between the equator and the South Pole; according, therefore, to theory, the trades of the northern hemisphere will be stronger than those of the southern, and will consequently blow across the equator to some distance on the southern hemisphere. This state of things will tend to deflect equatorial currents southwards, impelling the warm water of the equatorial regions more into the southern or warm hemisphere than into the northern or cold hemisphere. The tendency of all this will be to exaggerate the difference of temperature already existing between the two hemispheres. If, on the other hand, the great ocean currents which convey the warm equatorial waters to temperate and polar regions be not produced by the impulse of the winds, but by difference of temperature, as Maury maintains, then in the case above supposed the equatorial waters would be deflected more into the northern or cold hemisphere than into the southern or warm hemisphere, because the difference of temperature between the equator and the poles would be greater on the cold than on the warm hemisphere. This, of course, would tend to neutralize or counteract that difference of temperature between the two hemispheres which had been previously produced by eccentricity. In short, this theory of circulation would effectually prevent eccentricity from seriously affecting climate.

Chapters VI. and VII. have been devoted to an examination of this form of the gravitation theory.

The above remarks apply equally to Dr. Carpenter's form of the theory; for according to a doctrine of General Oceanic Circulation resulting from difference of specific gravity between the water at the equator and at the poles, the equatorial water will be carried more to the cold than to the warm hemisphere. It is perfectly true that a belief in a general oceanic circulation may be held quite consistently with the theory of secular changes of climate, provided it be admitted that not this general circulation but ocean currents are the great agency employed in distributing heat over the globe. The advocates of the theory, however, admit no such thing, but regard ocean currents as of secondary importance. It may be stated that the existence of this general ocean circulation has never been detected by actual observation. It is simply assumed in order to account for certain facts, and it is asserted that such a circulation must take place as a physical necessity. I freely admit that were it not that the warm water of equatorial regions is being constantly carried off by means of ocean currents such as the Gulf-stream, it would accumulate till, in order to restoration of equilibrium, such a general movement as is supposed would be generated. But it will be shown that the warm water in equatorial regions is being drained off so rapidly by ocean currents that the actual density of an equatorial column differs so little from that of a polar column that the force of gravity resulting from that difference is so infinitesimal that it is doubtful whether it is sufficient to produce sensible motion. I have also shown in Chapter VIII. that all the facts which this theory is designed to explain are not only explained by the wind theory, but are deducible from it as necessary consequences. In Chapter XI. it is proved, by contrasting the quantity of heat conveyed by ocean currents from inter-tropical to temperate and polar regions with such an amount as could possibly be conveyed by means of a general oceanic circulation, that the latter sinks into insignificance before the former. In Chapters X. and XII. the various objections which have been advanced by Dr. Carpenter and Mr. Findlay are discussed at considerable length, and in Chapter IX. I have entered somewhat minutely into an examination of the mechanics of the gravitation theory. A statement of the wind theory is given in Chapter XIII.; and in Chapter XIV. is shown the relation of this theory to the theory of Secular changes of climate. This terminates the part of the inquiry relating to oceanic circulation.

We now come to the *crucial test* of the theories respecting the cause of the glacial epoch, viz., Warm Inter-glacial Periods. In Chapters XV. and XVI. I have given a statement of the geological facts which go to prove that that long epoch known as the Glacial was not one of continuous cold, but consisted of a succession of cold and warm periods. This condition of things is utterly inexplicable on every theory of the cause of the glacial epoch which has hitherto been advanced; but, according to the physical theory of secular changes of climate under consideration, it follows as a necessary consequence. In fact, the amount of geological evidence which has already been accumulated in reference to inter-glacial periods may now be regarded as perfectly sufficient to establish the truth of that theory.

If the glacial epoch resulted from some accidental distribution of sea and land, then there may or may not have been more than one glacial epoch, but if it resulted from the cause which we have assigned, then there must have been during the geological history of the globe a succession of glacial epochs corresponding to the secular variations in the eccentricity of the earth's orbit. A belief in the existence of recurring glacial epochs has been steadily gaining ground for many years past. I have, in Chapter XVIII., given at some length the facts on which this belief rests. It is true that the geological evidence of glacial epochs in prior ages is meagre in comparison with that of the glacial epoch of Post-tertiary times; but there is a reason for this in the nature of geological evidence itself. Chapter XVII. deals with the geological records of former glacial epochs, showing that they are not only imperfect, but that there is good reason why they should be so, and that the imperfection of the records in reference to them cannot be advanced as an argument against their existence.

If the glacial epoch resulted from a high condition of eccentricity, we have not only a means of determining the positive date of that epoch, but we have also a means of determining geological time in absolute measure. For if the glacial epochs of prior ages correspond to periods of high eccentricity, then the intervals between those periods of high eccentricity become the measure of the intervals between the glacial epochs. The researches of Lagrange and Leverrier into the secular variations of the elements of the orbits of the planets enable us to determine with tolerable accuracy the values of the eccentricity of the earth's orbit for, at least, four millions of years past and future. With the view of determining those values, I several years ago computed from Leverrier's formula the eccentricity of the earth's orbit and longitude of the perihelion, at intervals of ten thousand and fifty thousand years during a period of three millions of years in the past, and one million of years in the future. The tables containing these values will be found in Chapter XIX. These tables not only give us the date of the glacial epoch, but they afford, as will be seen from Chapter XXI., evidence as to the probable date of the Eocene and Miocene periods.

Ten years ago, when the theory was first advanced, it was beset by a very formidable difficulty, arising from the opinions which then prevailed in reference to geological time. One or two glacial epochs in the course of a million of years was a conclusion which at that time scarcely any geologist would admit, and most would have felt inclined to have placed the last glacial epoch at least one million of years back. But then if we assume that the glacial epoch was due to a high state of eccentricity, we should be compelled to admit of at least two glacial epochs during that lapse of time. It was the modern doctrine that the great changes undergone by the earth's crust were produced, not by convulsions of nature, but by the slow and almost imperceptible action of rain, rivers, snow, frost, ice, &c., which impressed so strongly on the mind of the geologist the vast duration of geological periods. When it was considered that the rocky face of our globe had been carved into hills and dales, and ultimately worn down to the sea-level by means of those apparently trifling agents, not only once or twice, but many times, during past ages, it was not surprising that the views entertained by geologists regarding the immense antiquity of our globe should not have harmonised with the deductions of physical science on the subject. It had been shown by Sir William Thomson and others, from physical considerations relating to the age of the sun's heat and the secular cooling of our globe, that the geological history of our earth's crust must be limited to a period of something like one hundred millions of years. But these speculations had but little weight when pitted against the stern and undeniable facts of sub-aërial denudation. How, then, were the two to be reconciled? Was it the physicist who had under-estimated geological time, or the geologist who had over-estimated it? Few familiar with modern physics, and who have given special attention to the subject, would admit that the sun could have been dissipating his heat at the present enormous rate for a period much beyond one hundred millions of years. The probability was that the amount of work performed on the earth's crust by the denuding agents in a period so immense as a million of years was, for reasons stated in Chapter XX., very much under-estimated. But the difficulty was how to prove this. How was it possible to measure the rate of operation of agents so numerous and diversified acting with such extreme slowness and irregularity over so immense areas? In other words, how was it possible to measure the rate of sub-aërial denudation? Pondering over this problem about ten years ago, an extremely simple and obvious method of solving it suggested itself to my mind. This method—the details of which will be found in Chapter XX.—showed that the rate of sub-aërial denudation is enormously greater than had been supposed. The method is now pretty generally accepted, and the result has already been to bring about a complete reconciliation between physics and geology in reference to time.

Chapter XXI. contains an account of the gravitation theories of the origin of the sun's heat. The energy possessed by the sun is generally supposed to have been derived from gravitation, combustion being totally inadequate as a source. But something more than gravitation is required before we can account for even one hundred millions of years' heat. Gravitation could not supply even one-half that amount. There must be some other and greater source than that of gravitation. There is, however, as is indicated, an obvious source from which far more energy may have been derived than could have been obtained from gravitation.

The method of determining the rate of sub-aërial denudation enables us also to arrive at a rough estimate of the actual mean thickness of the stratified rocks of the globe. It will be seen from Chapter XXII. that the mean thickness is far less than is generally supposed.

The physical cause of the submergence of the land during the glacial epoch, and the influence of change in the obliquity of the ecliptic on climate, are next considered. In Chapter XXVI. I have given the reasons which induce me to believe that coal is an inter-glacial formation.

The next two chapters—the one on the path of the ice in north-western Europe, the other on the north of England ice-sheet—are reprints of papers which appeared a few years ago in the *Geological Magazine*. Recent observations have confirmed the truth of the views advanced in these two chapters, and they are rapidly gaining acceptance among geologists.

I have given, at the conclusion, a statement of the molecular theory of glacier motion—a theory which I have been led to modify considerably on one particular point.

There is one point to which I wish particularly to direct attention—viz., that I have studiously avoided introducing into the theories propounded anything of a hypothetical nature. There is not, so far as I am aware, from beginning to end of this volume, a single hypothetical element: nowhere have I attempted to give a hypothetical explanation. The conclusions are in every case derived either from facts or from what I believe to be admitted principles. In short, I have aimed to prove that the theory of secular changes of climate follows, as a necessary consequence, from the admitted principles of physical science.

CHAPTER II.

OCEAN-CURRENTS IN RELATION TO THE DISTRIBUTION OF HEAT OVER THE GLOBE.

The absolute Heating-power of Ocean-currents.—Volume of the Gulf-stream.—Absolute Amount of Heat conveyed by it.—Greater Portion of Moisture in inter-tropical Regions falls as Rain in those Regions.—Land along the Equator tends to lower the Temperature of the Globe.—Influence of Gulf-stream on Climate of Europe.—Temperature of Space.—Radiation of a Particle.—Professor Dove on Normal Temperature.—Temperature of Equator and Poles in the Absence of Ocean-currents.—Temperature of London, how much due to Ocean-currents.

The absolute Heating-power of Ocean-currents.—There is perhaps no physical agent concerned in the distribution of heat over the surface of the globe the influence of which has been so much underrated as that of ocean-currents. This is, no doubt, owing to the fact that although their surface-temperature, direction, and general influence have obtained considerable attention, yet little or nothing has been done towards determining the absolute amount of heat or of cold conveyed by them or the resulting absolute increase or decrease of temperature.

The modern method of determining the amount of heat-effects in absolute measure is, doubtless, destined to cast new light on all questions connected with climate, as it has done, and is still doing, in every department of physics where energy, under the form of heat, is being studied. But this method has hardly as yet been attempted in questions of meteorology; and owing to the complicated nature of the phenomena with which the meteorologist has generally to deal, its application will very often prove practically impossible. Nevertheless, it is particularly suitable to all questions relating to the direct thermal effects of currents, whatever the nature of these currents may happen to be.

In the application of the method to an ocean-current, the two most important elements required as data are the volume of the stream and its mean temperature. But although we know something of the temperature of most of the great ocean-currents, yet, with the exception of the Gulf-stream, little has been ascertained regarding their volume.

The breadth, depth, and temperature of the Gulf-stream have formed the subject of extensive and accurate observations by the United States Coast Survey. In the memoirs and charts of that survey cross-sections of the stream at various places are given, showing its breadth and depth, and also the temperature of the water from the surface to the bottom. We are thus enabled to determine with some precision the mean temperature of the stream. And knowing its mean velocity at any given section, we have likewise a means of determining the number of cubic feet of water passing through that section in a given time. But although we can obtain with tolerable accuracy the mean temperature, yet observations regarding the velocity of the water at all depths have unfortunately not been made at any particular section. Consequently we have no means of estimating as accurately as we could wish the volume of the current. Nevertheless, since we know the surface-velocity of the water at places where some of the sections were taken, we are enabled to make at least a rough estimate of the volume.

From an examination of the published sections, I came to the conclusion some years ago* that the total quantity of water conveyed by the stream is probably equal to that of a stream fifty miles broad and 1,000 feet deep,† flowing at the rate of

* Phil. Mag. for February, 1867, p. 127.

† The Gulf-stream at the narrowest place examined by the Coast Survey, and where also its velocity was greatest, was found to be over 30 statute miles broad and 1,950 feet deep. But we must not suppose that this represents all the warm water which is received by the Atlantic from the equator; a great mass flows into the Atlantic without passing through the Straits of Florida.

four miles an hour, and that the mean temperature of the entire mass of moving water is not under 65° at the moment of leaving the Gulf. But to prevent the possibility of any objections being raised on the grounds that I may have over-estimated the volume of the stream, I shall take the velocity to be *two* miles instead of four miles an hour. We are warranted, I think, in concluding that the stream before it returns from its northern journey is on an average cooled down to at least 40°,* consequently it loses 25° of heat. Each cubic foot of water, therefore, in this case carries from the tropics for distribution upwards of 1,158,000 foot-pounds of heat. According to the above estimate of the size and velocity of the stream, which in Chapter XI. will be shown to be an under-estimate, 2,787,840,000,000 cubic feet of water are conveyed from the Gulf per hour, or 66,908,160,000,000 cubic feet daily. Consequently the total quantity of heat thus transferred per day amounts to 77,479,650,000,000,000,000 foot-pounds.

This estimate of the volume of the stream is considerably less by one-half than that given both by Captain Maury and by Sir John Herschel. Captain Maury considers the Gulf-stream equal to a stream thirty-two miles broad and 1,200 feet deep, flowing at the rate of five knots an hour.† This gives 6,165,700,000,000 cubic feet per hour as the quantity of water conveyed by this stream. Sir John Herschel's estimate is still greater. He considers it equal to a stream thirty miles broad and 2,200 feet deep, flowing at the rate of four miles an hour. ‡ This makes the quantity 7,359,900,000,000 cubic feet per hour. Dr. Colding,in his elaborate memoir on the Gulf-stream, estimates the volume at 5,760,000,000,000 cubic feet per hour, while Mr. Laughton's estimate is nearly double that of mine.

* It is probable that a large proportion of the water constituting the south-eastern branch of the Gulf-stream is never cooled down to 40°; but, on the other hand, the north-eastern branch, which passes into the arctic regions, will be cooled far below 40°, probably below 30°. Hence I cannot be over-estimating the extent to which the water of the Gulf-stream is cooled down in fixing upon 40° as the average minimum temperature.

† "Physical Geography of the Sea," § 24, 6th edition.

‡ "Physical Geography," § 54.

From observations made by Sir John Herschel and by M. Pouillet on the direct heat of the sun, it is found that, were no heat absorbed by the atmosphere, about eighty-three foot-pounds per second would fall upon a square foot of surface placed at right angles to the sun's rays.* Mr. Meech estimates that the quantity of heat cut off by the atmosphere is equal to about twenty-two per cent. of the total amount received from the sun. M. Pouillet estimates the loss at twenty-four per cent. Taking the former estimate, 64·74 foot-pounds per second will therefore be the quantity of heat falling on a square foot of the earth's surface when the sun is in the zenith. And were the sun to remain stationary in the zenith for twelve hours, 2,796,768 foot-pounds would fall upon the surface.

It can be shown that the total amount of heat received upon a unit surface on the equator, during the twelve hours from sunrise till sunset at the time of the equinoxes, is to the total amount which would be received upon that surface, were the sun to remain in the zenith during those twelve hours, as the diameter of a circle to half its circumference, or as 1 to 1·5708. It follows, therefore, that a square foot of surface on the equator receives from the sun at the time of the equinoxes 1,780,474 foot-pounds daily, and a square mile 49,636,750,000,000 foot-pounds daily. But this amounts to only 1-1560935th part of the quantity of heat daily conveyed from the tropics by the Gulf-stream. In other words, the Gulf-stream conveys as much heat as is received from the sun by 1,560,935 square miles at the equator. The amount thus conveyed is equal to all the heat which falls upon the globe within thirty-two miles on each side of the equator. According to calculations made by Mr. Meech,† the annual quantity of heat received by a unit surface on the frigid zone, taking the mean of the whole zone, is 5·45-12th of that received at the equator; consequently the quantity of heat conveyed by the Gulf-stream in one year is

* Trans. of Roy. Soc. of Edin., vol. xxi., p. 57. Phil. Mag., § 4, vol. ix., p. 36.

† "Smithsonian Contributions to Knowledge," vol. ix.

equal to the heat which falls on an average on 3,436,900 square miles of the arctic regions. The frigid zone or arctic regions contain 8,130,000 square miles. There is actually, therefore, nearly one-half as much heat transferred from tropical regions by the Gulf-stream as is received from the sun by the entire arctic regions, the quantity conveyed from the tropics by the stream to that received from the sun by the arctic regions being nearly as two to five.

But we have been assuming in our calculations that the percentage of heat absorbed by the atmosphere is no greater in polar regions than it is at the equator, which is not the case. If we make due allowance for the extra amount absorbed in polar regions in consequence of the obliqueness of the sun's rays, the total quantity of heat conveyed by the Gulf-stream will probably be nearly equal to one-half the amount received from the sun by the entire arctic regions.

If we compare the quantity of heat conveyed by the Gulf-stream with that conveyed by means of aërial currents, the result is equally startling. The density of air to that of water is as 1 to 770, and its specific heat to that of water is as 1 to 4·2; consequently the same amount of heat that would raise 1 cubic foot of water 1° would raise 770 cubic feet of air 4°·2, or 3,234 cubic feet 1°. The quantity of heat conveyed by the Gulf-stream is therefore equal to that which would be conveyed by a current of air 3,234 times the volume of the Gulf-stream, at the same temperature and moving with the same velocity. Taking, as before, the width of the stream at fifty miles, and its depth at 1,000 feet, and its velocity at two miles an hour, it follows that, in order to convey an equal amount of heat from the tropics by means of an aërial current, it would be necessary to have a current about $1\frac{1}{4}$ mile deep, and at the temperature of 65°, blowing at the rate of two miles an hour from every part of the equator over the northern hemisphere towards the pole. If its velocity were equal to that of a good sailing-breeze, which Sir John Herschel states to be about twenty-one miles an hour, the current would require to be above 600 feet deep. A greater quantity of heat is probably conveyed by the Gulf-stream alone from the tropical to the temperate and arctic regions than by all the aërial currents which flow from the equator.

We are apt, on the other hand, to over-estimate the amount of the heat conveyed from tropical regions to us by means of aërial currents. The only currents which flow from the equatorial regions are the upper currents, or anti-trades as they are called. But it is not possible that much heat can be conveyed directly by them. The upper currents of the trade-winds, even at the equator, are nowhere below the snow-line; they must therefore lie in a region of which the temperature is actually below the freezing-point. In fact, if those currents were warm, they would elevate the snow-line above themselves. The heated air rising off the hot burning ground at the equator, after ascending a few miles, becomes exposed to the intense cold of the upper regions of the atmosphere; it then very soon loses all its heat, and returns from the equator much colder than it went thither. It is impossible that we can receive any heat directly from the equatorial regions by means of aërial currents. It is perfectly true that the south-west wind, to which we owe so much of our warmth in this country, is a continuation of the anti-trade; but the heat which this wind brings to us is not derived from the equatorial regions. This will appear evident, if we but reflect that, before the upper current descends to the snow-line after leaving the equator, it must traverse a space of at least 2,000 miles; and to perform this long journey several days will be required. During all this time the air is in a region below the freezing-point; and it is perfectly obvious that by the time it begins to descend it must have acquired the temperature of the region in which it has been travelling.

If such be the case, it is evident that a wind whose temperature is below 32° could never warm a country such as ours, where the temperature does not fall below 38° or 39°. The heat of our south-west winds is derived, not directly from the equator, but from the warm water of the Atlantic—in fact, from the Gulf-stream. The upper current acquires its heat after it descends to the earth. There is one way, however, whereby heat is indirectly conveyed from the equator by the anti-trades; that is, in the form of aqueous vapour. In the formation of one pound of water from aqueous vapour, as Professor Tyndall strikingly remarks, a quantity of heat is given out sufficient to melt five pounds of cast iron.* It must, however, be borne in mind that the greater part of the moisture of the south-west and west winds is derived from the ocean in temperate regions. The upper current receives the greater part of its moisture after it descends to the earth, whilst the moisture received at the equator is in great part condensed, and falls as rain in those regions.

This latter assertion has been so frequently called in question that I shall give my reasons for making it. According to Dr. Keith Johnston ("Physical Atlas") the mean rainfall of the torrid regions is ninety-six inches per annum, while that of the temperate regions amounts to only thirty-seven inches. If the greater part of the moisture of the torrid regions does not fall as rain in those regions, it must fall as such beyond them. Now the area of the torrid to that of the two temperate regions is about as 39·3 to 51. Consequently ninety-six inches of rain spread over the temperate regions would give seventy-four inches; but this is double the actual rainfall of the temperate regions. If, again, it were spread over both temperate and polar regions this would yield sixty-four inches, which, however, is nearly double the mean rainfall of the temperate and polar regions. If we add to this the amount of moisture derived from the ocean within temperate and polar regions, we should have a far greater rainfall for these latitudes than for the torrid region, and we know, of course, that it is actually far less. This proves the truth of the assertion that by far the greater part of the moisture of the torrid regions falls in those regions as rain. It will hardly do to object that the above may

* "Heat as a Mode of Motion," art. 240.

probably be an over-estimate of the amount of rainfall in the torrid zone, for it is not at all likely that any error will ever be found which will affect the general conclusion at which we have arrived.

Dr. Carpenter, in proof of the small rainfall of the torrid zone, adduces the case of the Red Sea, where, although evaporation is excessive, almost no rain falls. But the reason why the vapour raised from the Red Sea does not fall in that region as rain, is no doubt owing to the fact that this sea is only a narrow strip of water in a dry and parched land, the air above which is too greedy of moisture to admit of the vapour being deposited as rain. Over a wide expanse of ocean, however, where the air above is kept to a great extent in a constant state of saturation, the case is totally different.

Land at the Equator tends to Lower the Temperature of the Globe.—The foregoing considerations, as well as many others which might be stated, lead to the conclusion that, in order to raise the mean temperature of the whole earth, *water* should be placed along the equator, and not *land*, as is supposed by Sir Charles Lyell and others. For if land is placed at the equator, the possibility of conveying the sun's heat from the equatorial regions by means of ocean-currents is prevented. The transference of heat could then be effected only by means of the upper currents of the trades; for the heat conveyed by *conduction* along the solid crust, if any, can have no sensible effect on climate. But these currents, as we have just seen, are ill-adapted for conveying heat.

The surface of the ground at the equator becomes intensely heated by the sun's rays. This causes it to radiate its heat more rapidly into space than a surface of water heated under the same conditions. Again, the air in contact with the hot ground becomes also more rapidly heated than in contact with water, and consequently the ascending current of air carries off a greater amount of heat. But were the heat thus carried away transferred by means of the upper currents to high latitudes and there employed to warm the earth, then it might to a considerable extent compensate for the absence of ocean-currents, and in this case land at the equator might be nearly as well adapted as water for raising the temperature of the whole earth. But such is not the case; for the heat carried up by the ascending current at the equator is not employed in warming the earth, but is thrown off into the cold stellar space above. This ascending current, instead of being employed in warming the globe, is in reality one of the most effectual means that the earth has of getting quit of the heat received from the sun, and of thus maintaining a much lower temperature than it would otherwise possess. It is in the equatorial regions that the earth loses as well as gains the greater part of its heat; so that, of all places, here ought to be placed the substance best adapted for preventing the dissipation of the earth's heat into space, in order to raise the general temperature of the earth. Water, of all substances in nature, seems to possess this quality to the greatest extent; and, besides, it is a fluid, and therefore adapted by means of currents to carry the heat which it receives from the sun to every region of the globe.

These results show (although they have reference to only one stream) that the general influence of ocean-currents on the distribution of heat over the surface of the globe must be very great. If the quantity of heat transferred from equatorial regions by the Gulf-stream alone is nearly equal to all the heat received from the sun by the arctic regions, then how enormous must be the quantity conveyed from equatorial regions by all the ocean-currents together!

Influence of the Gulf-stream on the Climate of Europe.—In a paper read before the British Association at Exeter, Mr. A. G. Findlay objects to the conclusions at which I have arrived in former papers on the subject, that I have not taken into account the great length of time that the water requires in order to circulate, and the interference it has to encounter in its passage.

The objection is, that a stream so comparatively small as the Gulf-stream, after spreading out over such a large area of the Atlantic, and moving so slowly across to the shores of Europe, losing heat all the way, would not be able to produce any very sensible influence on the climate of Europe.

I am unable to perceive the force of this objection. Why, the very efficiency of the stream as a heating agent necessarily depends upon the slowness of its motion. Did the Gulf-stream move as rapidly along its whole course as it does in the Straits of Florida, it could produce no sensible effect on the climate of Europe. It does not require much consideration to perceive this. (1) If the stream during its course continued narrow, deep, and rapid, it would have little opportunity of losing its heat, and the water would carry back to the tropics the heat which it ought to have given off in the temperate and polar regions. (2) The Gulf-stream does not heat the shores of Europe by direct radiation. Our island, for example, is not heated by radiation from a stream of warm water flowing along its shores. The Gulf-stream heats our island *indirectly* by heating the winds which blow over it to our shores.

The anti-trades, or upper return-currents, as we have seen, bring no heat from the tropical regions. After traversing some 2,000 miles in a region of extreme cold they descend on the Atlantic as a cold current, and there absorb the heat and moisture which they carry to north-eastern Europe. Those aërial currents derive their heat from the Gulf-stream, or if it is preferred, from the warm water poured into the Atlantic by the Gulf-stream.

How, then, are these winds heated by the warm water? The air is heated in two ways, viz., by direct *radiation* from the water, and by *contact* with the water. Now, if the Gulf-stream continued a narrow and deep current during its entire course similar to what it is at the Straits of Florida, it could have little or no opportunity of communicating its heat to the air either by radiation or by contact. If the stream were only about forty or fifty miles in breadth, the aërial particles in their passage across it would not be in contact with the warm water more than an hour or two. Moreover, the number of particles in contact with the water, owing to the narrowness of the stream, would be small, and there would therefore be little opportunity for the air becoming heated by contact. The same also holds true in regard to radiation. The more we widen the stream and increase its area, the more we increase its radiating surface; and the greater the radiating surface, the greater is the quantity of heat thrown off. But this is not all; the number of aërial particles heated by radiation increases in proportion to the area of the radiating surface; consequently, the wider the area over which the waters of the Gulf-stream are spread, the more effectual will the stream be as a heating agent. And, again, in order that a very wide area of the Atlantic may be covered with the warm waters of the stream, slowness of motion is essential.

Mr. Findlay supposes that fully one-half of the Gulf-stream passes into the south-eastern branch, and that it is only the north-eastern branch of the current that can be effectual in raising the temperature of Europe. But it appears to me that it is to this south-eastern portion of the current, and not to the north-eastern, that we, in this country, are chiefly indebted for our heat. The south-west winds, to which we owe our heat, derive their temperature from this south-eastern portion which flows away in the direction of the Azores. The south-west winds which blow over the northern portion of the current which flows past our island up into the arctic seas cannot possibly cross this country, but will go to heat Norway and northern Europe. The north-eastern portion of the stream, no doubt, protects us from the ice of Greenland by warming the north-west winds which come to us from that cold region.

Mr. Buchan, Secretary of the Scottish Meteorological Society, has shown* that in a large tract of the Atlantic between latitudes 20° and 40° N., the mean pressure of the atmosphere is greater than in any other place on the globe. To the west of Madeira, between longitude 10° and 40° W., the mean annual pressure amounts to 30·2 inches, while between Iceland and Spitzbergen it is only 29·6, a lower mean pressure than is found

* Trans. Roy. Soc. of Edin., vol. xxv., part 2.

in any other place on the northern hemisphere. There must consequently, he concludes, be a general tendency in the air to flow from the former to the latter place along the earth's surface. Now, the air in moving from the lower to the higher latitudes tends to take a north-easterly direction, and in this case will pass over our island in its course. This region of high pressure, however, is situated in the very path of the south-eastern branch of the Gulf-stream, and consequently the winds blowing therefrom will carry directly to Britain the heat of the Gulf-stream.

As we shall presently see, it is as essential to the heating of our island as to that of the southern portion of Europe, that a very large proportion of the waters of the Gulf-stream should spread over the surface of the Atlantic and never pass up into the arctic regions.

Even according to Mr. Findlay's own theory, it is to the south-west wind, heated by the warm waters of the Atlantic, that we are indebted for the high temperature of our climate. But he seems to be under the impression that the Atlantic would be able to supply the necessary heat independently of the Gulf-stream. This, it seems to me, is the fundamental error of all those who doubt the efficiency of the stream. It is a mistake, however, into which one is very apt to fall who does not adopt the more rigid method of determining heat-results in absolute measure. When we apply this method, we find that the Atlantic, without the aid of such a current as the Gulf-stream, would be wholly unable to supply the necessary amount of heat to the south-west winds.

The quantity of heat conveyed by the Gulf-stream, as we have seen, is equal to all the heat received from the sun by 1,560,935 square miles at the equator. The mean annual quantity of heat received from the sun by the temperate regions per unit surface is to that received by the equator as 9·08 to 12.* Consequently, the quantity of heat conveyed by the stream is equal to all the heat received from the sun by 2,062,960 square

* See "Smithsonian Contributions to Knowledge," vol. ix.

miles of the temperate regions. The total area of the Atlantic from the latitude of the Straits of Florida, 200 miles north of the tropic of Cancer, up to the Arctic Circle, including also the German Ocean, is about 8,500,000 square miles. In this case the quantity of heat carried by the Gulf-stream into the Atlantic through the Straits of Florida, is to that received by this entire area from the sun as 1 to 4·12, or in round numbers as 1 to 4. It therefore follows that one-fifth of all the heat possessed by the waters of the Atlantic over that area, even supposing that they absorb every ray that falls upon them, is derived from the Gulf-stream. Would those who call in question the efficiency of the Gulf-stream be willing to admit that a decrease of one-fourth in the total amount of heat received from the sun, over the entire area of the Atlantic from within 200 miles of the tropical zone up to the arctic regions, would not sensibly affect the climate of northern Europe? If they would not willingly admit this, why, then, contend that the Gulf-stream does not affect climate? for the stoppage of the Gulf-stream would deprive the Atlantic of 77,479,650,000,000,000,000 foot-pounds of energy in the form of heat per day, a quantity equal to one-fourth of all the heat received from the sun by that area.

How much, then, of the temperature of the south-west winds derived from the water of the Atlantic is due to the Gulf-stream?

Were the sun extinguished, the temperature over the whole earth would sink to *nearly* that of stellar space, which, according to the investigations of Sir John Herschel * and of M. Pouillet,† is not above — 239° F. Were the earth possessed of no atmosphere, the temperature of its surface would sink to exactly that of space, or to that indicated by a thermometer exposed to no other heat-influence than that of radiation from the stars. But the presence of the atmospheric envelope would slightly modify

* "Meteorology," section 36.

† *Comptes-Rendus*, July 9, 1838. Taylor's "Scientific Memoirs," vol. iv., p. 44 (1846).

the conditions of things; for the heat from the stars (which of course constitutes what is called the temperature of space) would, like the sun's heat, pass more freely through the atmosphere than the heat radiated back from the earth, and there would in consequence of this be an accumulation of heat on the earth's surface. The temperature would therefore stand a little higher than that of space; or, in other words, it would stand a little higher than it would otherwise do were the earth exposed in space to the direct radiation of the stars without the atmospheric envelope. But, for reasons which will presently be stated, we may in the meantime, till further light is cast upon this matter, take $-239°$ F. as probably not far from what would be the temperature of the earth's surface were the sun extinguished.

Suppose now that we take the mean annual temperature of the Atlantic at, say, 56°.* Then $239° + 56° = 295°$ represents the number of degrees of rise due to the heat which it receives. In other words, it takes all the heat that the Atlantic receives to maintain its temperature 295° above the temperature of space. Stop the Gulf-stream, and the Atlantic would be deprived of one-fifth of the heat which it possesses. Then, if it takes five parts of heat to maintain a temperature of 295° above that of space, the four parts which would remain after the stream was stopped would only be able to maintain a temperature of four-fifths of 295°, or 236° above that of space: the stoppage of the Gulf-stream would therefore deprive the Atlantic of an amount of heat which would be sufficient to maintain its temperature 59° above what it would otherwise be, did it depend alone upon the heat received directly from the sun. It does not, of course, follow that the Gulf-stream actually maintains the temperature 59° above what it would otherwise be were there no ocean-currents, because the actual heating-effect of the stream is neutralized to a very considerable extent by cold currents from

* The mean temperature of the Atlantic between the tropics and the arctic circle, according to Admiral Fitzroy's chart, is about 60°. But he assigns far too high a temperature for latitudes above 50°. It is probable that 56° is not far from the truth.

the arctic regions. But 59° of rise represents its actual power; consequently 59°, minus the lowering effect of the cold currents, represents the actual rise. What the rise may amount to at any particular place must be determined by other means.

This method of calculating how much the temperature of the earth's surface would rise or fall from an increase or a decrease in the absolute amount of heat received is that adopted by Sir John Herschel in his "Outlines of Astronomy," § 369[a].

About three years ago, in an article in the *Reader*, I endeavoured to show that this method is not rigidly correct. It has been shown from the experiments of Dulong and Petit, Dr. Balfour Stewart, Professor Draper, and others, that the rate at which a body radiates its heat off into space is not directly proportionate to its absolute temperature. The rate at which a body loses its heat as its temperature rises increases more rapidly than the temperature. As a body rises in temperature the rate at which it radiates off its heat increases; the *rate* of this increase, however, is not uniform, but increases with the temperature. Consequently the temperature is not lowered in proportion to the decrease of the sun's heat. But at the comparatively low temperature with which we have at present to deal, the error resulting from assuming the decrease of temperature to be proportionate to the decrease of heat would not be great.

It may be remarked, however, that the experiments referred to were made on solids; but, from certain results arrived at by Dr. Balfour Stewart, it would seem that the radiation of a material particle may be proportionate to its absolute temperature.* This physicist found that the radiation of a thick plate of glass increases more rapidly than that of a thin plate as the temperature rises, and that, if we go on continually diminishing the thickness of the plate whose radiation at different temperatures we are ascertaining, we find that as it grows thinner and thinner the rate at which it radiates off its heat as its temperature rises becomes less and less. In other words, as the

* The probable physical cause of this will be considered in the Appendix.

plate grows thinner and thinner its rate of radiation becomes more and more proportionate to its absolute temperature. And we can hardly resist the conviction that if we could possibly go on diminishing the thickness of the plate till we reached a film so thin as to embrace but only one particle in its thickness, its rate of radiation would be proportionate to its temperature. Dr. Balfour Stewart has very ingeniously suggested the probable reason why the rate of radiation of thick plates increases with rise of temperature more rapidly than that of thin. It is this: all substances are more diathermanous for heat of high temperatures than for heat of low temperatures. When a body is at a low temperature, we may suppose that only the exterior rows of particles supply the radiation, the heat from the interior particles being all stopped by the exterior ones, the substance being very opaque for heat of low temperature; while at a high temperature we may imagine that part of the heat from the interior particles is allowed to pass, thereby swelling the total radiation. But as the plate becomes thinner and thinner, the obstructions to interior radiation become less and less, and as these obstructions are greater for radiation at low temperatures than for radiation at high temperatures, it necessarily follows that, by reducing the thickness of the plate, we assist radiation at low temperatures more than we do at high.

In a gas, where each particle may be assumed to radiate by itself, and where the particles stand at a considerable distance from one another, the obstruction to interior radiation must be far less than in a solid. In this case the rate at which a gas radiates off its heat as its temperature rises must increase more slowly than that of a solid substance. In other words, its rate of radiation must correspond more nearly to its absolute temperature than that of a solid. If this be the case, a reduction in the amount of heat received from the sun, owing to an increase of his distance, should tend to produce a greater lowering effect on the temperature of the air than it does on the temperature of the solid ground. But as the temperature of our climate is determined by the temperature of the air, it must follow that the error of assuming that the decrease of temperature would be proportionate to the decrease in the intensity of the sun's heat may not be great.

It may be observed here, although it does not bear directly on this point, that although the air in a room, for example, or at the earth's surface is principally cooled by convection rather than by radiation, yet it is by radiation alone that the earth's atmosphere parts with its heat to stellar space; and this is the chief matter with which we are at present concerned. Air, like all other gases, is a bad radiator; and this tends to protect it from being cooled to such an extent as it would otherwise be, were it a good radiator like solids. True, it is also a bad absorber; but as it is cooled by radiation into space, and heated, not altogether by absorption, but to a very large extent by convection, it on the whole gains its heat more easily than it loses it, and consequently must stand at a higher temperature than it would do were it heated by absorption alone.

But, to return; the error of regarding the decrease of temperature as proportionate to the decrease in the amount of heat received, is probably neutralized by one of an opposite nature, viz., that of taking space at too high a temperature; for by so doing we make the result too small.

We know that absolute zero is at least 493° below the melting-point of ice. This is 222° below that of space. Consequently, if the heat derived from the stars is able to maintain a temperature of —239°, or 222° of absolute temperature, then nearly as much heat is derived from the stars as from the sun. But if so, why do the stars give so much heat and so very little light? If the radiation from the stars could maintain a thermometer 222° above absolute zero, then space must be far more transparent to heat-rays than to light-rays, or else the stars give out a great amount of heat, but very little light, neither of which suppositions is probably true. The probability is, I venture to presume, that the temperature of space is not very much above absolute zero. At the time when these investigations into the probable temperature of space were made, at least as regards the labours of Pouillet, the modern science of heat had no existence, and little or nothing was then known with certainty regarding absolute zero. In this case the whole matter would require to be reconsidered. The result of such an investigation in all probability would be to assign a lower temperature to stellar space than —239°.

Taking all these various considerations into account, it is probable that if we adopt —239° as the temperature of space, we shall not be far from the truth in assuming that the absolute temperature of a place above that of space is proportionate to the amount of heat received from the sun.

We may, therefore, in this case conclude that 59° of rise is probably not very far from the truth, as representing the influence of the Gulf-stream. The Gulf-stream, instead of producing little or no effect, produces an effect far greater than is generally supposed.

Our island has a mean annual temperature of about 12° above the normal due to its latitude. This excess of temperature has been justly attributed to the influence of the Gulf-stream. But it is singular how this excess should have been taken as the measure of the *rise resulting from the influence of the stream.* These figures only represent the number of degrees that the mean normal temperature of our island stands above what is called the normal temperature of the latitude.

The mode in which Professor Dove constructed his Tables of normal temperature was as follows:—He took the temperature of thirty-six equidistant points on every ten degrees of latitude. The mean temperature of these thirty-six points he calls in each case the *normal* temperature of the parallel. The excess above the normal merely represents how much the stream raises our temperature above the mean of all places on the same latitude, but it affords us no information regarding the absolute rise produced. In the Pacific, as well as in the Atlantic, there are immense masses of water flowing from the tropical to the temperate regions. Now, unless we know how much of the normal temperature of a latitude is due to ocean-currents, and how much to the direct heat of the sun, we could not possibly, from Professor Dove's Tables, form the most distant conjecture as to how much of our temperature is derived from the Gulf-stream. The overlooking of this fact has led to a general misconception regarding the positive influence of the Gulf-stream on temperature. The 12° marked in Tables of normal temperature do not represent the absolute effect of the stream, but merely show how much the stream raises the temperature of our country above the mean of all places on the same latitude. Other places have their temperature raised by ocean-currents as well as this country; only the Gulf-stream produces a rise of several degrees over and above that produced by other streams in the same latitude.

At present there is a difference merely of 80° between the mean temperature of the equator and the poles;* but were each part of the globe's surface to depend only upon the direct heat which it receives from the sun, there ought, according to theory, to be a difference of more than 200°. The annual quantity of heat received at the equator is to that received at the poles (supposing the proportionate quantity absorbed by the atmosphere to be the same in both cases) as 12 to 4·98, or, say, as 12 to 5. Consequently, if the temperatures of the equator and the poles be taken as proportionate to the absolute amount of heat received from the sun, then the temperature of the equator above that of space must be to that of the poles above that of space as 12 to 5. What ought, therefore, to be the temperatures of the equator and the poles, did each place depend solely upon the heat which it receives directly from the sun? Were all ocean and aërial currents stopped, so that there could be no transference of heat from one part of the earth's surface to another, what ought to be the temperatures of the equator and the poles? We can at least arrive at a rough estimate on this

* The mean temperature of the equator, according to Dove, is 79°·7, and that of the north pole 2°·3. But as there is, of course, some uncertainty regarding the actual mean temperature of the poles, we may take the difference in round numbers at 80°.

point. If we diminish the quantity of warm water conveyed from the equatorial regions to the temperate and arctic regions, the temperature of the equator will begin to rise, and that of the poles to sink. It is probable, however, that this process would affect the temperature of the poles more than it would that of the equator; for as the warm water flows from the equator to the poles, the area over which it is spread becomes less and less. But as the water from the tropics has to raise the temperature of the temperate regions as well as the polar, the difference of effect at the equator and poles might not, on that account, be so very great. Let us take a rough estimate. Say that, as the temperature of the equator rises one degree, the temperature of the poles sinks one degree and a half. The mean annual temperature of the globe is about 58°. The mean temperature of the equator is 80°, and that of the poles 0°. Let ocean and aërial currents now begin to cease, the temperature of the equator commences to rise and the temperature of the poles to sink. For every degree that the temperature of the equator rises, that of the poles sinks $1\frac{1}{2}$°; and when the currents are all stopped and each place becomes dependent solely upon the direct rays of the sun, the mean annual temperature of the equator above that of space will be to that of the poles, above that of space, as 12 to 5. When this proportion is reached, the equator will be 374° above that of space, and the poles 156°; for 374 is to 156 as 12 is to 5. The temperature of space we have seen to be —239°, consequently the temperature of the equator will in this case be 135°, reckoned from the zero of the Fahrenheit thermometer, and the poles 83° below zero. The equator would therefore be 55° warmer than at present, and the poles 83° colder. The difference between the temperature of the equator and the poles will in this case amount to 218°.

Now, if we take into account the quantity of positive energy in the form of heat carried by warm currents from the equator to the temperate and polar regions, and also the quantity of negative energy (cold) carried by cold currents from the polar regions to the equator, we shall find that they are sufficient to reduce the difference of temperature between the poles and the equator from 218° to 80°.

The quantity of heat received in the latitude of London, for example, is to that received at the equator nearly as 12 to 8. This, according to theory, should produce a difference of about 125°. The temperature of the equator above that of space, as we have seen, would be 374°. Therefore 249° above that of space would represent the temperature of the latitude of London. This would give 10° as its temperature. The stoppage of all ocean and aërial currents would thus increase the difference between the equator and the latitude of London by about 85°. The stoppage of ocean-currents would not be nearly so much felt, of course, in the latitude of London as at the equator and the poles, because, as has been already noticed, in all latitudes midway between the equator and the poles the two sets of currents to a considerable extent compensate each other—the warm currents from the equator raise the temperature, while the cold ones from the poles lower it; but as the warm currents chiefly keep on the surface and the cold return-currents are principally under-currents, the heating effect very greatly exceeds the cooling effect. Now, as we have seen, the stoppage of all currents would raise the temperature of the equator 55°; that is to say, the rise at the equator alone would increase the difference of temperature between the equator and that of London by 55°. But the actual difference, as we have seen, ought to be 85°; consequently the temperature of London would be lowered 30° by the stoppage of the currents. For if we raise the temperature of the equator 55° and lower the temperature of London 30°, we then increase the difference by 85°. The normal temperature of the latitude of London being 40°, the stoppage of all ocean and aërial currents would thus reduce it to 10°. But the Gulf-stream raises the actual mean temperature of London 10° above the normal. Consequently 30° + 10° = 40° represents the actual rise at London due to the influence of the Gulf-stream over and above all the lowering effects resulting from arctic currents. On some parts of the American shores on the latitude of London, the temperature is 10° below the normal. The stoppage of all ocean and aërial currents would therefore lower the temperature there only 20°.

It is at the equator and the poles that the great system of ocean and aërial currents produces its maximum effects. The influence becomes less and less as we recede from those places, and between them there is a point where the influence of warm currents from the equator and of cold currents from the poles exactly neutralize each other. At this point the stoppage of ocean-currents would not sensibly affect temperature. This point, of course, is not situated on the same latitude in all meridians, but varies according to the position of the meridian in relation to land, and ocean-currents, whether cold or hot, and other circumstances. A line drawn round the globe through these various points would be very irregular. At one place, such as on the western side of the Atlantic, where the arctic current predominates, the neutral line would be deflected towards the equator, while on the eastern side, where warm currents predominate, the line would be deflected towards the north. It is a difficult problem to determine the mean position of this line; it probably lies somewhere not far north of the tropics.

CHAPTER III.

OCEAN-CURRENTS IN RELATION TO THE DISTRIBUTION OF HEAT OVER THE GLOBE.—(*Continued.*)

Influence of the Gulf-stream on the Climate of the Arctic Regions.—Absolute Amount of Heat received by the Arctic Regions from the Sun.—Influence of Ocean-currents shown by another Method.—Temperature of a Globe all Water or all Land according to Professor J. D. Forbes.—An important Consideration overlooked.—Without Ocean-currents the Globe would not be habitable.—Conclusions not affected by Imperfection of Data.

Influence of the Gulf-stream on the Climate of the Arctic Regions.—Does the Gulf-stream pass into the arctic regions? Are the seas around Spitzbergen and North Greenland heated by the warm water of the stream?

Those who deny this nevertheless admit the existence of an arctic current. They admit that an immense mass of cold water is continually flowing south from the polar regions around Greenland into the Atlantic. If it be admitted, then, that a mass of water flows across the arctic circle from north to south, it must also be admitted that an equal mass flows across from south to north. It is also evident that the water crossing from south to north must be warmer than the water crossing from north to south; for the temperate regions are warmer than the arctic, and the ocean in temperate regions warmer than the ocean in the arctic; consequently the current which flows into the arctic seas, to compensate for the cold arctic current, must be a warmer current.

Is the Gulf-stream this warm current? Does this compensating warm current proceed from the Atlantic or from the Pacific? If it proceeds from the Atlantic, it is simply the warm water of the Gulf-stream. We may call it the warm water of the Atlantic if we choose; but this cannot materially affect the question at issue, for the heat which the waters of the Atlantic possess is derived, as we have seen, to an enormous extent from the water brought from the tropics by the Gulf-stream. If we deny that the warm compensating current comes from the Atlantic, then we must assume that it comes from the Pacific. But if the cold current flows from the arctic regions into the Atlantic, and the warm compensating current from the Pacific into the arctic regions, the highest temperature should be found on the Pacific side of the arctic regions and not on the Atlantic side; the reverse, however, is the case. In the Atlantic, for example, the 41° isothermal line reaches to latitude 65° 30′, while in the Pacific it nowhere goes beyond latitude 57°. The 27° isotherm reaches to latitude 75° in the Atlantic, but in the Pacific it does not pass beyond 64°. And the 14° isotherm reaches the north of Spitzbergen in latitude 80°, whereas on the Pacific side of the arctic regions it does not reach to latitude 72°.

On no point of the earth's surface does the mean annual temperature rise so high above the normal as in the northern Atlantic, just at the arctic circle, at a spot believed to be in the middle of the Gulf-stream. This place is no less than 22°·5 above the normal, while in the northern Pacific the temperature does not anywhere rise more than 9° above the normal. These facts prove that the warm current passes up the Atlantic into the arctic regions and not up the Pacific, or at least that the larger amount of warm water must pass into the arctic regions through the Atlantic. In other words, the Gulf-stream is the warm compensating current. Not only must there be a warm stream, but one of very considerable magnitude, in order to compensate for the great amount of cold water that is constantly flowing from the arctic regions, and also to maintain the temperature of those regions so much above the temperature of space as they actually are.

No doubt, when the results of the late dredging expedition

are published, they will cast much additional light on the direction and character of the currents forming the north-eastern branch of the Gulf-stream.

The average quantity of heat received by the arctic regions as a whole per unit surface to that received at the equator, as we have already seen, is as 5·45 to 12, assuming that the percentage of rays cut off by the atmosphere is the same at both places. In this case the mean annual temperature of the arctic regions, taken as a whole, would be about—69°, did those regions depend entirely for their temperature upon the heat received directly from the sun. But the temperature would not even reach to this; for the percentage of rays cut off by the atmosphere in arctic regions is generally believed to be greater than at the equator, and consequently the actual mean quantity of heat received by the arctic regions will be less than 5·45-12ths of what is received at the equator.

In the article on Climate in the "Encyclopædia Britannica" there is a Table calculated upon the principle that the quantity of heat cut off is proportionate to the number of aërial particles which the rays have to encounter before reaching the surface of the earth—that, as a general rule, if the tracts of the rays follow an arithmetical progression, the diminished force with which the rays reach the ground will form a decreasing geometrical progression. According to this Table about 75 per cent. of the sun's rays are cut off by the atmosphere in arctic regions. If 75 per cent. of the rays were cut off by the atmosphere in arctic regions, then the direct rays of the sun could not maintain a mean temperature 100° above that of space. But this is no doubt much too high a percentage for the quantity of heat cut off; for recent discoveries in regard to the absorption of radiant heat by gases and vapours prove that Tables computed on this principle must be incorrect. The researches of Tyndall and Melloni show that when rays pass through any substance, the absorption is rapid at first: but the rays are soon "sifted," as it is called, and they then pass onwards with but little further obstruction. Still, however, owing to the dense fogs which prevail in arctic regions, the quantity of heat cut off must be considerable. If as much as 50 per cent. of the sun's rays are cut off by the atmosphere in arctic regions, the amount of heat received directly from the sun would not be sufficient to maintain a mean annual temperature of — 100°. Consequently the arctic regions must depend to an enormous extent upon ocean-currents for their temperature.

Influence of Ocean-currents shown by another Method.—That the temperature of the arctic regions would sink enormously, and the temperature of the equator rise enormously, were all ocean-currents stopped, can be shown by another method—viz., by taking the mean annual temperature from the equator to the pole along a meridian passing through the ocean, say, the Atlantic, and comparing it with the mean annual temperature taken along a meridian passing through a great continent, say, the Asiatic.

Professor J. D. Forbes, in an interesting memoir,* has endeavoured by this method to determine what would be the temperature of the equator and the poles were the globe all water or all land. He has taken the temperature of the two meridians from the tables and charts of Professor Dove, and ascertained the exact proportion of land and water on every 10° of latitude from the equator to the poles, with the view of determining what proportion of the average temperature of the globe in each parallel is due to the land, and what to the water which respectively belongs to it. He next endeavours to obtain a formula for expressing the mean temperature of a given parallel, and thence arrives at "an approximate answer to the inquiry as to what would have been the equatorial or polar temperature of the globe, or that of any latitude, had its surface been entirely composed of land or of water."

The result at which he arrived is this: that, were the surface of the globe all water, 71°·7 would be the temperature of the equator, and 12°·5 the temperature of the poles; and were the

* Trans. of Roy. Soc. Edin., vol. xxii., p. 75.

surface all land, 109°·8 would be the temperature of the equator, and — 25°·6 the temperature of the poles.

But in Professor Forbes's calculations no account whatever is taken of the influence of currents, whether of water or of air, and the difference of temperature is attributed wholly to difference of latitude and the physical properties of land and water in relation to their powers in absorbing and detaining the sun's rays, and to the laws of conduction and of convection which regulate the internal motion of heat in the one and in the other. He considers that the effects of currents are all compensatory.

"If a current of hot water," he says, "moderates the cold of a Lapland winter, the counter-current, which brings the cold of Greenland to the shores of the United States, in a great measure restores the balance of temperature, so far as it is disturbed by this particular influence. The prevalent winds, in like manner, including the trade-winds, though they render some portions of continents, on the average, hotter or colder than others, produce just the contrary effect elsewhere. Each continent, if it has a cold eastern shore, has likewise a warm western one; and even local winds have for the most part established laws of compensation. In a given parallel of latitude all these secondary causes of local climate may be imagined to be mutually compensatory, and the outstanding gradation of mean or normal temperature will mainly depend, 1st, upon the effect of latitude simply; 2nd, on the distribution of land and water considered in their primary or *statical* effect."

It is singular that a physicist so acute as Professor Forbes should, in a question such as this, leave out of account the influence of currents, under the impression that their effects were compensatory.

If there is a constant transference of hot water from the equatorial regions to the polar, and of cold water from the polar regions to the equatorial (a thing which Professor Forbes admitted), then there can only be one place between the equator and the pole where the two sets of currents compensate each other. At all places on the equatorial side of this point a cooling effect is the result. Starting from this neutral point, the preponderance of the cooling effect over the heating increases as we approach towards the equator, and the preponderance of the heating effect over the cooling increases as we recede from this point towards the pole—the cooling effect reaching a maximum at the equator, and the heating effect a maximum at the pole.

Had Professor Forbes observed this important fact, he would have seen at once that the low temperature of the land in high latitudes, in comparison with that of the sea, was no index whatever as to how much the temperature of those regions would sink were the sea entirely removed and the surface to become land; for the present high temperature of the sea is not due wholly to the mere physical properties of water, but to a great extent is due to the heat brought by currents from the equator. Now, unless it is known how much of the absolute temperature of the ocean in those latitudes is due to currents, we cannot tell how much the removal of the sea would lower the absolute temperature of those places. Were the sea removed, the continents in high latitudes would not simply lose the heating advantages which they presently derive from the mere fact of their proximity to so much sea, but the removal would, in addition to this, deprive them of an enormous amount of heat which they at present receive from the tropics by means of ocean-currents. And, on the other hand, at the equator, were the sea removed, the continents there would not simply lose the cooling influences which result from their proximity to so much water, but, in addition to this, they would have to endure the scorching effects which would result from the heat which is at present carried away from the tropics by ocean-currents.

We have already seen that Professor Forbes concluded that the removal of the sea would raise the mean temperature of the equator 30°, and lower the temperature of the poles 28°; it is therefore perfectly certain that, had he added to his result the effect due to ocean-currents, and had he been aware that about one-fifth of all the heat possessed by the Atlantic is actually derived from the equator by means of the Gulf-stream, he would have assigned a temperature to the equator and the poles, of a globe all land, differing not very far from what I have concluded would be the temperature of those places were all ocean and aërial currents stopped, and each place to depend solely upon the heat which it received directly from the sun.

Without Ocean-currents the Globe would not be habitable.—All these foregoing considerations show to what an extent the climatic condition of our globe is due to the thermal influences of ocean-currents.

As regards the northern hemisphere, we have two immense oceans, the Pacific and the Atlantic, extending from the equator to near the north pole, or perhaps to the pole altogether. Between these two oceans lie two great continents, the eastern and the western. Owing to the earth's spherical form, far too much heat is received at the equator and far too little at high latitudes to make the earth a suitable habitation for sentient beings. The function of these two great oceans is to remove the heat from the equator and carry it to temperate and polar regions. Aërial currents could not do this. They might remove the heat from the equator, but they could not, as we have already seen, carry it to the temperate and polar regions; for the greater portion of the heat which aërial currents remove from the equator is dissipated into stellar space: the ocean alone can convey the heat to distant shores. But aërial currents have a most important function; for of what avail would it be, though ocean-currents should carry heat to high latitudes, if there were no means of distributing the heat thus conveyed over the land? The function of aërial currents is to do this. Upon this twofold arrangement depends the thermal condition of the globe. Exclude the waters of the Pacific and the Atlantic from temperate and polar regions and place them at the equator, and nothing now existing on the globe could live in high latitudes.

Were these two great oceans placed beside each other on one side of the globe, and the two great continents placed beside each other on the other side, the northern hemisphere would not then be suitable for the present order of things: the land on the central and on the eastern side of the united continent would be far too cold.

The foregoing Conclusions not affected by the Imperfection of the Data.—The general results at which we have arrived in reference to the influence of ocean-currents on the climatic condition of the globe are not affected by the imperfection of the data employed. It is perfectly true that considerable uncertainty prevails regarding some of the data; but, after making the fullest allowance for every possible error, the influence of currents is so enormous that the general conclusion cannot be materially affected. I can hardly imagine that any one familiar with the physics of the subject will be likely to think that, owing to possible errors in the data, the effects have probably been doubled. Even admitting, however, that this were proved to be the case, still that would not materially alter the general conclusion at which we have arrived. The influence of ocean-currents in the distribution of heat over the surface of the globe would still be admittedly enormous, whether we concluded that owing to them the present temperature of the equator is 55° or 27° colder than it would otherwise be, or the poles 83° or 41° hotter than they would be did no currents exist.

Nay, more, suppose we should again halve the result; even in that case we should have to admit that, owing to ocean-currents, the equator is about 14° colder and the poles about 21° hotter than they would otherwise be; in other words, we should have to admit that, were it not for ocean-currents, the mean temperature of the equator would be about 100° and the mean temperature of the poles about — 21°.

If the influence of ocean-currents in reducing the difference between the temperature of the equator and poles amounted to only a few degrees, it would of course be needless to put much weight on any results arrived at by the method of calculation which I have adopted; but when it is a matter of two hundred degrees, it is not at all likely that the general results will be very much affected by any errors which may ever be found in the data.

Objections of a palæontological nature have frequently been urged against the opinion that our island is much indebted for its mild climate to the influence of the Gulf-stream; but, from what has already been stated, it must be apparent that all objections of that nature are of little avail. The palæontologist may detect, from the character of the flora and fauna brought up from the sea-bottom by dredging and other means, the presence of a warm or of a cold current; but this can never enable him to prove that the temperate and polar regions are not affected to an enormous extent by warm water conveyed from the equatorial regions. For anything that palæontology can show to the contrary, were ocean-currents to cease, the mean annual temperature of our island might sink below the present mid-winter temperature of Siberia. What would be the thermal condition of our globe were there no ocean-currents is a question for the physicist; not for the naturalist.

CHAPTER IV.

OUTLINE OF THE PHYSICAL AGENCIES WHICH LEAD TO SECULAR CHANGES OF CLIMATE.

Eccentricity of the Earth's Orbit; its Effect on Climate.—Glacial Epoch not the direct Result of an Increase of Eccentricity.—An important Consideration overlooked.—Change of Eccentricity affects Climate only indirectly.—Agencies which are brought into Operation by an Increase of Eccentricity.—How an Accumulation of Snow is produced.—The. Effect of Snow on the Summer Temperature.—Reason of the low Summer Temperature of Polar Regions.—Deflection of Ocean-currents the chief Cause of secular Changes of Climate.—How the foregoing Causes deflect Ocean-currents.—Nearness of the Sun in Perigee a Cause of the Accumulation of Ice.—A remarkable Circumstance regarding the Causes which lead to secular Changes of Climate.—The primary Cause an Increase of Eccentricity.—Mean Temperature of whole Earth should be greater in Aphelion than in Perihelion.—Professor Tyndall on the Glacial Epoch.—A general Reduction of Temperature will not produce a Glacial Epoch.—Objection from the present Condition of the Planet Mars.

Primary cause of Change of Eccentricity of the Earth's Orbit.—There are two causes affecting the position of the earth in relation to the sun, which must, to a very large extent, influence the earth's climate; viz., the precession of the equinoxes and the change in the eccentricity of the earth's orbit. If we duly examine the combined influence of these two causes, we shall find that the northern and southern portions of the globe are subject to an excessively slow secular change of climate, consisting in a slow periodic change of alternate warmer and colder cycles.

According to the calculations of Leverrier, the superior limit of the earth's eccentricity is 0·07775.* The eccentricity is at

* *Connaissance des Temps* for 1863 (Additions). Lagrange's determination makes the superior limit 0·07641 (Memoirs of the Berlin Academy for 1782, p. 273). Recently the laborious task of re-investigating the whole subject of the secular variations of the elements of the planetary orbits was undertaken by Mr.

present diminishing, and will continue to do so during 23,980 years, from the year 1800 A.D., when its value will be then ·00314.

The change in the eccentricity of the earth's orbit may affect the climate in two different ways; viz., by either increasing or diminishing the mean annual amount of heat received from the sun, or by increasing or diminishing the difference between summer and winter temperature.

Let us consider the former case first. The total quantity of heat received from the sun during one revolution is inversely proportional to the minor axis.

The difference of the minor axis of the orbit when at its maximum and its minimum state of eccentricity is as 997 to 1000. This small amount of difference cannot therefore sensibly affect the climate. Hence we must seek for our cause in the second case under consideration.

There is of course as yet some little uncertainty in regard to the exact mean distance of the sun. I shall, however, in the present volume assume it to be 91,400,000 miles. When the eccentricity is at its superior limit, the distance of the sun from the earth, when the latter is in the aphelion of its orbit, is no less than 98,506,350 miles; and when in the perihelion it is only 84,293,650 miles. The earth is therefore 14,212,700 miles further from the sun in the former position than in the latter. The direct heat of the sun being inversely as the square of the distance, it follows that the amount of heat received by the earth when in these two positions will be as 19 to 26. Taking the present eccentricity to be ·0168, the earth's distance during winter, when nearest to the sun, is 89,864,480 miles. Suppose now that, according to the precession of the equinoxes, winter in our northern hemisphere should happen when the earth is in

Stockwell, of the United States. He has taken into account the disturbing influence of the planet Neptune, the existence of which was not known when Leverrier's computations were made; and he finds that the eccentricity of the earth's orbit will always be included within the limits of 0 and 0·0693888. Mr. Stockwell's elaborate Memoir, extending over no fewer than two hundred pages, will be found in the eighteenth volume of the "Smithsonian Contributions to Knowledge."

the aphelion of its orbit, at the time when the orbit is at its greatest eccentricity; the earth would then be 8,641,870 miles further from the sun in winter than at present. The direct heat of the sun would therefore be one-fifth less during that season than at present; and in summer one-fifth greater. This enormous difference would affect the climate to a very great extent. But if winter under these circumstances should happen when the earth is in the perihelion of its orbit, the earth would then be 14,212,700 miles nearer the sun in winter than in summer. In this case the difference between winter and summer in the latitude of this country would be almost annihilated. But as the winter in the one hemisphere corresponds with the summer in the other, it follows that while the one hemisphere would be enduring the greatest extremes of summer heat and winter cold, the other would be enjoying a perpetual summer.

It is quite true that whatever may be the eccentricity of the earth's orbit, the two hemispheres must receive equal quantities of heat per annum; for proximity to the sun is exactly compensated by the effect of swifter motion—the total amount of heat received from the sun between the two equinoxes is the same in both halves of the year, whatever the eccentricity of the earth's orbit may be. For example, whatever extra heat the southern hemisphere may at present receive from the sun during its summer months owing to greater proximity to the sun, is exactly compensated by a corresponding loss arising from the shortness of the season; and, on the other hand, whatever deficiency of heat we in the northern hemisphere may at present have during our summer half year in consequence of the earth's distance from the sun, is also exactly compensated by a corresponding length of season.

It has been shown in the introductory chapter that a simple change in the sun's distance would not alone produce a glacial epoch, and that those physicists who confined their attention to purely astronomical effects were perfectly correct in affirming that no increase of eccentricity of the earth's orbit could account for that epoch. But the important fact was overlooked that although the glacial epoch could not result directly from an increase of eccentricity, it might nevertheless do so indirectly. The glacial epoch, as I hope to show, was not due directly to an increase in the eccentricity of the earth's orbit, but to a number of physical agents that were brought into operation as a result of an increase.

I shall now proceed to give an outline of what these physical agents were, how they were brought into operation, and the way in which they led to the glacial epoch.

When the eccentricity is about its superior limit, the combined effect of all those causes to which I allude is to lower to a very great extent the temperature of the hemisphere whose winters occur in aphelion, and to raise to nearly as great an extent the temperature of the opposite hemisphere, where winter of course occurs in perihelion.

With the eccentricity at its superior limit and the winter occurring in the aphelion, the earth would be 8,641,870 miles further from the sun during that season than at present. The reduction in the amount of heat received from the sun owing to his increased distance would, upon the principle we have stated in Chapter II., lower the midwinter temperature to an enormous extent. In temperate regions the greater portion of the moisture of the air is at present precipitated in the form of rain, and the very small portion which falls as snow disappears in the course of a few weeks at most. But in the circumstances under consideration, the mean winter temperature would be lowered so much below the freezing-point that what now falls as rain during that season would then fall as snow. This is not all; the winters would then not only be colder than now, but they would also be much longer. At present the winters are nearly eight days shorter than the summers; but with the eccentricity at its superior limit and the winter solstice in aphelion, the length of the winters would exceed that of the summers by no fewer than thirty-six days. The lowering of the temperature and the lengthening of the winter would both tend to the same effect, viz., to increase the amount of snow accumulated during the winter; for, other things being equal, the larger the snow-accumulating period the greater the accumulation. I may remark, however, that the absolute quantity of heat received during winter is not affected by the decrease in the sun's heat,* for the additional length of the season compensates for this decrease. As regards the absolute amount of heat received, increase of the sun's distance and lengthening of the winter are compensatory, but not so in regard to the amount of snow accumulated.

The consequence of this state of things would be that, at the commencement of the short summer, the ground would be covered with the winter's accumulation of snow.

Again, the presence of so much snow would lower the summer temperature, and prevent to a great extent the melting of the snow.

There are three separate ways whereby accumulated masses of snow and ice tend to lower the summer temperature, viz.:—

First. By means of direct radiation. No matter what the intensity of the sun's rays may be, the temperature of snow and ice can never rise above 32°. Hence the presence of snow and ice tends by direct radiation to lower the temperature of all surrounding bodies to 32°.

In Greenland, a country covered with snow and ice, the pitch has been seen to melt on the side of a ship exposed to the direct rays of the sun, while at the same time the surrounding air was far below the freezing-point; a thermometer exposed to the direct radiation of the sun has been observed to stand above 100°, while the air surrounding the instrument was actually 12° below the freezing-point.† A similar experience has been recorded by travellers on the snow-fields of the Alps.‡

These results, surprising as they no doubt appear, are what

* When the eccentricity is at its superior limit, the absolute quantity of heat received by the earth during the year is, however, about one three-hundredth part greater than at present. But this does not affect the question at issue.

† Scoresby's "Arctic Regions," vol. ii., p. 379. Daniell's "Meteorology," vol. ii., p. 123.

‡ Tyndall, "On Heat," article 364.

we ought to expect under the circumstances. The diathermancy of air has been well established by the researches of Professor Tyndall on radiant heat. Perfectly dry air seems to be nearly incapable of absorbing radiant heat. The entire radiation passes through it almost without any sensible absorption. Consequently the pitch on the side of the ship may be melted, or the bulb of the thermometer raised to a high temperature by the direct rays of the sun, while the surrounding air remains intensely cold. "A joint of meat," says Professor Tyndall, "might be roasted before a fire, the air around the joint being cold as ice."* The air is cooled by *contact* with the snow-covered ground, but is not heated by the radiation from the sun.

When the air is humid and charged with aqueous vapour, a similar cooling effect also takes place, but in a slightly different way. Air charged with aqueous vapour is a good absorber of radiant heat, but it can only absorb those rays which agree with it in *period*. It so happens that rays from snow and ice are, of all others, those which it absorbs best. The humid air will absorb the total radiation from the snow and ice, but it will allow the greater part of, if not nearly all, the sun's rays to pass unabsorbed. But during the day, when the sun is shining, the radiation from the snow and ice to the air is negative; that is, the snow and ice cool the air by radiation. The result is, the air is cooled by radiation from the snow and ice (or rather, we should say, *to* the snow and ice) more rapidly than it is heated by the sun; and, as a consequence, in a country like Greenland, covered with an icy mantle, the temperature of the air, even during summer, seldom rises above the freezing-point. Snow is a good reflector, but as simple reflection does not change the character of the rays they would not be absorbed by the air, but would pass into stellar space.

Were it not for the ice, the summers of North Greenland, owing to the continuance of the sun above the horizon, would be as warm as those of England; but, instead of this, the

* Tyndall, "On Heat," article 364.

Greenland summers are colder than our winters. Cover India with an ice sheet, and its summers would be colder than those of England.

Second. Another cause of the cooling effect is that the rays which fall on snow and ice are to a great extent reflected back into space.* But those that are not reflected, but absorbed, do not raise the temperature, for they disappear in the mechanical work of melting the ice. The latent heat of ice is about 142° F.; consequently in the melting of every pound of ice a quantity of heat sufficient to raise one pound of water 142° disappears, and is completely lost, so far as temperature is concerned. This quantity of heat is consumed, not in raising the temperature of the ice, but in the mechanical work of tearing the molecules separate against the forces of cohesion binding them together into the solid form. No matter what the intensity of the sun's heat may be, the surface of the ground will remain permanently at 32° so long as the snow and ice con tinue unmelted.

Third. Snow and ice lower the temperature by chilling the air and condensing the vapour into thick fogs. The great strength of the sun's rays during summer, due to his nearness at that season, would, in the first place, tend to produce an increased amount of evaporation. But the presence of snow-clad mountains and an icy sea would chill the atmosphere and condense the vapour into thick fogs. The thick fogs and cloudy sky would effectually prevent the sun's rays from reaching the earth, and the snow, in consequence, would remain unmelted during the entire summer. In fact, we have this very condition of things exemplified in some of the islands of the Southern Ocean at the present day. Sandwich Land, which is in the same parallel of latitude as the north of Scotland, is covered with ice and snow the entire summer; and in the island of South Georgia, which is in the same parallel as the centre of England, the perpetual snow descends to the very sea-beach. The following is Captain Cook's description of this dismal place:—"We thought it very extraordinary," he says,

* See Phil. Mag., March, 1870, p.

"that an island between the latitudes of 54° and 55° should, in the very height of summer, be almost wholly covered with frozen snow, in some places many fathoms deep. . . . . The head of the bay was terminated by ice-cliffs of considerable height; pieces of which were continually breaking off, which made a noise like a cannon. Nor were the interior parts of the country less horrible. The savage rocks raised their lofty summits till lost in the clouds, and valleys were covered with seemingly perpetual snow. Not a tree nor a shrub of any size were to be seen. The only signs of vegetation were a strong-bladed grass growing in tufts, wild burnet, and a plant-like moss seen on the rocks. . . . . We are inclined to think that the interior parts, on account of their elevation, never enjoy heat enough to melt the snow in such quantities as to produce a river, nor did we find even a stream of fresh water on the whole coast."*

Captain Sir James Ross found the perpetual snow at the sea-level at Admiralty Inlet, South Shetland, in lat. 64°; and while near this place the thermometer in the very middle of summer fell at night to 23° F.; and so rapidly was the young ice forming around the ship that he began, he says, "to have serious apprehensions of the ships being frozen in." † At the comparatively low latitude of 59° S., in long. 171° E. (the corresponding latitude of our Orkney Islands), snow was falling on the longest day, and the surface of the sea at 32°.‡ And during the month of February (the month corresponding to August in our hemisphere) there were only three days in which they were not assailed by snow-showers.§

In the Straits of Magellan, in 53° S. lat., where the direct heat of the sun ought to be as great as in the centre of England, MM. Churrca and Galcano have seen snow fall in the middle of summer; and though the day was eighteen hours long, the thermometer seldom rose above 42° or 44°, and never above 51°.‖

* Captain Cook's "Second Voyage," vol. ii., pp. 232, 235.
† "Antarctic Regions," vol. ii., pp. 345—349.
‡ Ibid., vol. i., p. 167. § Ibid., vol. ii., p. 362.
‖ Edinburgh Philosophical Journal, vol. iv., p. 266.

This rigorous condition of climate chiefly results from the rays of the sun being intercepted by the dense fogs which envelope those regions during the entire summer; and the fogs again are due to the air being chilled by the presence of the snow-clad mountains and the immense masses of floating ice which come from the antarctic seas. The reduction of the sun's heat and lengthening of the winter, which would take place when the eccentricity is near to its superior limit and the winter in aphelion, would in this country produce a state of things perhaps as bad as, if not worse than, that which at present exists in South Georgia and South Shetland.

If we turn our attention to the polar regions, we shall find that the cooling effects of snow and ice are even still more marked. The coldness of the summers in polar regions is owing almost solely to this cause. Captain Scoresby states that, in regard to the arctic regions, the general obscurity of the atmosphere arising from fogs or clouds is such that the sun is frequently invisible during several successive days. At such times, when the sun is near the northern tropic, there is scarcely any sensible quantity of light from noon till midnight.* "And snow," he says, "is so common in the arctic regions, that it may be boldly stated that in nine days out of ten during the months of April, May, and June more or less falls." †

On the north side of Hudson's Bay, for example, where the quantity of floating ice during summer is enormous, and dense fogs prevail, the mean temperature of June does not rise above the freezing-point, being actually 13°·5 below the normal temperature; while in some parts of Asia under the same latitude, where there is comparatively little ice, the mean temperature of June is as high as 60°.

The mean temperature of Van Rensselaer Harbour, in lat. 78° 37′ N., long. 70° 53′ W., was accurately determined from hourly observations made day and night over a period of two years by Dr. Kane. It was found to be as follows:—

* Scoresby's "Arctic Regions," vol. i., p. 378. † Ibid., p. 425.

	°
Winter	−28·59
Spring	−10·59
Summer	+33·38
Autumn	− 4·03

But although the quantity of heat received from the sun at that latitude ought to have been greater during the summer than in England,* yet nevertheless the temperature is only 1°·38 above the freezing-point.

The temperature of Port Bowen, lat. 73° 14′ N., was found to be as follows:—

	°
Winter	−25·09
Spring	− 5·77
Summer	+34·40
Autumn	+10·58

Here the summer is only 2°·4 above the freezing-point.

The condition of things in the antarctic regions is even still worse than in the arctic. Captain Sir James Ross, when between lat. 66° S. and 77° 5′ S., during the months of January and February, 1841, found the mean temperature to be only 26°·5; and there were only two days when it rose even to the freezing-point. When near the ice-barrier on the 8th of February, 1841, a season of the year equivalent to August in England, he had the thermometer at 12° at noon; and so rapidly was the young ice forming around the ships, that it was with difficulty that he escaped being frozen in for the winter. "Three days later," he says, "the thick falling snow prevented our seeing to any distance before us; the waves as they broke over the ships froze as they fell on the decks and rigging, and covered our clothes with a thick coating of ice." † On visiting the barrier next year about the same season, he again ran the risk of being frozen in. He states that the surface of the sea presented one unbroken sheet of young ice as far as the eye could discover from the masthead.

Lieutenant Wilkes, of the American Exploring Expedition,

* See Meech's memoir "On the Intensity of the Sun's Heat and Light," "Smithsonian Contributions," vol. ix.

† "Antarctic Regions," vol. i., p. 240.

says that the temperature they experienced in the antarctic regions surprised him, for they seldom, if ever, had it above 30°, even at midday. Captain Nares, when in latitude 64° S., between the 13th and 25th February last (1874), found the mean temperature of the air to be 31°·5; a lower temperature than is met with in the arctic regions, in August, ten degrees nearer the pole.*

These extraordinarily low temperatures during summer, which we have just been detailing, were due solely to the presence of snow and ice. In South Georgia, Sandwich Land, and some other places which we have noticed, the summers ought to be about as warm as those of England; yet to such an extent is the air cooled by means of floating ice coming from the antarctic regions, and the rays of the sun enfeebled by the dense fogs which prevail, that there is actually not heat sufficient even in the very middle of summer to melt the snow lying on the sea-beach.

We read with astonishment that a country in the latitude of England should in the very middle of summer be covered with snow down to the sea-shore—the thermometer seldom rising much above the freezing-point. But we do not consider it so surprising that the summer temperature of the polar regions should be low, for we are accustomed to regard a low temperature as the normal condition of things there. We are, however, mistaken if we suppose that the influence of ice on climate is less marked at the poles than at such places as South Georgia or Sandwich Land.

It is true that a low summer temperature is the normal state of matters in very high latitudes, but it is so only in consequence of the perpetual presence of snow and ice. When we speak of the normal temperature of a place we mean, of course, as we have already seen, the normal temperature under the present condition of things. But were the ice removed from those regions, our present Tables of normal summer temperature would be valueless. These Tables give us the normal June temperature while the ice remains, but they do not afford us

* *Challenger* Reports, No. 2, p. 10.

the least idea as to what that temperature would be were the ice removed. The mere removal of the ice, all things else remaining the same, would raise the summer temperature enormously. The actual June temperature of Melville Island, for example, is 37°, and Port Franklin, Nova Zembla, 36°·5; but were the ice removed from the arctic regions, we should then find that the summer temperature of those places would be about as high as that of England. This will be evident from the following considerations:—

The temperature of a place, other things being equal, is proportionate to the quantity of heat received from the sun. If Greenland receives per given surface as much heat from the sun as England, its temperature ought to be as high as that of England. Now, from May 10 till August 3, a period of eighty-five days, the quantity of heat received from the sun in consequence of his remaining above the horizon is actually greater at the north pole than at the equator.

Column II. of the following Table, calculated by Mr. Meech,* represents the quantity of heat received from the sun on the 15th of June at every 10° of latitude. To simplify the Table, I have taken 100 as the unit quantity received at the equator on that day instead of the unit adopted by Mr. Meech:—

	I. Latitude.	II. Quantity of heat.	III. June temperature.
Equator	0°	100	80°·0
	10	111	81·1
	20	118	81·1
	30	123	77·3
	40	125	68·0
	50	125	58·8
	60	123	51·4
	70	127	39·2
	80	133	30·2
North Pole..	90	136	27·4

The calculations are, of course, made upon the supposition that the quantity of rays cut off in passing through the atmo-

* See "Smithsonian Contributions," vol. ix.

sphere is the same at the poles as at the equator, which, as we know, is not exactly the case. But, notwithstanding the extra loss of solar heat in high latitudes caused by the greater amount of rays that are cut off, still, if the temperature of the arctic summers were at all proportionate to the quantity of heat received from the sun, it ought to be very much higher than it actually is. Column III. represents the actual mean June temperature, according to Prof. Dove, at the corresponding latitudes. A comparison of these two columns will show the very great deficiency of temperature in high latitudes during summer. At the equator, for example, the quantity of heat received is represented by 100 and the temperature 80°; while at the pole the temperature is only 27°·4, although the amount of heat received is 136. This low temperature during summer, from what has been already shown, is due chiefly to the presence of snow and ice. If by some means or other we could remove the snow and ice from the arctic regions, they would then enjoy a temperate, if not a hot, summer. In Greenland, as we have already seen, snow falls even in the very middle of summer, more or less, nine days out of ten; but remove the snow from the northern hemisphere, and a snow-shower in Greenland during summer would be as great a rarity as it would be on the plains of India.

Other things being equal, the quantity of solar heat received in Greenland during summer is considerably greater than in England. Consequently, were it not for snow and ice, it would enjoy as warm a climate during summer as that of England. Conversely, let the polar snow and ice extend to the latitude of England, and the summers of that country would be as cold as those of Greenland. Our summers would then be as cold as our winters are at present, and snow in the very middle of summer would perhaps be as common as rain.

Mr. Murphy's Theory.—In a paper read before the Geological Society by Mr. Murphy* he admits that the glacial climate was due to an increase of eccentricity, but maintains in opposition to me that the glaciated hemisphere must be that in which the

* Quart. Journ. Geol. Soc., vol. xxv., p. 350.

summer occurs in *aphelion* during the greatest eccentricity of the earth's orbit.

I fear that Mr. Murphy must be resting his theory on the mistaken idea that a summer in aphelion ought to melt less snow and ice than one in perihelion. It is quite true that the longer summer in aphelion—other things being equal—is colder than the shorter one in perihelion, but the quantity of heat received from the sun is the same in both cases. Consequently the quantity of snow and ice melted ought also to be the same; for the amount melted is in proportion to the quantity of energy in the form of heat received.

It is true that with us at present less snow and ice are melted during a cold summer than during a warm one. But this is not a case in point, for during a cold summer we have less heat than during a warm summer, the length of both being the same. The coldness of the summers in this case is owing chiefly to a portion of the heat which we ought to receive from the sun being cut off by some obstructing cause.

The reason why we have so little snow, and consequently so little ice, in temperate regions, is not, as Mr. Murphy seems to suppose, that the heat of summer melts it all, but that there is so little to melt. And the reason why we have so little to melt is that, owing to the warmth of our winters, we have generally rain instead of snow. But if you increase the eccentricity very much, and place the winter in perihelion, we should probably have no snow whatever, and, as far as glaciation is concerned, it would then matter very little what sort of summer we had.

But it is not correct to say that the perihelion summer of the glacial epôch must have been hot. There are physical reasons, as we have just seen, which go to prove that, notwithstanding the nearness of the sun at that season, the temperature would seldom, if ever, rise much above the freezing-point.

Besides, Mr. Murphy overlooks the fact that the nearness of the sun during summer was nearly as essential to the production of the ice, as we shall shortly see, as his great distance during winter.

We must now proceed to the consideration of an agency which is brought into operation by the foregoing condition of things, an agency far more potent than any which has yet come under our notice, viz., the *Deflection of Ocean-currents.*

Deflection of Ocean-currents the chief Cause of secular Changes of Climate.—The enormous extent to which the thermal condition of the globe is affected by ocean-currents seems to cast new light on the mystery of geological climate. What, for example, would be the condition of Europe were the Gulf-stream stopped, and the Atlantic thus deprived of one-fifth of the absolute amount of heat which it is now receiving above what it has in virtue of the temperature of space? If the results just arrived at be at all justifiable, it follows that the stoppage of the stream would lower the temperature of northern Europe to an extent that would induce a condition of climate as severe as that of North Greenland; and were the warm currents of the North Pacific also at the same time to be stopped, the northern hemisphere would assuredly be subjected to a state of general glaciation.

Suppose also that the warm currents, having been withdrawn from the northern hemisphere, should flow into the Southern Ocean: what then would be the condition of the southern hemisphere? Such a transference of heat would raise the temperature of the latter hemisphere about as much as it would lower the temperature of the former. It would consequently raise the mean temperature of the antarctic regions much above the freezing-point, and the ice under which those regions are at present buried would, to a great extent at least, disappear. The northern hemisphere, thus deprived of the heat from the equator, would be under a condition of things similar to that which prevailed during the glacial epoch; while the other hemisphere, receiving the heat from the equator, would be under a condition of climate similar to what we know prevailed in the northern hemisphere during a part of the Upper Miocene period, when North Greenland enjoyed a climate as mild as that of England at the present day.

This is no mere picture of the imagination, no mere hypothesis devised to meet a difficult case; for if what has already been stated be not completely erroneous, all this follows as a necessary consequence from physical principles. If the warm currents of the equatorial regions be all deflected into one hemisphere, such must be the condition of things. How then do the agencies which we have been considering deflect ocean-currents?

How the foregoing Causes deflect Ocean-currents.—A high condition of eccentricity tends, we have seen, to produce an accumulation of snow and ice on the hemisphere whose winters occur in aphelion. This accumulation tends in turn to lower the summer temperature, to cut off the sun's rays, and so to retard the melting of the snow. In short, it tends to produce on that hemisphere a state of glaciation. Exactly opposite effects take place on the other hemisphere, which has its winter in perihelion. There the shortness of the winters and the highness of the temperature, owing to the sun's nearness, combine to prevent the accumulation of snow. The general result is that the one hemisphere is cooled and the other heated. This state of things now brings into play the agencies which lead to the deflection of the Gulf-stream and other great ocean-currents.

Owing to the great difference between the temperature of the equator and the poles, there is a constant flow of air from the poles to the equator. It is to this that the trade-winds owe their existence. Now as the strength of these winds, as a general rule, will depend upon the difference of temperature that may exist between the equator and higher latitudes, it follows that the trades on the cold hemisphere will be stronger than those on the warm. When the polar and temperate regions of the one hemisphere are covered to a large extent with snow and ice, the air, as we have just seen, is kept almost at the freezing-point during both summer and winter. The trades on that hemisphere will, of necessity, be exceedingly powerful; while on the other hemisphere, where there is comparatively little snow and ice, and the air is warm, the trades will, as a consequence, be weak. Suppose now the northern hemisphere to be the cold one. The north-east trade-winds of this hemisphere will far exceed in strength the south-east trade-winds of the southern hemisphere. The *median-line* between the trades will consequently lie to a very considerable distance to the south of the equator. We have a good example of this at the present day. The difference of temperature between the two hemispheres at present is but trifling to what it would be in the case under consideration; yet we find that the south-east trades of the Atlantic blow with greater force than the north-east trades, and the result is that the south-east trades sometimes extend to 10° or 15° N. lat., whereas the north-east trades seldom blow south of the equator. The effect of the northern trades blowing across the equator to a great distance will be to impel the warm water of the tropics over into the Southern Ocean. But this is not all; not only would the median-line of the trades be shifted southwards, but the great equatorial currents of the globe would also be shifted southwards.

Let us now consider how this would affect the Gulf-stream. The South American continent is shaped somewhat in the form of a triangle, with one of its angular corners, called Cape St. Roque, pointing eastwards. The equatorial current of the Atlantic impinges against this corner; but as the greater portion of the current lies a little to the north of the corner, it flows westward into the Gulf of Mexico and forms the Gulf-stream. A considerable portion of the water, however, strikes the land to the south of the Cape and is deflected along the shores of Brazil into the Southern Ocean, forming what is known as the Brazilian current.

Now it is perfectly obvious that the shifting of the equatorial current of the Atlantic only a few degrees to the south of its present position—a thing which would certainly take place under the conditions which we have been detailing—would turn the entire current into the Brazilian branch, and instead of flowing chiefly into the Gulf of Mexico as at present, it would all flow into the Southern Ocean, and the Gulf-stream would consequently be stopped. The stoppage of the Gulf-stream, combined with all those causes which we have just been considering, would place Europe under glacial conditions; while, at the same time, the temperature of the Southern Ocean would, in consequence of the enormous quantity of warm water received, have its temperature (already high from other causes) raised enormously.

Deflection of the Gulf-stream during the Glacial Epoch indicated by the Difference between the Clyde and Canadian Shell-beds.—That the glaciation of north-western Europe resulted to a great extent from the stoppage of the Gulf-stream may, I think, be inferred from a circumstance pointed out by the Rev. Mr. Crosskey, several years ago, in a paper read before the Glasgow Geological Society.* He showed that the difference between the glacial shells of Canada and those now existing in the Gulf of St. Lawrence is much less marked than the difference between the glacial shells of the Clyde beds and those now existing in the Firth. And from this he justly infers that the change of climate in Canada since the glacial epoch has been far less complete than in Scotland.

The return of the Gulf-stream has raised the mean annual temperature of our island no less than 15° above the normal, while Canada, deprived of its influence and exposed to a cold stream from polar regions, has been kept nearly as much below the normal.

Let us compare the present temperature of the two countries. In making our comparison we must, of course, compare places on the same latitude. It will not do, for example, to compare Glasgow with Montreal or Quebec, places on the latitude of the south of France and north of Italy. It will be found that the difference of temperature between the two countries is so enormous as to appear scarcely credible to those who have not examined the matter. The temperatures have all been taken from Professor Dove's work on the "Distribution of Heat over the Surface of the Globe," and his Tables published in the Report of the British Association for 1847.

* Trans. of Glasgow Geol. Soc. for 1866.

The mean temperature of Scotland for January is about 38° F., while in some parts of Labrador, on the same latitude, and all along the central parts of North America lying to the north of Upper Canada, it is actually 10°, and in many places 13° below zero. The January temperature at the Cumberland House, which is situated on the latitude of the centre of England, is more than 13° below zero. Here is a difference of no less than 51°. The normal temperature for the month of January in the latitude of Glasgow, according to Professor Dove, is 10°. Consequently, owing to the influence of the Gulf-stream, we are 28° warmer during that month than we would otherwise be, while vast tracts of country in America are 23° colder than they should be.

The July temperature of Glasgow is 61°, while on the same latitude in Labrador and places to the west it is only 49°. Glasgow during that month is 3° above the normal temperature, while America, owing to the influence of the cold polar stream, is 9° below it. The mean annual temperature of Glasgow is nearly 50°, while in America, on the same latitude, it is only 30°, and in many places as low as 23°. The mean normal temperature for the whole year is 35°. Our mean annual temperature is therefore 15° above the normal, and that of America from 5° to 12° below it. The American winters are excessively cold, owing to the continental character of the climate, and the absence of any benefit from the Gulf-stream, while the summers, which would otherwise be warm, are, in the latitude of Glasgow, cooled down to a great extent by the cold ice from Greenland; and the consequence is, that the mean annual temperature is about 20° or 27° below that of ours. The mean annual temperature of the Gulf of St. Lawrence is as low as that of Lapland or Iceland. It is no wonder, then, that the shells which flourished in Canada during the glacial epoch have not left the gulf and the neighbouring seas.

We have good reason to believe that the climate of America during the glacial epoch was even then somewhat more severe than that of Western Europe, for the erratics of America extend as far south as latitude 40°, while on the old continent they are not found much beyond latitude 50°. This difference may have resulted from the fact that the western side of a continent is always warmer than the eastern.

In order to determine whether the cold was as great in America during the glacial epoch as in Western Europe, we must not compare the fossils found in the glacial beds about Montreal, for example, with those found in the Clyde beds, for Montreal lies much further to the south than the Clyde. The Clyde beds must be compared with those of Labrador, while the beds of Montreal must be compared with those of the south of France and the north of Italy, if any are to be found there.

On the whole, it may be concluded that had the Gulf-stream not returned to our shores at the close of the glacial epoch, and had its place been supplied by a cold stream from the polar regions, similar to that which washes the shores of North America, it is highly probable that nearly every species found in our glacial beds would have had their representatives flourishing in the British seas at the present day.

It is no doubt true that when we compare the places in which the Canadian shell-beds referred to by Mr. Crosskey are situated with places on the same latitude in Europe, the difference of climate resulting from the influence of the Gulf-stream is not so great as between Scotland and those places which we have been considering; but still the difference is sufficiently great to account for why the change of climate in Canada has been less complete than in Scotland.

And what holds true in regard to the currents of the Atlantic holds also true, though perhaps not to the same extent, of the currents of the Pacific.

Nearness of the Sun in Perigee a Cause of the Accumulation of Ice.—But there is still another cause which must be noticed :—A strong under current of air *from* the north implies an equally strong upper current *to* the north. Now if the effect of the under current would be to impel the warm water at the equator to the south, the effect of the upper current would be to carry the aqueous vapour formed at the equator to the north; the upper current, on reaching the snow and ice of temperate regions, would deposit its moisture in the form of snow; so that, notwithstanding the great cold of the glacial epoch, it is probable that the quantity of snow falling in the northern regions would be enormous. This would be particularly the case during summer, when the earth would be in the perihelion and the heat at the equator great. The equator would be the furnace where evaporation would take place, and the snow and ice of temperate regions would act as a condenser.

Heat to produce *evaporation* is just as essential to the accumulation of snow and ice as cold to produce *condensation*. Now at Midsummer, on the supposition of the eccentricity being at its superior limit, the sun would be 8,641,870 miles nearer than at present during that season. The effect would be that the intensity of the sun's rays would be one-fifth greater than now. That is to say, for every five rays received by the ocean at present, six rays would be received then, consequently the evaporation during summer would be excessive. But the ice-covered land would condense the vapour into snow. It would, no doubt, be during summer that the greatest snowfall would take place. In fact, the nearness of the sun during that season was as essential to the production of the glacial epoch as was his distance during winter.

The direct effect of eccentricity is to produce on one of the hemispheres a long and cold winter. This alone would not lead to a condition of things so severe as that which we know prevailed during the glacial epoch. But the snow and ice thus produced would bring into operation, as we have seen, a host of physical agencies whose combined efforts would be quite sufficient to do this.

A remarkable Circumstance regarding those Causes which lead to Secular Changes of Climate.—There is one remarkable circumstance connected with those physical causes which deserves special notice. They not only all lead to one result, viz., an accumulation of snow and ice, but they react on one another.

It is quite a common thing in physics for the effect to react on the cause. In electricity and magnetism, for example, cause and effect in almost every case mutually act and react upon each other. But it is usually, if not universally, the case that the reaction of the effect tends to weaken the cause. The weakening influences of this reaction tend to impose a limit on the efficiency of the cause. But, strange to say, in regard to the physical causes concerned in the bringing about of the glacial condition of climate, cause and effect mutually reacted so as to strengthen each other. And this circumstance had a great deal to do with the extraordinary results produced.

We have seen that the accumulation of snow and ice on the ground resulting from the long and cold winters tended to cool the air and produce fogs which cut off the sun's rays. The rays thus cut off diminished the melting power of the sun, and so increased the accumulation. As the snow and ice continued to accumulate, more and more of the rays were cut off; and on the other hand, as the rays continued to be cut off, the *rate* of accumulation increased, because the quantity of snow and ice melted became thus annually less and less.

Again, during the long and dreary winters of the glacial epoch the earth would be radiating off its heat into space. Had the heat thus lost simply gone to lower the temperature, the lowering of the temperature would have tended to diminish the rate of loss; but the necessary result of this was the formation of snow and ice rather than the lowering of temperature.

And, again, the formation of snow and ice facilitated the rate at which the earth lost its heat; and on the other hand, the more rapidly the earth parted with its heat, the more rapidly were the snow and ice formed.

Further, as the snow and ice accumulated on the one hemisphere, they at the same time continued to diminish on the other. This tended to increase the strength of the trade-winds on the cold hemisphere, and to weaken those on the warm. The effect of this on ocean currents would be to impel the warm water of the tropics more to the warm hemisphere than to the cold. Suppose the northern hemisphere to be the cold one, then as the snow and ice began gradually to accumulate there, the ocean currents of that hemisphere would begin to decrease in volume, while those on the southern, or warm, hemisphere, would *pari passu* increase. This withdrawal of heat from the northern hemisphere would tend, of course, to lower the temperature of that hemisphere and thus favour the accumulation of snow and ice. As the snow and ice accumulated the ocean currents would decrease, and, on the other hand, as the ocean currents diminished the snow and ice would accumulate,—the two effects mutually strengthening each other.

The same must have held true in regard to aërial currents. The more the polar and temperate regions became covered with snow and ice, the stronger would become the trades and anti-trades of the hemisphere; and the stronger those winds became, the greater would be the amount of moisture transferred from the tropical regions by the anti-trades to the temperate regions; and on the other hand, the more moisture those winds brought to temperate regions, the greater would be the quantity of snow produced.

The same process of mutual action and reaction would take place among the agencies in operation on the warm hemisphere, only the result produced would be diametrically opposite of that produced in the cold hemisphere. On this warm hemisphere action and reaction would tend to raise the mean temperature and diminish the quantity of snow and ice existing in temperate and polar regions.

Had it been possible for each of those various physical agents which we have been considering to produce its direct effects without influencing the other agents or being influenced by them, its real efficiency in bringing about either the glacial condition of climate or the warm condition of climate would not have been so great.

The primary cause that set all those various physical agencies in operation which brought about the glacial epoch, was a high state of eccentricity of the earth's orbit. When the eccentricity is at a high value, snow and ice begin to accumulate, owing to the increasing length and coldness of the winter on that hemisphere whose winter solstice is approaching toward the aphelion. The accumulating snow then begins to bring into operation all the various agencies which we have been describing; and, as we have just seen, these, when once in full operation, mutually aid one another. As the eccentricity increases century by century, the temperate regions become more and more covered with snow and ice, first by reason of the continued increase in the coldness and length of the winters, and secondly, and chiefly, owing to the continued increase in the potency of those physical agents which have been called into operation. This glacial state of things goes on at an increasing rate, and reaches a maximum when the solstice point arrives at the aphelion. After the solstice passes the aphelion, a contrary process commences. The snow and ice gradually begin to diminish on the cold hemisphere and to make their appearance on the other hemisphere. The glaciated hemisphere turns, by degrees, warmer and the warm hemisphere colder, and this continues to go on for a period of ten or twelve thousand years, until the winter solstice reaches the perihelion. By this time the conditions of the two hemispheres have been reversed; the formerly glaciated hemisphere has now become the warm one, and the warm hemisphere the glaciated. The transference of the ice from the one hemisphere to the other continues as long as the eccentricity remains at a high value. This will, perhaps, be better understood from an inspection of the frontispiece.

The Mean Temperature of the whole Earth should be greater in Aphelion than in Perihelion.—When the eccentricity becomes reduced to about its present value, its influence on climate is but little felt. It is, however, probable that the present extension of ice on the southern hemisphere may, to a considerable extent, be the result of eccentricity. The difference in the climatic conditions of the two hemispheres is just what should be according to theory:—(1) The mean temperature of that hemisphere is less than that of the northern. (2) The winters of the southern hemisphere are colder than those of the northern. (3) The summers, though occurring in perihelion, are also comparatively cold; this, as we have seen, is what ought to be according to theory. (4) The mean temperature of the whole earth is greater in June, when the earth is in aphelion, than in December, when it is in perihelion. This, I venture to affirm, is also what ought to follow according to theory, although this very fact has been adduced as a proof that eccentricity has at present but little effect on the climatic condition of our globe.

That the mean temperature of the whole earth would, during the glacial epoch, be greater when the earth was in aphelion than when in perihelion will, I think, be apparent from the following considerations:—When the earth was in the perihelion, the sun would be over the hemisphere nearly covered with snow and ice. The great strength of the sun's rays would in this case have little effect in raising the temperature; it would be spent in melting the snow and ice. But when the earth was in the aphelion, the sun would be over the hemisphere comparatively free, or perhaps wholly free, from snow and ice. Consequently, though the intensity of the sun's rays would be less than when the earth was in perihelion, still it ought to have produced a higher temperature, because it would be chiefly employed in heating the ground and not consumed in melting snow and ice.

Professor Tyndall on the Glacial Epoch.—"So natural," says Professor Tyndall, "was the association of ice and cold, that even celebrated men assumed that all that is needed to produce a great extension of our glaciers is a diminution of the sun's temperature. Had they gone through the foregoing reflections and calculations, they would probably have demanded *more* heat instead of less for the production of a glacial epoch. What they really needed were *condensers* sufficiently powerful to congeal the vapour generated by the heat of the sun." (*The Forms of Water*, p. 154. See also, to the same effect, *Heat Considered as a Mode of Motion*, chap. vi.)

I do not know to whom Professor Tyndall here refers, but certainly his remarks have no application to the theory under consideration, for according to it, as we have just seen, the ice of the glacial epoch was about as much due to the nearness of the sun in perigee as to his great distance in apogee.

There is one theory, however, to which his remarks justly apply, viz., the theory that the great changes of climate during geological ages resulted from the passage of our globe through different temperatures of space. What Professor Tyndall says shows plainly that the glacial epoch was not brought about by our earth passing through a cold part of space. A general reduction of temperature over the whole globe certainly would not produce a glacial epoch. Suppose the sun were extinguished and our globe exposed to the temperature of stellar space (−239° F.), this would certainly freeze the ocean solid from its surface to its bottom, but it would not cover the land with ice.

Professor Tyndall's conclusions are, of course, equally conclusive against Professor Balfour Stewart's theory, that the glacial epoch may have resulted from a general diminution in the intensity of the sun's heat.

Nevertheless it would be in direct opposition to the well-established facts of geology to assume that the ice periods of the glacial epoch were warm periods. We are as certain from palæontological evidence that the cold was then much greater than now, as we are from physical evidence that the accumulation of ice was greater than now. Our glacial shell-beds and remains of the mammoth, the reindeer, and musk-ox, tell of cold as truly as the markings on the rocks do of ice.

Objection from the Present Condition of the Planet Mars.—It has been urged as an objection by Professor Charles Martins* and others, that if a high state of eccentricity could produce a glacial epoch, the planet Mars ought to be at present under a glacial condition. The eccentricity of its orbit amounts to 0·09322, and one of its southern winter solstices is, according to

* *Revue des Deux Mondes* for 1867.

Dr. Oudemans, of Batavia,* within 17° 41′ 8″ of aphelion. Consequently, it is supposed that one of the hemispheres should be in a glacial state and the other free from snow and ice. But it is believed that the snow accumulates around each pole during its winter and disappears to a great extent during its summer.

There would be force in this objection were it maintained that eccentricity alone can produce a glacial condition of climate, but such is not the case, and there is no good ground for concluding that those physical agencies which led to the glacial epoch of our globe exist in the planet Mars. It is perfectly certain that either water must be different in constitution in that planet from what it is in our earth, or else its atmospheric envelope must be totally different from ours. For it is evident from what has been stated in Chapter II., that were our globe to be removed to the distance of Mars from the sun, the lowering of the temperature resulting from the decrease in the sun's heat would not only destroy every living thing, but would convert the ocean into solid ice.

But it must be observed that the eccentricity of Mars' orbit is at present far from its superior limit of 0·14224, and it may so happen in the economy of nature that when it approaches to that limit a glacial condition of things may supervene.

The truth is, however, that very little seems to be known with certainty regarding the climatic condition of Mars. This is obvious from the fact that some astronomers believe that the planet possesses a dense atmosphere which protects it from cold; while others maintain that its atmosphere is so exceedingly thin that its mean temperature is below the freezing-point. Some assert that the climatic condition of Mars resembles very much that of our earth, while others affirm that its seas are actually frozen solid to the bottom, and the poles covered with ice thirty or forty miles in thickness. For reasons which will be explained in the Appendix, Mars, notwithstanding its greater distance from the sun, may enjoy a climate as warm as that of our earth.

* Letter to the author, February, 1870.

CHAPTER V.

REASON WHY THE SOUTHERN HEMISPHERE IS COLDER THAN THE NORTHERN.

Adhémar's Explanation.—Adhémar's Theory founded upon a physical Mistake in regard to Radiation.—Professor J. D. Forbes on Underground Temperature.—Generally accepted Explanation.—Low Temperature of Southern Hemisphere attributed to Preponderance of Sea.—Heat transferred from Southern to Northern Hemisphere by Ocean-current the true Explanation. —A large Portion of the Heat of the Gulf-stream derived from the Southern Hemisphere.

Adhémar's Explanation.—It has long been known that on the southern hemisphere the temperature is lower and the accumulation of ice greater than on the northern. This difference has usually been attributed to the great preponderance of sea on the southern hemisphere. M. Adhémar, on the other hand, attempts to explain this difference by referring it to the difference in the amount of heat lost by the two hemispheres in consequence of the difference of seven days in the length of their respective winters. As the northern winter is shorter than the summer, he concludes that there is an accumulation of heat on that hemisphere, while, on the other hand, the southern winter being longer than the summer, there is therefore a loss of heat on the southern hemisphere. "The south pole," he says, "loses in one year more heat than it receives, because the total duration of its night surpasses that of its day by 168 hours; and the contrary takes place for the north pole. If, for example, we take for unity the mean quantity of heat which the sun sends off in one hour, the heat accumulated at the end of the year at the north pole will be expressed by 168, while the heat lost by the south pole will be equal to 168 times what the radiation lessens it by in one hour, so that at the end of the year the difference in the heat of the two hemispheres will be represented by 336 times what the earth receives from the sun or loses in an hour by radiation."*

Adhémar supposes that about 10,000 years hence, when our northern winter will occur in aphelion and the southern in perihelion, the climatic conditions of the two hemispheres will be reversed; the ice will melt at the south pole, and the northern hemisphere will become enveloped in one continuous mass of ice, leagues in thickness, extending down to temperate regions.

This theory seems to be based upon an erroneous interpretation of a principle, first pointed out, so far as I am aware, by Humboldt in his memoir "On Isothermal Lines and Distribution of Heat over the Globe."† This principle may be stated as follows:—

Although the total quantity of heat received by the earth from the sun in one revolution is inversely proportional to the minor axis of the orbit, yet this amount, as was proved by D'Alembert, is equally distributed between the two hemispheres, whatever the eccentricity may be. Whatever extra heat the southern hemisphere may at present receive from the sun daily during its summer months owing to greater proximity to the sun, is exactly compensated by a corresponding loss arising from the shortness of the season; and, on the other hand, whatever daily deficiency of heat we in the northern hemisphere may at present have during our summer half-year, in consequence of the earth's distance from the sun, is also exactly compensated by a corresponding length of season.

But the surface temperature of our globe depends as much upon the amount of heat radiated into space as upon the amount derived from the sun, and it has been thought by some that this compensating principle holds true only in regard to the

* "Révolutions de la Mer," p. 37 (second edition).
† Edin. Phil. Journ., vol. iv., p. 262 (1821).

latter. In the case of the heat lost by radiation the reverse is supposed to take place. The southern hemisphere, it is asserted, has not only a colder winter than the northern in consequence of the sun's greater distance, but it has also a longer winter; and the extra loss of heat from radiation during winter is not compensated by its nearness to the sun during summer, for it gains no additional heat from this proximity. And in the same way it is argued that as our winter in the northern hemisphere, owing to the less distance of the sun, is not only warmer than that of the southern hemisphere, but is also at the same time shorter, so our hemisphere is not cooled to such an extent as the southern. And thus the mean temperature of the winter half-year, as well as the intensity of the sun's heat, is affected by a change in the sun's distance.

Although I always regarded this cause of Humboldt's to be utterly inadequate to produce such effects as those attributed to it by Adhémar, still, in my earlier papers * I stated it to be a *vera causa* which ought to produce some sensible effect on climate. But shortly afterwards on a more careful consideration of the whole subject, I was led to suspect that the circumstance in question can, according to theory, produce little or no effect on the climatic condition of our globe.

As there appears to be a considerable amount of misapprehension in reference to this point, which forms the basis of Adhémar's theory, I may here give it a brief consideration.†

The rate at which the earth radiates into space the heat received from the sun depends upon the temperature of its surface; and the temperature of its surface (other things being equal) depends upon the rate at which the heat is received. The greater the rate at which the earth receives heat from the sun, the greater will therefore be the rate at which it will lose that heat by radiation. Now the total quantity of heat received during winter by the southern hemisphere is exactly

* Phil. Mag., § 4, vol. xxviii., p. 131. *Reader*, December 2nd, 1865.

† This point will be found discussed at considerable length in the Phil. Mag. for September, 1869.

equal to that received during winter by the northern. But as the southern winter is longer than the northern the rate at which the heat is received, and consequently the rate of radiation, during that season must be less on the southern hemisphere than on the northern. Thus the southern hemisphere loses heat during a longer period than the northern, and therefore the less rate of radiation (were it not for a circumstance presently to be noticed) would wholly compensate for the longer period, and the total quantity of heat lost during winter would be the same on both hemispheres. The southern summer is shorter than the northern, but the heat is more intense, and the surface of the ground kept at a higher temperature; consequently the rate of radiation into space is greater.

When the rate at which a body receives heat is increased, the temperature of the body rises till the rate of radiation equals the rate of absorption, after which equilibrium is restored; and when the rate of absorption is diminished, the temperature falls till the rate of radiation equals that of absorption.

But notwithstanding all this, owing to the slow conductivity of the ground for heat, more heat will pass into it during the longer summer of aphelion than during the shorter one of perihelion; for the amount of heat which passes into the ground depends on the length of time during which the earth is receiving heat, as well as upon the amount received. In like manner, more heat will pass out of the ground during the longer winter in aphelion than during the shorter one in perihelion. Suppose the length of the days on the one hemisphere (say the northern) to be 23 hours, and the length of the nights, say one hour; while on the other hemisphere the days are one hour and the nights 23 hours. Suppose also that the quantity of heat received from the sun by the southern hemisphere during the day of one hour to be equal to that received by the northern hemisphere during the day of 23 hours. It is evident that although the surface of the ground on the southern hemisphere would receive as much heat from the sun during the short day of one hour as the surface of the northern hemisphere during the long day of 23 hours, yet, owing to the slow conductivity of the ground for heat, the amount absorbed would not be nearly so much on the southern hemisphere as on the northern. The temperature of the surface during the day, it is true, would be far higher on the southern hemisphere than on the northern, and consequently the rate at which the heat would pass into the ground would be greater on that hemisphere than on the northern; but, notwithstanding the greater rate of absorption resulting from the high temperature of the surface, it would not compensate for the shortness of the day. On the other hand, the surface of the ground on the southern hemisphere would be colder during the long night of 23 hours than it would be on the northern during the short night of only one hour; and the low temperature of the ground would tend to lessen the rate of radiation into space. But the decrease in the rate of radiation would not compensate fully for the great length of the night. The general and combined result of all those causes would be that a slight accumulation of heat would take place on the northern hemisphere and a slight loss on the southern. But this loss of heat on the one hemisphere and gain on the other would not go on accumulating at a uniform rate year by year, as Adhémar supposes.

Of course we are at present simply considering the earth as an absorber and radiator of heat, without taking into account the effects of distribution of sea and land and other modifying causes, and are assuming that everything is the same in both hemispheres, with the exception that the winter of the one hemisphere is longer than that of the other.

What, then, is the amount of heat stored up by the one hemisphere and lost by the other? Is it such an amount as to sensibly affect climate?

The experiments and observations which have been made on underground temperature afford us a means of making at least a rough estimate of the amount. And from these it will be seen that the influence of an excess of seven or eight days in the length of the southern winter over the northern could hardly produce an effect that would be sensible.

Observations were made at Edinburgh by Professor J. D. Forbes on three different substances; viz., sandstone, sand, and trap-rock. By calculation, we find from the data afforded by those observations that the total quantity of heat accumulated in the ground during the summer above the mean temperature was as follows :—In the sandstone-rock, a quantity sufficient to raise the temperature of the rock 1° C. to a depth of 85 feet 6 inches; in the sand a quantity sufficient to raise the temperature 1° C. to a depth of 72 feet 6 inches; and in the trap-rock a quantity only sufficient to raise the temperature 1° C. to a depth of 61 feet 6 inches.

Taking the specific heat of the sandstone per unit volume, as determined by Regnault, at ·4623, and that of sand at ·3006, and trap at 5283, and reducing all the results to one standard, viz., that of water, we find that the quantity of heat stored up in the sandstone would, if applied to water, raise its temperature 1° C. to a depth of 39 feet 6 inches; that stored up in the sand would raise the temperature of the water 1° C. to a depth of 21 feet 8 inches, and that stored up in the trap would raise the water 1° C. to the depth of 32 feet 6 inches. We may take the mean of these three results as representing pretty accurately the quantity stored up in the general surface of the country. This would be equal to 31 feet 3 inches depth of water raised 1° C. The quantity of heat lost by radiation during winter below the mean was found to be about equal to that stored up during summer.

The total quantity of heat per square foot of surface received by the equator from sunrise till sunset at the time of the equinoxes, allowing 22 per cent. for the amount cut off in passing through the atmosphere, is 1,780,474 foot-pounds. In the latitude of Edinburgh about 938,460 foot-pounds per square foot of surface is received, assuming that not more than 22 per cent. is cut off by the atmosphere. At this rate a quantity of heat would be received from the sun in two days ten hours (say, three days) sufficient to raise the temperature of the water 1° C. to the required depth of 31 feet 3 inches. Consequently the total quantity of heat stored up during summer in the latitude of Edinburgh is only equal to what we receive from the sun during three days at the time of the equinoxes. Three days' sunshine during the middle of March or September, if applied to raise the temperature of the ground, would restore all the heat lost during the entire winter; and another three days' sunshine would confer on the ground as much heat as is stored up during the entire summer. But it must be observed that the total duration of sunshine in winter is to that of summer in the latitude of Edinburgh only about as 4 to 7. Here is a difference of two months. But this is not all; the quantity of heat received during winter is scarcely one-third of that received during summer; yet, notwithstanding this enormous difference between summer and winter, the ground during winter loses only about six days' sun-heat below the maximum amount possessed by it in summer.

But if what has already been stated is correct, this loss of heat sustained by the earth during winter is not chiefly owing to radiation during the longer absence of the sun, but to the decrease in the quantity of heat received in consequence of his longer absence combined with the obliquity of his rays during that season. Now in the case of the two hemispheres, although the southern winter is longer than the northern, yet the quantity of heat received by each is the same. But supposing it held true, which it does not, that the loss of heat sustained by the earth in winter is as much owing to radiation resulting from the excess in the length of the winter nights over those of the summer as to the deficiency of heat received in winter from that received in summer, three days' heat would then in this case be the amount lost by radiation in consequence of this excess in the length of the winter nights. The total length of the winter nights to those of the summer is, as we have seen, about as 7 to 4. This is a difference of nearly 1200 hours. But the excess of the south polar winter over the north amounts to only about 184 hours. Now if 1200 hours give a loss of three days' sun-heat, 184 hours will give a loss of scarcely $5\frac{1}{2}$ hours.

It is no doubt true that the two cases are not exactly analogous; but it is obvious that any error which can possibly arise from regarding them as such cannot materially alter the conclusion to which we have arrived. Supposing the effect were double, or even quadruple, what we have concluded it to be, still it would not amount to a loss of two days' heat, which could certainly have little or no influence on climate.

But even assuming all the preceding reasoning to be incorrect, and that the southern hemisphere, in consequence of its longer winter, loses heat to the extravagant extent of 168 hours, supposed by Adhémar, still this could not materially affect climate. The climate is influenced by the mere *temperature* of the *surface* of the ground, and not by the quantity of heat or cold that may be stored up under the surface. The climate is determined, so far as the ground is concerned, by the temperature of the surface, and is wholly independent of the temperature which may exist under the surface. Underground temperature can only affect climate through the surface. If the surface could, for example, be kept covered with perpetual snow, we should have a cold and sterile climate, although the temperature of the ground under the snow was actually at the boiling-point. Let the ground to a depth of, say 40 or 50 feet, be deprived of an amount of heat equal to that received from the sun in 168 hours. This could produce little or no sensible effect on climate; for, owing to the slow conductivity of the ground for heat, this loss would not sensibly affect the temperature of the surface, as it would take several months for the sun's heat to penetrate to that depth and restore the lost heat. The cold, if I may be allowed to use the expression, would come so slowly out to the surface that its effect in lowering the temperature of the surface would scarcely be sensible. And, again, if we suppose the 168 hours' heat to be lost by the mere surface of the ground, the effect would certainly be sensible, but it would only be so for a few days. We might in this case have a week's frozen soil, but that would be all. Before the air had time to become very sensibly affected by the low temperature of the surface the frozen soil would be thawed.

The storing up of heat or cold in the ground has in reality very little to do with climate. Some physicists explain, for example, why the month of July is warmer than June by referring it to the fact that by the month of July the ground has become possessed of a larger accumulation of heat than it possessed in June. This explanation is evidently erroneous. The ground in July certainly possesses a greater store of heat than it did in June; but this is not the reason why the former month is hotter than the latter. July is hotter than June because the *air* (not the *ground*) has become possessed of a larger store of heat than it had in June. Now the air is warmer in July than in June because, receiving little increase of temperature from the direct rays of the sun, it is heated chiefly by radiation from the earth and by contact with its warm surface. Consequently, although the sun's heat is greater in June than it is in July, it is near the middle of July before the air becomes possessed of its maximum store of heat. We therefore say that July is hotter than June because the air is hotter, and consequently the temperature in the shade is greater in the former month than in the latter.

It is therefore, I presume, quite apparent that Adhémar's theory fails to explain why the southern hemisphere is colder than the northern.

The generally accepted Explanation.—The difference in the mean temperature of the two hemispheres is usually attributed to the proportion of sea to land in the southern hemisphere and of land to sea in the northern hemisphere. This, no doubt, will account for the greater *annual range* of temperature on the northern hemisphere, but it seems to me that it will not account for the excess of *mean* temperature possessed by that hemisphere over the southern.

The general influence of land on climate is to exaggerate the variation of temperature due to the seasons. On continents the summers are hotter and the winters colder than on the ocean. The days are also hotter and the nights colder on land than on sea. This is a result which follows from the mere physical properties of land and water, independently of currents, whether of ocean or of air. But it nevertheless follows, according to theory (and this is a point which has been overlooked), that the mean annual temperature of the ocean ought to be greater than that of the land in equatorial regions as well as in temperate and polar regions. This will appear obvious for the following reasons:—(1) The ground stores up heat only by the slow process of conduction, whereas water, by the mobility of its particles and its transparency for heat-rays, especially those from the sun, becomes heated to a considerable depth rapidly. The quantity of heat stored up in the ground is thus comparatively small, while the quantity stored up in the ocean is great. (2) The air is probably heated more rapidly by contact with the ground than with the ocean; but, on the other hand, it is heated far more rapidly by radiation from the ocean than from the land. The aqueous vapour of the air is to a great extent diathermanous to radiation from the ground, while it absorbs the rays from water and thus becomes heated. (3) The air radiates back a considerable portion of its heat, and the ocean absorbs this radiation from the air more readily than the ground does. The ocean will not reflect the heat from the aqueous vapour of the air, but absorbs it, while the ground does the opposite. Radiation from the air, therefore, tends more readily to heat the ocean than it does the land. (4) The aqueous vapour of the air acts as a screen to prevent the loss by radiation from water, while it allows radiation from the ground to pass more freely into space; the atmosphere over the ocean consequently throws back a greater amount of heat than is thrown back by the atmosphere over the land. The sea in this case has a much greater difficulty than the land has in getting quit of the heat received from the sun; in other words, the land tends to lose its heat more rapidly than the sea. The consequence of all these circumstances is that the ocean must stand at a higher mean temperature than the land. A state of equilibrium is never gained until the rate at which a body is receiving heat is equal to the rate at which it is losing it; but as equal surfaces of sea and land receive from the sun the same amount of heat, it therefore follows that, in order that the sea may get quit of its heat as rapidly as the land, it *must stand at a higher temperature* than the land. The temperature of the sea must continue to rise till the amount of heat thrown off into space equals that received from the sun; when this point is reached, equilibrium is established and the temperature remains stationary. But, owing to the greater difficulty that the sea has in getting rid of its heat, the mean temperature of equilibrium of the ocean must be higher than that of the land; consequently the mean temperature of the ocean, and also of the air immediately over it, in tropical regions should be higher than the mean temperature of the land and the air over it.

The greater portion of the southern hemisphere, however, is occupied by water, and why then, it may be asked, is this water hemisphere colder than the land hemisphere? Ought it uot also to follow that the sea in inter-tropical regions should be warmer than the land under the same parallels; yet, as we know, the reverse is actually found to be the case. How then is all this to be explained, if the foregoing reasoning be correct? We find when we examine Professor Dove's charts of mean annual temperature, that the ocean in inter-tropical regions has a mean annual temperature below the normal, and the land a mean annual temperature above the normal. Both in the Pacific and in the Atlantic the mean temperature sinks to $2^{\circ}\cdot3$ below the normal, while on the land it rises $4^{\circ}\cdot6$ above the normal. The explanation in this case is obviously this: the temperature of the ocean in inter-tropical regions, as we have already seen, is kept much lower than it would otherwise be by the enormous amount of *heat* that is being constantly carried away from those regions into temperate and polar regions, and of *cold* that is being constantly carried from temperate and polar regions to the tropical regions by means of ocean-currents. The same principle which explains why the sea in inter-tropical regions has a lower mean annual temperature than the land, explains also why the southern hemisphere has a lower mean annual temperature than the northern. The temperature of the southern hemisphere is lowered by the transference of heat by means of ocean-currents.

Heat transferred from the Southern to the Northern Hemisphere by Ocean-currents the true Explanation.—The great ocean-currents of the globe take their rise in three immense streams from the Southern Ocean, which, on reaching the tropical regions, become deflected in a westerly direction and flow along the southern side of the equator for thousands of miles. Perhaps more than one half of this mass of moving water returns into the Southern Ocean without ever crossing the equator, but the quantity which crosses over to the northern hemisphere is enormous. This constant flow of water from the southern hemisphere to the northern in the form of surface currents must be compensated by *under currents* of equal magnitude from the northern hemisphere to the southern. The currents, however, which cross the equator are far higher in temperature than their compensating under currents; consequently there is a constant transference of heat from the southern hemisphere to the northern. Any currents taking their rise in the northern hemisphere and flowing across into the southern are comparatively trifling, and the amount of heat transferred by them is also trifling. There are one or two currents of considerable size, such as the Brazilian branch of the great equatorial current of the Atlantic, and a part of the South Equatorial Drift-current of the Pacific, which cross the equator from north to south; but these cannot be regarded as northern currents; they are simply southern currents deflected back after crossing over to the northern hemisphere. The heat which these currents possess is chiefly obtained on the southern hemisphere before crossing over to the northern; and although the northern hemisphere may not gain much heat by means of them, it, on the other hand, does not lose much, for the heat which they give out in their progress along the southern hemisphere does not belong to the northern hemisphere.

But, after making the fullest allowance for the amount of heat carried across the equator from the northern hemisphere to the southern, we shall find, if we compare the mean temperature of the currents from south to north with that of the great compensating under currents and the one or two small surface currents, that the former is very much higher than the latter. The mean temperature of the water crossing the equator from south to north is probably not under 65°, that of the under currents is probably not over 39°. But to the under currents we must add the surface currents from north to south; and assuming that this will raise the mean temperature of the entire mass of water flowing south to, say, 45°, we have still a difference of 20° between the temperature of the masses flowing north and south. Each cubic foot of water which crosses the equator will in this case transfer about 965,000 foot-pounds of heat from the southern hemisphere to the northern. If we had any means of ascertaining the volume of those great currents crossing the equator, we should then be able to make a rough estimate of the total amount of heat transferred from the southern hemisphere to the northern; but as yet no accurate estimate has been made on this point. Let us assume, what is probably below the truth, that the total amount of water crossing the equator is at least double that of the Gulf-stream as it passes through the Straits of Florida, which amount we have already found to be equal to 66,908,160,000,000 cubic feet daily. Taking the quantity of heat conveyed by each cubic foot of water of the Gulf-stream as 1,158,000 foot-pounds, it is found, as we have seen, that an amount of heat is conveyed by this current equal to all the heat that falls within 32 miles on each side of the equator. Then, if each cubic foot of water crossing the equator transfers 965,000 foot-pounds, and the quantity of water be double that of the Gulf-stream, it follows that the amount of heat transferred from the southern hemisphere to the northern is equal to all the heat falling within 52 miles on each side of the equator, or equal to all the heat falling on the southern hemisphere within 104 miles of the equator. This quantity taken from the southern hemisphere and added to the northern will therefore make a difference in the amount of heat possessed by the two hemispheres equal to all the heat which falls on the southern hemisphere within somewhat more than 208 miles of the equator.

A large Portion of the Heat of the Gulf-stream derived from the Southern Hemisphere.—It can be proved that a very large portion of the heat conveyed by the Gulf-stream comes from the southern hemisphere. The proof is as follows:—

If all the heat came from the northern hemisphere, it could only come from that portion of the Atlantic, Caribbean Sea, and Gulf of Mexico which lies to the north of the equator. The entire area of these seas, extending to the Tropic of Cancer, is about 7,700,000 square miles. But this area is not sufficient to supply the current passing through the "Narrows" with the necessary heat. Were the heat which passes through the Straits of Florida derived exclusively from this area, the following table would then represent the relative quantity per unit surface possessed by the Atlantic in the three zones, assuming that one half of the heat of the Gulf-stream passes into the arctic regions and the other half remains to warm the temperate regions* :—

From the equator to the Tropic of Cancer	773
From the Tropic of Cancer to the Arctic Circle . . .	848
From the Arctic Circle to the North Pole	610

These figures show that the Atlantic, from the equator to the Tropic of Cancer, would be as cold as from the Tropic of Cancer to the North Pole, were it not that a large proportion of the heat possessed by the Gulf-stream is derived from the southern hemisphere.

* See Phil. Mag. for October, 1870, p. 259.

CHAPTER VI.

EXAMINATION OF THE GRAVITATION THEORY OF OCEANIC CIRCULATION.—LIEUT. MAURY'S THEORY.

Introduction.—Ocean-currents, according to Maury, due to Difference of Specific Gravity.—Difference of Specific Gravity resulting from Difference of Temperature.—Difference of Specific Gravity resulting from Difference of Saltness.—Maury's two Causes neutralize each other.—How, according to him, Difference in Saltness acts as a Cause.

Introduction.—Few subjects have excited more interest and attention than the cause of ocean circulation; and yet few are in a more imperfect and unsatisfactory condition, nor is there any question regarding which a greater diversity of opinion has prevailed. Our incomplete acquaintance with the facts relating to the currents of the ocean and the modes of circulation actually in operation, is no doubt one reason for this state of things. But doubtless the principal cause of such diversity of opinion lies in the fact that the question is one which properly belongs to the domain of physics and mechanics, while as yet no physicist of note (if we except Dr. Colding, of Copenhagen) has given, as far as I know, any special attention to the subject. It is true that in works of meteorology and physical geography reference is continually made to such eminent physicists as Herschel, Pouillet, Buff, and others; but when we turn to the writings of these authors we find merely a few remarks expressive of their opinions on the subject, and no special discussion or investigation of the matter, nor anything which could warrant us in concluding that such investigations have ever been made. At present the question cannot be decided by a reference to authorities.

The various theories on the subject may be classed under two divisions; the first of these attributes the motion of the water to the *impulse of the wind,* and the second to the *force of gravity* resulting from difference of density. But even amongst those who adopt the former theory, it is generally held that the winds are not the sole cause, but that, to a certain extent at least, difference of specific gravity contributes to produce motion of the waters. This is a very natural conclusion; and in the present state of physical geography on this subject one can hardly be expected to hold any other view.

The supporters of the latter theory may be subdivided into two classes. The first of these (of which Maury may be regarded as the representative) attributes the Gulf-stream, and other sensible currents of the ocean, to difference of specific gravity. The other class (at present the more popular of the two, and of which Dr. Carpenter may be considered the representative) denies altogether that such currents can be produced by difference of specific gravity,* and affirms that there is a general movement of the upper portion of the ocean from the equator to the poles, and a counter-movement of the under portion from the poles to the equator. This movement is attributed to difference of specific gravity between equatorial and polar water, resulting from difference of temperature.

The widespread popularity of the gravitation theory is no doubt, to a great extent, owing to the very great prominence given to it by Lieut. Maury in his interesting and popular work, "The Physical Geography of the Sea." Another cause which must have favoured the reception of this theory is the ease with which it is perceived how, according to it, circulation of the waters of the ocean is supposed to follow. One has no difficulty, for example, in perceiving that if the inter-tropical waters of the ocean are expanded by heat, and the waters around the poles contracted by cold, the surface of the ocean will stand at a higher level at the equator than at the poles. Equilibrium being thus disturbed, the water at the equator

* Proceedings of the Royal Society, No. 138, p. 596, foot-note.

will tend to flow towards the poles as a surface current, and the water at the poles towards the equator as an under current. This, at first sight, looks well, especially to those who take but a superficial view of the matter.

We shall examine this theory at some length, for two reasons: 1, because it lies at the root of a great deal of the confusion and misconception which have prevailed in regard to the whole subject of ocean-currents: 2, because, if the theory is correct, it militates strongly against the physical theory of secular changes of climate advanced in this volume. We have already seen (Chapter IV.) that when the eccentricity of the earth's orbit reaches a high value, a combination of physical circumstances tends to lower the temperature of the hemisphere which has its winter solstice in aphelion, and to raise the temperature of the opposite hemisphere, whose winter solstice will, of course, be in perihelion. The direct result of this state of things, as was shown, is to strengthen the force of the trade-winds on the cold hemisphere, and to weaken their strength on the warm hemisphere: and this, in turn, we also saw, tends to impel the warm water of the inter-tropical region on to the warm hemisphere, and to prevent it, in a very large degree, from passing into the cold hemisphere. This deflection of the ocean-currents tends to an enormous extent to increase the difference of temperature previously existing between the two hemispheres. In other words, the warm and equable condition of the one hemisphere, and the cold and glacial condition of the other, are, to a great extent, due to this deflection of ocean-currents. But if the theory be correct which attributes the motion of ocean-currents to a difference in density between the sea in inter-tropical and polar regions, then it follows that these currents (other things being equal) ought to be stronger on the cold hemisphere than on the warm, because there is a greater difference of temperature and, consequently, a greater difference of density, between the polar seas of the cold hemisphere and the equatorial seas, than between the polar seas of the warm hemisphere and the equatorial seas. And this being the case, notwithstanding the influence of the trade-winds of the cold hemisphere blowing over upon the warm, the currents will, in all probability, be stronger on the cold hemisphere than on the warm. In other words, the influence of the powerful trade-winds of the cold hemisphere to transfer the warm water of the equator to the warm hemisphere will probably be more than counterbalanced by the tendency of the warm and buoyant waters of the equator to flow towards the dense and cold waters around the pole of the cold hemisphere. But if ocean-currents are due not to difference in specific gravity, but to the influence of the winds, then it is evident that the waters at the equator will be impelled, not into the cold hemisphere, but into the warm.

For this reason I have been the more anxious to prove that inter-tropical heat is conveyed to temperate and polar regions by ocean-currents, and not by means of any general movement of the ocean resulting from difference of gravity. I shall therefore on this account enter more fully into this part of the subject than I otherwise would have done. Irrespective of all this, however, the important nature of the whole question, and the very general interest it excites, warrant a full consideration of the subject.

I shall consider first that form of the gravitation theory advocated by Maury in his work on the "Physical Geography of the Sea," which attributes the motion of the Gulf-stream and other sensible currents of the ocean to differences of specific gravity. One reason which has induced me to select Maury's work is, that it not only contains a much fuller discussion on the cause of the motion of ocean-currents than is to be found anywhere else, but also that it has probably passed through a greater number of editions than any other book of a scientific character in the English language in the same length of time.

Examination of Lieut. Maury's Gravitation Theory.—Although Lieut. Maury has expounded his views on the cause of ocean-currents at great length in the various editions of his work, yet it is somewhat difficult to discover what they really are. This arises chiefly from the generally confused and sometimes contradictory nature of his hydrodynamical conceptions. After a repeated perusal of several editions of his book, the following, I trust, will be found to be a pretty accurate representation of his theory :—

Ocean-currents, according to Maury, due to Difference of Specific Gravity.—Although Maury alludes to a number of causes which, he thinks, tend to produce currents, yet he deems their influence so small that, practically, all currents may be referred to difference of specific gravity.

"If we except," he says, "the tides, and the partial currents of the sea, such as those that may be created by the wind, we may lay it down as a rule that all the currents of the ocean owe their origin to the differences of specific gravity between sea-water at one place and sea-water at another ; for wherever there is such a difference, whether it be owing to difference of temperature or to difference of saltness, &c., it is a difference that disturbs equilibrium, and currents are the consequence" (§ 467) *. To the same effect see §§ 896, 37, 512, 520, and 537.

Notwithstanding the fact that he is continually referring to difference of specific gravity as the great cause of currents, it is difficult to understand in what way he conceives this difference to act as a cause.

Difference of specific gravity between the waters of the ocean at one place and another can give rise to currents only through the influence of the earth's gravity. All currents resulting from difference of specific gravity can be ultimately resolved into the general principle that the molecules that are specifically heavier *descend* and displace those that are specifically lighter. If, for example, the ocean at the equator be expanded by heat or by any other cause, it will be forced by the denser waters in temperate and polar regions to rise so that its surface shall stand at a higher level than the surface of the ocean in

* The edition from which I quote, unless the contrary is stated, is the one published by Messrs. T. Nelson and Sons, 1870, which is a reprint of the new edition published in 1859 by Messrs. Sampson Low and Co.

these regions. The surface of the ocean will become an inclined plane, sloping from the equator to the poles. Hydrostatically, the ocean, considered as a mass, will then be in a state of equilibrium; but the individual molecules will not be in equilibrium. The molecules at the surface in this case may be regarded as lying on an inclined plane sloping from the equator down to the poles, and as these molecules are at liberty to move they will not remain at rest, but will descend the incline towards the poles. When the waters at the equator are expanded, or the waters at the poles contracted, gravitation makes, as it were, a twofold effort to restore equilibrium. It in the first place sinks the waters at the poles, and raises the waters at the equator, in order that the two masses may balance each other; but this very effort of gravitation to restore equilibrium to the mass destroys the equilibrium of the molecules by disturbing the level of the ocean. It then, in the second place, endeavours to restore equilibrium to the molecules by pulling the lighter surface water at the equator down the incline towards the poles. This tends not only to restore the level of the ocean, but to bring the lighter water to occupy the surface and the denser water the bottom of the ocean; and when this is done, complete equilibrium is restored, both to the mass of the ocean and to its individual molecules, and all further motion ceases. But if heat be constantly applied to the waters of the equatorial regions, and cold to those of the polar regions, and a permanent disturbance of equilibrium maintained, then the continual effort of gravitation to restore equilibrium will give rise to a constant current. In this case, the heat and the cold (the agents which disturb the equilibrium of the ocean) may be regarded as causes of the current, inasmuch as without them the current would not exist; but the real efficient cause, that which impels the water forward, is the force of gravity. But the force of gravity, as has already been noticed, cannot produce motion (perform work) unless the thing acted upon *descend*. Descent is implied in the very conception of a current produced by difference of specific gravity.

But Maury speaks as if difference of specific gravity could give rise to a current without any descent.

"It is not necessary," he says, "to associate with oceanic currents the idea that they must of necessity, as on land, run from a higher to a lower level. So far from this being the case, some currents of the sea actually run up hill, while others run on a level. The Gulf-stream is of the first class" (§ 403). "The top of the Gulf-stream runs on a level with the ocean; therefore we know it is not a descending current" (§ 18). And in § 9 he says that between the Straits of Florida and Cape Hatteras the waters of the Gulf-stream "are actually forced up an inclined plane, whose submarine ascent is not less than 10 inches to the mile." To the same effect see §§ 25, 59.

It is perfectly true that "it is not necessary to associate with ocean-currents the idea that they must of necessity, as on land, run from a higher to a lower level." But the reason of this is that ocean-currents do not, like the currents on land, owe their motion to the force of gravitation. If ocean-currents result from difference of specific gravity between the waters in tropical and polar regions, as Maury maintains, then it is necessary to assume that they are descending currents. Whatever be the cause which may give rise to a difference of specific gravity, the motion which results from this difference is due wholly to the force of gravity; but gravity can produce no motion unless the water *descend.*

This fact must be particularly borne in mind while we are considering Maury's theory that currents are the result of difference of specific gravity.

Ocean-currents, then, according to that writer, owe their existence to the difference of specific gravity between the waters of inter-tropical and polar regions. This difference of specific gravity he attributes to two causes—(1) to difference as to *temperature,* (2) to difference as to saltness. There are one or two causes of a minor nature affecting the specific gravity of the sea, to which he alludes; but these two determine the general result. Let us begin with the consideration of the first of these two causes, viz.:—

Difference of Specific Gravity resulting from Difference of Temperature.—Maury explains his views on this point by means of an illustration. "Let us now suppose," he says, "that all the water within the tropics, to the depth of one hundred fathoms, suddenly becomes oil. The aqueous equilibrium of the planet would thereby be disturbed, and a general system of currents and counter currents would be immediately commenced—the oil, in an unbroken sheet on the surface, running toward the poles, and the water, in an under current, toward the equator. The oil is supposed, as it reaches the polar basin, to be reconverted into water, and the water to become oil as it crosses Cancer and Capricorn, rising to the surface in inter-tropical regions, and returning as before" (§ 20). "Now," he says (§ 22), "do not the cold waters of the north, and the warm waters of the Gulf, made specifically lighter by tropical heat, and which we see actually preserving such a system of counter-currents, hold, at least in some degree, the relation of the supposed water and oil?"

In § 24 he calculates that at the Narrows of Bemini the difference in weight between the volume of the Gulf-water that crosses a section of the stream in one second, and an equal volume of water at the ocean temperature of the latitude, supposing the two volumes to be equally salt, is fifteen millions of pounds. Consequently the force per second operating to propel the waters of the Gulf towards the pole would in this case, he concludes, be the "equilibrating tendency due to fifteen millions of pounds of water in the latitude of Bemini." In §§ 511 and 512 he states that the effect of expanding the waters at the torrid zone by heat, and of contracting the waters at the frigid zone by cold, is to produce a set of surface-currents of warm and light water from the equator towards the poles, and another set of under currents of cooler and heavy water from the poles towards the equator. (See also to the same effect §§ 513, 514, 896.)

There can be no doubt that his conclusion is that the waters in inter-tropical regions are expanded by heat, while those in polar regions are contracted by cold, and that this tends to produce a surface current from the equator to the poles, and an under current from the poles to the equator.

We shall now consider his second great cause of ocean currents, viz. :—

Difference of Specific Gravity resulting from Difference in Degree of Saltness.—Maury maintains, and that correctly, that saltness increases the density of water—that, other things being equal, the saltest water is the densest. He suggests "that one of the purposes which, in the grand design, it was probably intended to accomplish by having the sea salt and not fresh, was to impart to its waters the forces and powers necessary to make their circulation complete" (§ 495).

Now it is perfectly obvious that if difference in saltness is to co-operate with difference in temperature in the production of ocean-currents, the saltest waters, and consequently the densest, must be in the polar regions, and the waters least salt, and consequently lightest, must be in equatorial and inter-tropical regions. Were the saltest waters at the equator, and the freshest at the poles, it would tend to neutralize the effect due to heat, and, instead of producing a current, would simply tend to prevent the existence of the currents which otherwise would result from difference of temperature.

A very considerable portion of his work, however, is devoted to proving that the waters of equatorial and inter-tropical regions are salter and heavier than those of the polar regions; and yet, notwithstanding this, he endeavours to show that this difference in respect to saltness between the waters of the equatorial and the polar regions is one of the chief causes, if not the chief cause, of ocean-currents. In fact, it is for this special end that so much labour is bestowed in proving that the saltest water is in the equatorial and inter-tropical regions, and the freshest in the polar.

"In the present state of our knowledge," he says, "concerning this wonderful phenomenon (for the Gulf-stream is one of the most marvellous things in the ocean) we can do little more than conjecture. But we have two causes in operation which we may safely assume are among those concerned in producing the Gulf-stream. One of these is the increased saltness of its water after the trade-winds have been supplied with vapour from it, be it much or little; and the other is the diminished quantum of salt which the Baltic and the Northern Seas contain" (§ 37). "Now here we have, on one side, the Caribbean Sea and Gulf of Mexico, with their waters of brine; on the other, the great Polar Basin, the Baltic, and the North Sea, the two latter with waters that are but little more than brackish. In one set of these sea-basins the water is heavy, in the other it is light. Between them the ocean intervenes; but water is bound to seek and to maintain its level; and here, therefore, we unmask one of the agents concerned in causing the Gulf-stream" (§ 38). To the same effect see §§ 52, 522, 523, 524, 525, 526, 528, 530, 554, 556.

Lieut. Maury's *two causes neutralize each other.* Here we have two theories put forth regarding the cause of ocean-currents, the one in direct opposition to the other. According to the one theory, ocean-currents exist because the waters of equatorial regions, in consequence of their higher temperature, are *less dense* than the waters of the polar regions; but according to the other theory, ocean-currents exist because the waters of equatorial regions, in consequence of their greater saltness, are *more dense* than the waters of the polar regions. If the one cause be assigned as a reason why ocean-currents exist, then the other can be equally assigned as a reason why they should not exist. According to both theories it is the difference of density between the equatorial and polar waters that gives rise to currents; but while the one theory maintains that the equatorial waters are *lighter* than the polar, the other holds that they are *heavier.* Either the one theory or the other may be true, or neither; but it is logically impossible that both of them can. Let it be observed that it is not two currents, the one contrary to the other, with which we have at present to do; it is not temperature producing currents in one direction, and saltness producing currents in the contrary direction. We have two theories regarding the origin of currents, the one diametrically opposed to the other. The tendency of the one cause assigned is to prevent the action of the other. If temperature is allowed to act, it will make the inter-tropical waters lighter than the polar, and then, according to theory, a current will result. But if we bring saltness into play (the other cause) it will do the reverse: it will increase the density of the inter-tropical waters and diminish the density of the polar; and so far as it acts it will diminish the currents produced by temperature, because it will diminish the difference of specific gravity between the inter-tropical and polar regions which had been previously caused by temperature. And when the effects of saltness are as powerful as those of temperature, the difference of specific gravity produced by temperature will be completely effaced, or, in other words, the waters of the equatorial and polar seas will be of the same density, and consequently no current will exist. And so long as the two causes continue in action, no current can arise, unless the energy of the one cause should happen to exceed that of the other; and even then a current will only exist to the extent by which the strength of the one exceeds that of the other.

The contrary nature of the two theories will be better seen by considering the way in which it is supposed that difference in saltness is produced and acts as a cause.

If there is a constant current resulting from the difference in saltness between the equatorial and polar waters, then there must be a cause which maintains this difference. The current is simply the effort to restore the equilibrium lost by the difference; and the current would very soon do this, and then all motion would cease, were there not a constantly operating cause maintaining the disturbance. What, then, according to Maury, is the cause of this disturbance, or, in other words, what is it that keeps the equatorial waters salter than the polar?

The agencies in operation are stated by him to be heat, radiation, evaporation, precipitation, and secretion of solid matter in the form of shells, &c. The two most important, however, are evaporation and precipitation.

The trade-winds enter the equatorial regions as relatively dry winds thirsting for vapour; consequently they absorb far more moisture than they give out; and the result is that in inter-tropical regions, evaporation is much in excess of precipitation; and as fresh water only is taken up, the salt being left behind, the process, of course, tends to increase the saltness of the inter-tropical seas. Again, in polar and extra-tropical regions the reverse is the case; precipitation is in excess of evaporation. This tends in turn to diminish the saltness of the waters of those regions. (See on these points §§ 31, 33, 34, 37, 179, 517, 526, and 552.)

In the system of circulation produced by difference of temperature, as we have already seen, the surface-currents flow from the equator to the poles, and the under or return currents from the poles to the equator; but in the system produced by difference of saltness, the surface currents flow from the poles to the equator, and the return under currents from the equator to the poles. That the surface currents produced by difference of saltness flow from the poles to the equator, Maury thinks is evident for the two following reasons:—

(1) As evaporation is in excess of precipitation in inter-tropical regions, more water is taken off the surface of the ocean in those regions than falls upon it in the form of rain. This excess of water falls in the form of rain on temperate and polar regions, where, consequently, precipitation is in excess of evaporation. The lifting of the water off the equatorial regions and its deposit on the polar tend to lower the level of the ocean in equatorial regions and to raise the level in polar; consequently, in order to restore the level of the ocean, the surface water at the polar regions flows towards the equatorial regions.

(2) As the water taken up at the equator is fresh, and the

salt is left behind, the ocean, in inter-tropical regions, is thus made salter and consequently denser. This dense water, therefore, sinks and passes away as an under current. This water, evaporated from inter-tropical regions, falls as fresh and lighter water in temperate and polar regions; and therefore not only is the level of the ocean raised, but the waters are made lighter. Hence, in order to restore equilibrium, the waters in temperate and polar regions will flow as a surface current towards the equator. Under currents will flow from the equator to the poles, and surface or upper currents from the poles to the equator. Difference in temperature and difference in saltness, therefore, in every respect tend to produce opposite effects.

That the above is a fair representation of the way in which Maury supposes difference in saltness to act as a cause in the production of ocean-currents will appear from the following quotations:—

"In those regions, as in the trade-wind region, where evaporation is in excess of precipitation, the general level of this supposed sea would be altered, and immediately as much water as is carried off by evaporation would commence to flow in from north and south toward the trade-wind or evaporation region, to restore the level" (§ 509). "On the other hand, the winds have taken this vapour, borne it off to the extra-tropical regions, and precipitated it, we will suppose, where precipitation is in excess of evaporation. Here is another alteration of sea-level, by elevation instead of by depression; and hence we have the motive power for a *surface current from each pole towards the equator*, the object of which is only to supply the demand for evaporation in the trade-wind regions" (§ 510).

The above result would follow, supposing the ocean to be fresh. He then proceeds to consider an additional result that follows in consequence of the saltness of the ocean.

"Let evaporation now commence in the trade-wind region, as it was supposed to do in the case of the fresh-water seas, and as it actually goes on in nature—and what takes place? Why a lowering of the sea-level as before. But as the vapour of salt water is fresh, or nearly so, fresh water only is taken up from the ocean; that which remains behind is therefore more salt. Thus, while the level is lowered in the salt sea, the equilibrium is destroyed because of the saltness of the water; for the water that remains after evaporation takes place is, on account of the solid matter held in solution, specifically heavier than it was before any portion of it was converted into vapour" (§ 517).

"The vapour is taken from the surface-water; the surface-water thereby becomes more salt, and, under certain conditions, heavier. When it becomes heavier, it sinks; and hence we have, due to the salts of the sea, a vertical circulation, namely, a descent of heavier—because salter and cooler—water from the surface, and an ascent of water that is lighter—because it is not so salt—from the depths below" (§ 518).

In section 519 he goes on to show that this vapour removed from the inter-tropical region is precipitated in the polar regions, where precipitation is in excess of evaporation. "In the precipitating regions, therefore, the level is destroyed, as before explained, by elevation, and in the evaporating regions by depression; which, as already stated, gives rise to a system of *surface* currents, moved by gravity alone, from the *poles towards the equator*" (§ 520).

"This fresh water being emptied into the Polar Sea and agitated by the winds, becomes mixed with the salt; but as the agitation of the sea by the winds is supposed to extend to no great depth, it is only the upper layer of salt water, and that to a moderate depth, which becomes mixed with the fresh. The specific gravity of this upper layer, therefore, is diminished just as much as the specific gravity of the sea-water in the evaporating regions was increased. *And thus we have a surface current of saltish water from the poles towards the equator, and an under current of water salter and heavier from the equator to the poles*" (§ 522).

"This property of saltness imparts to the waters of the ocean another peculiarity, by which the sea is still better adapted for the regulation of climates, and it is this : by evaporating fresh water from the salt in the tropics, the surface water becomes heavier than the average of sea-water. This heavy water is also warm water ; it sinks, and being a good retainer, but a bad conductor, of heat, this water is employed in transporting through *under currents* heat for the mitigation of climates in far distant regions" (§ 526).

"For instance, let us suppose the waters in a certain part of the torrid zone to be 90°, but by reason of the fresh water which has been taken from them in a state of vapour, and consequently, by reason of the proportionate increase of salts, these waters are heavier than waters that may be cooler, but not so salt. This being the case, the tendency would be for this warm but salt and heavy water to flow off as an *under current towards the polar or some other regions of lighter water*" (§ 554).

That Maury supposes the warm water at the equator to flow to the polar regions as an under current is further evident from the fact that he maintains that the climate of the arctic regions is mitigated by a warm under current, which comes from the equatorial regions, and passes up through Davis Straits (see §§ 534—544).

The question now suggests itself: to which of these two antagonistic causes does Maury really suppose ocean-currents must be referred? Whether does he suppose, difference in temperature or difference in saltness, to be the real cause? I have been unable to find anything from which we can reasonably conclude that he prefers the one cause to the other. It would seem that he regards both as real causes, and that he has failed to perceive that the one is destructive of the other. But it is difficult to conceive how he could believe that the sea in equatorial regions, by virtue of its higher temperature, *is* lighter than the sea in polar regions, while at the same time it *is not* lighter but heavier, in consequence of its greater saltness —how he could believe that the warm water at the equator flows to the poles as an upper current, and the cold water at the poles to the equator as an *under* current, while at the same time the warm water at the equator does not flow to the poles as a surface current, nor the cold water at the poles to the equator as an under current, but the reverse. And yet, unless these absolute impossibilities be possible, how can an ocean-current be the result of both causes?

The only explanation of the matter appears to be that Maury has failed to perceive the contradictory nature of his two theories. This fact is particularly seen when he comes to apply his two theories to the case of the Gulf-stream. He maintains, as has already been stated, that the waters of the Gulf-stream are salter than the waters of the sea through which they flow (see §§ 3, 28, 29, 30, 34, and several other places). And he states, as we have already seen (see p. 104), that the existence of the Gulf-stream is due principally to the difference of density of the water of the Caribbean Sea and the Gulf of Mexico as compared with that of the great Polar Basin and the North Sea. There can be no doubt whatever that it is the *density* of the waters of the Gulf-stream at its fountain-head, the Gulf of Mexico, resulting from its superior saltness, and the deficiency of density of the waters in polar regions and the North Sea, &c., that is here considered to be unmasked as one of the agents. If this be a cause of the motion of the Gulf-stream, how then can the difference of temperature between the waters of inter-tropical and polar regions assist as a cause? This difference of temperature will simply tend to undo all that has been done by difference of saltness: for it will tend to make the waters of the Gulf of Mexico lighter, and the waters of the polar regions heavier. But Maury maintains, as we have seen, that this difference of temperature is also a cause, which shows that he does not perceive the contradiction.

This is still further apparent. He holds, as stated, that "the waters of the Gulf-stream are salter than the waters of the sea through which they flow," and that this excess in saltness, by making the water heavier, is a cause of the motion of the stream. But he maintains that, notwithstanding the effect which greater saltness has in increasing the density of the waters of the Gulf-stream, yet, owing to their higher temperature, they are actually lighter than the water through which they flow; and as a proof that this is the case, he adduces the fact that the surface of the Gulf-stream is roof-shaped (§§ 39—41), which it could not be were its waters not actually lighter than the waters through which the stream flows. So it turns out that, in contradiction to what he had already stated, it is the lesser density of the waters of the Gulf-stream that is the real cause of their motion. The greater saltness of the waters, to which he attributes so much, can in no way be regarded as a cause of motion. Its effect, so far as it goes, is to stop the motion of the stream rather than to assist it.

But, again, although he asserts that difference of saltness and difference of temperature are both causes of ocean-currents, yet he appears actually to admit that temperature and saltness neutralize each other so as to prevent change in the specific gravity of the ocean, as will be seen from the following quotation:—

"It is the trade-winds, then, which prevent the thermal and specific-gravity curves from conforming with each other in intertropical seas. The water they suck up is fresh water; and the salt it contained, being left behind, is just sufficient to counterbalance, by its weight, the effect of thermal dilatation upon the specific gravity of sea-water between the parallels of 34° north and south. As we go from 34° to the equator, the water grows warmer and expands. It would become lighter; but the trade-winds, by taking up vapour without salt, make the water salter, and therefore heavier. The conclusion is, the proportion of salt in sea-water, its expansibility between 62° and 82°, and the thirst of the trade-winds for vapour are, where they blow, so balanced as to produce *perfect compensation;* and a more beautiful compensation cannot, it appears to me, be found in the mechanism of the universe than that which we have here stumbled upon. It is a triple adjustment; the power of the sun to expand, the power of the winds to evaporate, and the quantity of salts in the sea—these are so proportioned and adjusted that when both the wind and the sun have each played with its forces upon the inter-tropical waters of the ocean, *the residuum of heat and of salt should be just such as to balance each other in their effects; and so the aqueous equilibrium of the torrid zone is preserved*" (§ 436, eleventh edition).

"Between 35° or 40° and the equator evaporation is in excess of precipitation; and though, as we approach the equator on either side from these parallels, the solar ray warms and expands the surface-water of the sea, the winds, by the vapour they carry off, and the salt they leave behind, *prevent it from making that water lighter*" (§ 437, eleventh edition).

"Philosophers have admired the relations between the size of the earth, the force of gravity, and the strength of fibre in the flower-stalks of plants; but how much more exquisite is the system of counterpoises and adjustments here presented between the sea and its salts, the winds and the heat of the sun!" (§ 438, eleventh edition).

How can this be reconciled with all that precedes regarding ocean-currents being the result of difference of specific gravity caused by a difference of temperature and difference of saltness? Here is a distinct recognition of the fact that difference in saltness, instead of producing currents, tends rather to prevent the existence of currents, by counteracting the effects of difference in temperature. And so effectually does it do this, that for 40°, or nearly 3,000 miles, on each side of the equator there is absolutely no difference in the specific gravity of the ocean, and consequently nothing, either as regards difference of temperature or difference of saltness, that can possibly give rise to a current.

But it is evident that, if between the equator and latitude 40° the two effects completely neutralize each other, it is not at all likely that between latitude 40° and the poles they will not to a large extent do the same thing. And if so, how can ocean-currents be due either to difference in temperature or to difference in saltness, far less to both. If there be any difference of specific gravity of the ocean between latitude 40° and the poles, it must be only to the extent by which the one cause has failed to neutralize the other. If, for example, the waters in latitude 40°, by virtue of higher temperature, are less dense than the waters in the polar regions, they can be so only to the extent that difference in saltness has failed to neutralize the effect of difference in temperature. And if currents result, they can do so only to the extent that difference in saltness has thus fallen short of being able to produce complete compensation. Maury, after stating his views on compensation, seems to become aware of this; but, strangely, he does not appear to perceive, or, at least, he does not make any allusion to the fact, that all this is fatal to his theories about ocean-currents being the combined result of differences of temperature and of saltness. For, in opposition to all that he had previously advanced regarding the difficulty of finding a cause sufficiently powerful to account for such currents as the Gulf-stream, and the great importance that difference in saltness had in their production, he now begins to maintain that so great is the influence of difference in temperature that difference in saltness, and a number of other compensating causes are actually necessary to prevent the ocean-currents from becoming too powerful.

"If all the inter-tropical heat of the sun," he says, "were to pass into the seas upon which it falls, simply raising the temperature of their waters, it would create a thermo-dynamical force in the ocean capable of transporting water scalding hot from the torrid zone, and spreading it while still in the tepid state around the poles . . . . Now, suppose there were no trade-winds to evaporate and to counteract the dynamical force of the sun, this hot and light water, by becoming hotter and lighter, would flow off in currents with almost mill-tail velocity towards the poles, covering the intervening sea with a mantle of warmth as a garment. The cool and heavy water of the polar basin, coming out as under currents, would flow equatorially with equal velocity."

"Thus two antagonistic forces are unmasked, and, being unmasked, we discover in them a most exquisite adjustment—a compensation—by which the dynamical forces that reside in the sunbeam and the trade-wind are made to counterbalance each other, by which the climates of inter-tropical seas are regulated, and by which the set, force, and volume of oceanic currents are measured" (§§ 437 and 438, eleventh edition).

CHAPTER VII.

EXAMINATION OF THE GRAVITATION THEORY OF OCEANIC CIRCULATION.—LIEUT. MAURY'S THEORY (*continued*).

Methods of determining the Question.—The Force resulting from Difference of Specific Gravity.—Sir John Herschel's Estimate of the Force.—Maximum Density of Sea-Water.—Rate of Decrease of Temperature of Ocean at Equator.—The actual Amount of Force resulting from Difference of Specific Gravity.—M. Dubuat's Experiments.

How the Question may be Determined.—Whether the circulation of the ocean is due to difference in specific gravity or not may be determined in three ways: viz. (1) by direct experiment; (2) by ascertaining the absolute amount of *force* acting on the water to produce motion, in virtue of difference of specific gravity, and thereafter comparing it with the force which has been shown by experiment to be necessary to the production of sensible motion; or (3) by determining the greatest possible amount of *work* which gravity can perform on the waters in virtue of difference of specific gravity, and then ascertaining if the work of gravity does or does not equal the work of the resistances in the required motion. But Maury has not adopted either of these methods.

The Force resulting from Difference of Specific Gravity.—I shall consider first whether the force resulting from difference of specific gravity be sufficient to account for the motion of ocean-currents.

The inadequacy of this cause has been so clearly shown by Sir John Herschel, that one might expect that little else would be required than simply to quote his words on the subject, which are as follows:—

"First, then, if there were no atmosphere, there would be no Gulf-stream, or any other considerable ocean-current (as distinguished from a mere surface-drift) whatever. By the action of the sun's rays, the *surface* of the ocean becomes *most* heated, and the heated water will, therefore, neither directly tend to *ascend* (which it could not do without leaving the sea) nor to *descend*, which it cannot do, being rendered buoyant, nor to move laterally, no lateral impulse being given, and which it could only do by reason of a general declivity of surface, the dilated portion occupying a higher level. Let us see what this declivity would amount to. The equatorial surface-water has a temperature of 84°. At 7,200 feet deep the temperature is 39°, the level of which temperature rises to the surface in latitude 56°. Taking the dilatability of sea-water to be the same as that of fresh, a uniformly progressive increase of temperature, from 39° to 84° Fahr., would dilate a column of 7,200 feet by 10 feet, to which height, therefore, above the spheroid of equilibrium (or above the sea-level in lat. 56°), the equatorial surface is actually raised by dilatation. An arc of 56° on the earth's surface measures 3,360 geographical miles; so that we have a slope of 1-28th of an inch per geographical mile, or 1-32nd of an inch per statute mile for the water so raised to run down. As the accelerating force corresponding to such a slope (of 1-10th of a second, 0″·1) is less than one two-millionth part of gravity, we may dismiss this as a cause capable of creating only a very trifling surface-drift, and not worth considering, even were it in the proper direction to form, by concentration, a current from east to west, *which it would not be, but the very reverse.*" *

It is singular how any one, even though he regarded this conclusion as but a rough approximation to the truth, could entertain the idea that ocean-currents can be the result of difference in specific gravity. There are one or two reasons, however, which may be given for the above not having been generally received as conclusive. Herschel's calculations refer to the difference of gravity resulting from difference of tempe-

* "Physical Geography," article 57.

rature; but this is only one of the causes to which Maury appeals, and even not the one to which he most frequently refers. He insists so strongly on the effects of difference of saltness, that many might think that, although Herschel may have shown that difference in specific gravity arising from difference of temperature could not account for the motion of ocean-currents, yet nevertheless that this, combined with the effects resulting from difference in saltness, might be a sufficient explanation of the phenomena. Such, of course, would not be the case with those who perceived the contradictory nature of Maury's two causes; but probably many read the "Physical Geography of the Sea" without being aware that the one cause is destructive of the other. Again, a few plausible objections, which have never received due consideration, have been strongly urged by Maury and others against the theory that ocean-currents can be caused by the impulses of the winds; and probably these objections appear to militate as strongly against this theory as Herschel's arguments against Maury's.

There is one trifling objection to Herschel's result: he takes 39° as the temperature of maximum density. This, however, as we shall see, does not materially affect his conclusions.

Observations on the temperature of the maximum density of sea-water have been made by Erman, Despretz, Rossetti, Neumann, Marcet, Hubbard, Horner, and others. No two of them have arrived at exactly the same conclusion. This probably arises from the fact that the temperature of maximum density depends upon the amount of salt held in solution. No two seas, unless they are equal as to saltness, have the same temperature of maximum density. The following Table of Despretz will show how rapidly the temperature of both the freezing-point and of maximum density is lowered by additional amounts of salt:—

Amount of salt.	Temperature of freezing-point.	Temperature of Maximum density.
0·000123	−1°·21 C.	+1°·19 C.
0·0246	−2·24	− 1·69
0·0371	−2·77	− 4·75
0·0741	−5·28	−16·00

He found the temperature of maximum density of sea-water, whose density at 20° C. was 1·0273, to be —3°·67 C. (25°·4 F.), and the temperature of freezing-point —2°·55 (27°·4 F.).* Somewhere between 25° and 26° F. may therefore be regarded as the temperature of maximum density of sea-water of average saltness. We have no reason to believe that the ocean, from the surface to the bottom, even at the poles, is at 27°·4 F., the freezing-point.

The actual slope resulting from difference of specific gravity, as we shall presently see, does not amount to 10 feet. Herschel's estimate was, however, made on insufficient data, both as to the rate of expansion of sea-water and that at which the temperature of the ocean at the equator decreases from the surface downwards. We are happily now in the possession of data for determining with tolerable accuracy the amount of slope due to difference of temperature between the equatorial and polar seas. The rate of expansion of sea-water from 0° C. to 100° C. has been experimentally determined by Professor Muncke, of Heidelberg.† The valuable reports of Captain Nares, of H.M.S. *Challenger*, lately published by the Admiralty, give the rate at which the temperature of the Atlantic at the equator decreases from the surface downwards. These observations show clearly that the super-heating effect of the sun's rays does not extend to any great depth. They also prove that at the equator the temperature decreases as the depth increases so rapidly that at 60 fathoms from the surface the temperature is 62°·4, the same as at Madeira at the same depth; while at the depth of 150 fathoms it is only 51°, about the same as that in the Bay of Biscay (Reports, p. 11). Here at the very outset we have broad and important facts hostile to the theory of a flow of water resulting from difference of temperature between the ocean in equatorial and temperate and polar regions.

Through the kindness of Staff-Captain Evans, Hydrographer

* Philosophical Magazine, vol. xii. p. 1 (1838).

† "Mémoires par divers Savans," tom. i., p. 318, St. Petersburgh, 1831. See also twelfth number of Meteorological Papers, published by the Board of Trade, 1865, p. 16.

of the Admiralty, I have been favoured with a most valuable set of serial temperature soundings made by Captain Nares of the *Challenger*, close to the equator, between long. 14° 49′ W. and 32° 16′ W. The following Table represents the mean of the whole of these observations :—

Fathoms.	Temperature.	Fathoms.	Temperature.	Fathoms.	Temperature.
Surface.	77°·9	90	58°·0	800	39°·1
10	77·2	100	55·6	900	38·2
20	77·1	150	51·0	1000	36·9
30	76·9	200	46·6	1100	37·6
40	71·7	300	42·2	1200	36·7
50	64·0	400	40·3	1300	35·8
60	60·4	500	38·9	1400	36·4
70	59·4	600	39·2	1500	36·1
80	58·0	700	39·0	Bottom.	34·7

We have in this Table data for determining the height at which the surface of the ocean at the equator ought to stand above that of the poles. Assuming 32° F. to be the temperature of the ocean at the poles from the surface to the bottom and the foregoing to be the rate at which the temperature of the ocean at the equator decreases from the surface downwards, and then calculating according to Muncke's Table of the expansion of sea-water, we have only 4 feet 6 inches as the height to which the level of the ocean at the equator ought to stand above that at the poles in order that the ocean may be in static equilibrium. In other words, the equatorial column requires to be only 4 feet 6 inches higher than the polar in order that the two may balance each other.

Taking the distance from the equator to the poles at 6,200 miles, the force resulting from the slope of 4½ feet in 6,200 will amount to only 1-7,340,000th that of gravity, or about 1-1000th of a grain on a pound of water. But, as we shall shortly see, there can be no permanent current resulting from difference of temperature while the two columns remain in equilibrium, for the current is simply an effort to the retardation of equilibrium. In order to permanent circulation there must be a permanent disturbance of equilibrium. Or, in other words, the weight of the polar column must be kept in excess of that of the equatorial. Suppose, then, that the weight of the polar column exceeds that of the equatorial by 2 feet of water, the difference of level between the two columns will, in that case, amount to only 2 feet 6 inches. This would give a force of only 1-13,200,000th that of gravity, or not much over 1-1,900th of a grain on a pound of water, tending to draw the water down the slope from the equator to the poles, a force which does not much exceed the weight of a grain on a ton of water. But it must be observed that this force of a grain per ton would affect only the water at the surface; a very short distance below the surface the force, small as it is, would be enormously reduced. If water were a perfect fluid, and offered no resistance to motion, it would not only flow down an incline, however small it might be, but would flow down with an accelerated motion. But water is not a perfect fluid, and its molecules do offer considerable resistance to motion. Water flowing down an incline, however steep it may be, soon acquires a uniform motion. There must therefore be a certain inclination below which no motion can take place. Experiments were made by M. Dubuat with the view of determining this limit.* He found that when the inclination was 1 in 500,000, the motion of the water was barely perceptible; and he came to the conclusion that when the inclination is reduced to 1 in 1,000,000, all motion ceases. But the inclination afforded by the difference of temperature between the sea in equatorial and polar regions does not amount to one-seventh of this, and consequently it can hardly produce even that "trifling surface-drift" which Sir John Herschel is willing to attribute to it.

There is an error into which some writers appear to fall to which I may here refer. Suppose that at the equator we have to descend 10,000 feet before water equal in density to that at the poles is reached. We have in this case a plain with a slope

* Dubuat's "Hydraulique," tom. i., p. 64 (1816). See also British Association Report for 1834, pp. 422, 451.

of 10,000 feet in 6,200 miles, forming the upper surface of the water of maximum density. Now this slope exercises no influence in the way of producing a current, as some seem to think; for it is not a case of disturbed equilibrium, but the reverse. It is the condition of static equilibrium resulting from a difference between the temperature of the water at the equator and the poles. The only slope that has any tendency to produce motion is that which is formed by the surface of the ocean in the equatorial regions being higher than the surface at the poles; but this is an inclination of only 4 feet 6 inches, and is therefore wholly inadequate to produce such currents as the Gulf-stream.

CHAPTER VIII.

EXAMINATION OF THE GRAVITATION THEORY OF OCEANIC CIRCULATION.—DR. CARPENTER'S THEORY.

Gulf-stream according to Dr. Carpenter not due to Difference of Specific Gravity.—Facts to be Explained.—The Explanation of the Facts.—The Explanation hypothetical.—The Cause assigned for the hypothetical Mode of Circulation.—Under currents account for all the Facts better than the Gravitation Hypothesis.—Known Condition of the Ocean inconsistent with that Hypothesis.

DR. CARPENTER does not suppose, with Lieut. Maury, that the difference of temperature between the ocean in equatorial and polar regions can account for the Gulf-stream and other great currents of the ocean. He maintains, however, that this difference is quite sufficient to bring about a slow general interchange of water between the polar and inter-tropical areas—to induce a general movement of the upper portion of the ocean from the equator to the poles and a counter-movement of the under portion in a contrary direction. It is this general movement which, according to that author, is the great agent by which heat is distributed over the globe.*

In attempting to estimate the adequacy of this hypothesis as an explanation of the phenomena involved, there are obviously two questions to be considered: namely, (1) is the difference of temperature between the sea in inter-tropical and polar regions sufficiently great to produce the required movement? and (2) assuming that there is such a movement, does it convey the amount of heat which Dr. Carpenter supposes? I shall begin with the consideration of the first of these two points.

* See Proceedings of the Royal Society for December, 1868, November, 1869. Lecture delivered at the Royal Institute, *Nature*, vol. i., p. 490. Proceedings of the Royal Geographical Society, vol. xv.

But before doing so let us see what the facts are which this gravitation theory is intended to explain.

The Facts to be Explained.—Dr. Carpenter considers that the great mass of warm water proved during recent dredging expeditions to occupy the depths of the North Atlantic, must be referred, not to the Gulf-stream, but to a general movement of water from the equator. "The inference seems inevitable," he says, "that the bulk of the water in the warm area must have come thither from the south-west. The influence of the Gulf-stream proper (meaning by this the body of superheated water which issues through the 'Narrows' from the Gulf of Mexico), if it reaches this locality at all (which is very doubtful), could only affect the *most superficial* stratum; and the same may be said of the surface-drift caused by the prevalence of south-westerly winds, to which some have attributed the phenomena usually accounted for by the extension of the Gulf-stream to these regions. And the presence of the body of water which lies between 100 and 600 fathoms deep, and the range of whose temperature is from 48° to 42°, can scarcely be accounted for on any other hypothesis than that of a *great general movement of equatorial water towards the polar area,* of which movement the Gulf-stream constitutes a peculiar case modified by local conditions. In like manner the Arctic stream which underlies the warm superficial stratum in our cold area constitutes a peculiar case, modified by the local conditions to be presently explained, of *a great general movement of polar water towards the equatorial area,* which depresses the temperature of the deepest parts of the great oceanic basins nearly to the freezing-point."

It is well-known that, wherever temperature-observations have been made in the Atlantic, the bottom of that ocean has been found to be occupied by water of an ice-cold temperature. And this holds true not merely of the Atlantic, but also of the ocean in inter-tropical regions—a fact which has been proved by repeated observations, and more particularly of late by those of Commander Chimmo in the China Sea and Indian Ocean, where a temperature as low as 32° Fahr. was found at a depth of 2,656 fathoms. In short, the North Atlantic, and probably the inter-tropical seas also, may be regarded, Dr. Carpenter considers, as divided horizontally into two great layers or strata—an upper warm, and a lower cold stratum. All these facts I, of course, freely admit; nor am I aware that their truth has been called in question by any one, no matter what his views may have been as to the mode in which they are to be explained.

The Explanation of the Facts.—We have next the explanation of the facts, which is simply this:—The cold water occupying the bottom of the Atlantic and of inter-tropical seas is to be accounted for by the supposition that *it came from the polar regions.* This is obvious, because the cold possessed by the water could not have been derived from the crust of the earth beneath: neither could it have come from the surface; for the temperature of the bottom water is far below the normal temperature of the latitude in which it is found. Consequently "the inference seems irresistible that this depression must be produced and maintained by the convection of cold from the polar towards the equatorial area." Of course, if we suppose a flow of water from the poles towards the equator, we must necessarily infer a counter flow from the equator towards the poles; and while the water flowing from equatorial to polar regions will be *warm*, that flowing from polar to equatorial regions will be *cold*. The doctrine of a mutual interchange of equatorial and polar water is therefore a *necessary consequence* from the admission of the foregoing facts. With this *explanation of the facts* I need hardly say that I fully agree; nor am I aware that its correctness has ever been disputed. Dr. Carpenter surely cannot charge me with overlooking the fact of a mutual interchange of equatorial and polar water, seeing that my estimate of the thermal power of the Gulf-stream, from which it is proved that the amount of heat conveyed from equatorial to temperate and polar regions is enormously greater than had ever been anticipated, was made a considerable time before he began to write on the subject of oceanic circulation.* And in my paper "On Ocean-currents in relation to the Distribution of Heat over the Globe"† (the substance of which is reproduced in Chapters II. and III. of this volume), I have endeavoured to show that, were it not for the raising of the temperature of polar and high temperate regions and the lowering of the temperature of inter-tropical regions by means of this interchange of water, these portions of the globe would not be habitable by the present existing orders of beings.

The explanation goes further :—"It is along the surface and upper portion of the ocean that the equatorial waters flow towards the poles, and it is along the bottom and under portion of the ocean that polar waters flow towards the equator; or, in other words, the warm water keeps the *upper* portion of the ocean and the cold water the *under* portion." With this explanation I to a great extent agree. It is evident that, in reference to the northern hemisphere at least, the most of the water which flows from inter-tropical to polar regions (as, for example, the Gulf-stream) keeps to the surface and upper portion of the ocean; but for reasons which I have already stated, a very large proportion of this water must return in the form of *under* currents; or, which is the same thing, the return compensating current, whether it consist of the identical water which originally came from the equator or not, must flow towards the equator as an under current. That the cold water which is found at the bottom of the Atlantic and of inter-tropical seas must have come as under currents is perfectly obvious, because water which should come along the surface of the ocean from the polar regions would not be cold when it reached inter-tropical regions.

The Explanation hypothetical.—Here the general agreement between us in a great measure terminates, for Dr. Carpenter is not satisfied with the explanation generally adopted by the

* Trans. of Glasgow Geol. Soc. for April, 1867. Phil. Mag. for February, 1867, and June, 1867 (Supplement).

† Phil. Mag. for February, 1870.

advocates of the *wind theory*, viz., that the cold water found in temperate and inter-tropical areas comes from polar regions as compensating under currents, but advances a *hypothetical* form of circulation to account for the phenomenon. He assumes that there is a *general set* or flow of the surface and upper portion of the ocean from the equator to polar regions, and a *general set* or flow of the bottom and under portion of the ocean from polar regions to the equator. Mr. Ferrel (*Nature*, June 13, 1872) speaks of that "interchanging motion of the water between the equator and the pole *discovered* by Dr. Carpenter." In this, however, Mr. Ferrel is mistaken; for Dr. Carpenter not only makes no claim to any discovery of the kind, but distinctly admits that none such has yet been made. Although in some of his papers he speaks of a "*set* of warm surface-water in the southern oceans toward the Antarctic pole" as being well known to navigators, yet he nowhere affirms, as far as I know, that the existence of such a general oceanic circulation as he advocates has ever been directly determined from observations. This mode of circulation is *simply inferred* or *assumed* in order to account for the facts referred to above. "At present," Dr. Carpenter says, "I claim for it no higher character than that of a good working *hypothesis* to be used as a guide in further inquiry" (§ 16); and lest there should be any misapprehension on this point, he closes his memoir thus:—"At present, as I have already said, I claim for the doctrine of a general oceanic circulation no higher a character than that of a good working *hypothesis* consistent with our present knowledge of facts, and therefore entitled to be *provisionally* adopted for the purpose of stimulating and directing further inquiry."

I am unable to agree with him, however, on this latter point. It seems to me that there is no necessity for adopting any hypothetical mode of circulation to account for the facts, as they can be quite well accounted for by means of that mode of circulation which does *actually exist*. It has been determined from direct observation that surface-currents flow from equatorial to polar regions, and their paths have been actually mapped out.

But if it is established that currents flow from equatorial to polar regions, it is equally so that return currents flow from polar to equatorial regions; for if the one *actually* exists, the other of necessity *must* exist. We know also on physical grounds, to which I have already referred, and which fall to be considered more fully in a subsequent chapter, that a very large portion of the water flowing from polar to equatorial regions must be in the form of under currents. If there are cold under currents, therefore, flowing from polar to temperate and equatorial regions, this is all that we really require to account for the cold water which is found to occupy the bed of the ocean in those regions. It does not necessarily follow, because cold water may be found at the bottom of the ocean all along the equator, that there must be a direct flow from the polar regions to every point of the equator. Water brought constantly from the polar regions to various points along the equator by means of under currents will necessarily accumulate, and in course of time spread over the bottom of the intertropical seas. It must either do this, or the currents on reaching the equator must bend upwards and flow to the surface in an unbroken mass. Considerable portions of some of those currents may no doubt do so and join surface-currents; but probably the greater portion of the water coming from polar regions extends itself over the floor of the equatorial seas. In a letter in *Nature*, January 11, 1872, I endeavoured to show that the surface-currents of the ocean are not separate and independent of one another, but form one grand system of circulation, and that the impelling cause keeping up this system of circulation is not the *trade-winds* alone, as is generally supposed, but the *prevailing winds of the entire globe considered also as one grand system.* The evidence for this opinion, however, will be considered more fully in the sequel.

Although the under currents are parts of one general system of oceanic circulation produced by the impulse of the system of prevailing winds, yet their direction and position are nevertheless, to a large extent, determined by different laws. The water at the surface, being moved by the force of the wind, will follow the path of *greatest pressure and traction,*—the effects resulting from the general contour of the land, which to a great extent are common to both sets of currents, not being taken into account; while, on the other hand, the under currents from polar regions (which to a great extent are simply "indraughts" compensating for the water drained from equatorial regions by the Gulf-stream and other surface currents) will follow, as a general rule, the path of *least resistance.*

The Cause assigned for the Hypothetical Mode of Circulation.—Dr. Carpenter assigns a cause for his mode of circulation; and that cause he finds in the difference of specific gravity between equatorial and polar waters, resulting from the difference of temperature between these two regions. "Two separate questions," he says, "have to be considered, which have not, perhaps, been kept sufficiently distinct, either by Mr. Croll or by myself;—*first,* whether there is adequate evidence of the existence of a general vertical oceanic circulation; and *second,* whether, supposing its existence to be provisionally admitted, a *vera causa* can be found for it in the difference of temperature between the oceanic waters of the polar and equatorial areas" (§ 17). It seems to me that the facts adduced by Dr. Carpenter do not necessarily require the assumption of any such mode of circulation as that advanced by him. The phenomena can be satisfactorily accounted for otherwise; and therefore there does not appear to be any necessity for considering whether his hypothesis be sufficient to produce the required effect or not.

An important Consideration overlooked.—But there is one important consideration which seems to have been overlooked—namely, the fact that the sea is salter in inter-tropical than in polar regions, and that this circumstance, so far as it goes, must tend to neutralize the effect of difference of temperature. It is probable, indeed, that the effect produced by difference of temperature is thus entirely neutralized, and that no difference of density whatever exists between the sea in inter-tropical and polar regions, and consequently that there is no difference of level nor anything to produce such a general motion as Dr. Carpenter supposes. This, I am glad to find, is the opinion of Professor Wyville Thomson.

"I am greatly mistaken," says that author, "if the low specific gravity of the polar sea, the result of the condensation and precipitation of vapour evaporated from the inter-tropical area, do not fully counterbalance the contraction of the superficial film by arctic cold. . . . Speaking in the total absence of all reliable data, it is my general impression that if we were to set aside all other agencies, and to trust for an oceanic circulation to those conditions only which are relied upon by Dr. Carpenter, if there were any general circulation at all, which seems very problematical, the odds are rather in favour of a warm under current travelling northwards by virtue of its excess of salt, balanced by a surface return current of fresher though colder arctic water." *

This is what actually takes place on the west and north-west of Spitzbergen. There the warm water of the Gulf-stream flows underneath the cold polar current. And it is the opinion of Dr. Scoresby, Mr. Clements Markham, and Lieut. Maury that this warm water, in virtue of its greater saltness, is denser than the polar water. Mr. Leigh Smith found on the north-west of Spitzbergen the temperature at 500 fathoms to be 52°, and once even 64°, while the water on the surface was only a degree or two above freezing.† Mr. Aitken, of Darroch, in a paper lately read before the Royal Scottish Society of Arts, showed experimentally that the polar water in regions where the ice is melting is actually less dense than the warm and more salt tropical waters. Nor will it help the matter in the least to maintain that difference of specific gravity is not the reason why the warm water of the Gulf-stream passes under the polar stream—because if difference of specific gravity be not the cause of the warm water underlying the cold water in polar regions, then difference of specific gravity may likewise not

* "The Depths of the Sea," pp. 376 and 377.

† "The Threshold of the Unknown Region," p. 95.

be the cause of the cold water underlying the warm at the equator; and if so, then there is no necessity for the gravitation hypothesis of oceanic circulation.

There is little doubt that the super-heated stratum at the surface of the inter-tropical seas, which stratum, according to Dr. Carpenter, is of no great thickness, is less dense than the polar water: but if we take a column extending from the surface down to the bottom of the ocean, this column at the equator will be found to be as heavy as one of equal length in the polar area. And if this be the case, then there can be no difference of level between the equator and the poles, and no disturbance of static equilibrium nor anything else to produce circulation.

Under Currents account for all the Facts better than Dr. Carpenter's Hypothesis.—Assuming, for the present, the system of prevailing winds to be the true cause of oceanic currents, it necessarily follows (as will be shown hereafter) that a large quantity of Atlantic water must be propelled into the Arctic Ocean; and such, as we know, is actually the case. The Arctic Ocean, however, as Professor Wyville Thomson remarks, is a well-nigh closed basin, not permitting of a free outflow into the Pacific Ocean of the water impelled into it.

But it is evident that the water which is thus being constantly carried from the inter-tropical to the arctic regions must somehow or other find its way back to the equator; in other words, there must be a return current equal in magnitude to the direct current. Now the question to be determined is, what path must this return current take? It appears to me that it will take the *path of least resistance*, whether that path may happen to be at the surface or under the surface. But that the path of least resistance will, as a general rule, lie at a very considerable distance below the surface is, I think, evident from the following considerations. At the surface the general direction of the currents is opposite to that of the return current. The surface motion of the water in the Atlantic is from the equator to the pole; but the return current must be from the pole to the equator. Consequently the surface currents will oppose the motion of any return current unless that current lie at a considerable depth below the surface currents. Again, the winds, as a general rule, blow in an opposite direction to the course of the return current, because, according to supposition, the winds blow in the direction of the surface currents. From all these causes the path of least resistance to the return current will, as a general rule, not be at the surface, but at a very considerable depth below it.

A large portion of the water from the polar regions no doubt leaves those regions as surface currents; but a surface current of this kind, on meeting with some resistance to its onward progress along the surface, will dip down and continue its course as an under current. We have an example of this in the case of the polar current, which upon meeting the Gulf-stream on the banks of Newfoundland divides—a portion of it dipping down and pursuing its course underneath that stream into the Gulf of Mexico and the Caribbean Sea. And that this under current is a real and tangible current, in the proper sense of the term, and not an imperceptible movement of the water, is proved by the fact that large icebergs deeply immersed in it are often carried southward with considerable velocity against the united force of the wind and the Gulf-stream.

Dr. Carpenter refers at considerable length (§ 134) to Mr. Mitchell's opinion as to the origin of the polar current, which is the same as that advanced by Maury, viz., that the impelling cause is difference of specific gravity. But although Dr. Carpenter quotes Mr. Mitchell's opinion, he nevertheless does not appear to adopt it: for in §§ 90-93 and various other places he distinctly states that he does not agree with Lieut. Maury's view that the Gulf-stream and polar current are caused by difference of density. In fact, Dr. Carpenter seems particularly anxious that it should be clearly understood that he dissents from the theory maintained by Maury. But he does not merely deny that the Gulf-stream and polar current can be caused by difference of density; he even goes so far as to affirm that no sensible current whatever can be due to that cause, and adduces the authority of Sir John Herschel in support of that opinion:—"The doctrine of Lieut. Maury," he says, "was powerfully and convincingly opposed by Sir John Herschel; who showed, beyond all reasonable doubt, first, that the Gulf-stream really has its origin in the propulsive force of the trade-winds, and secondly, that the greatest disturbance of equilibrium which can be supposed to result from the agencies invoked by Lieut. Maury would be utterly inadequate to generate and maintain either the Gulf-stream or any other sensible current" (§ 92). This being Dr. Carpenter's belief, it is somewhat singular that he should advance the case of the polar current passing under the Gulf-stream as evidence in favour of his theory; for in reality he could hardly have selected a case more hostile to that theory. In short, it is evident that, if a polar current impelled by a force other than that of gravity can pass from the banks of Newfoundland to the Gulf of Mexico (a distance of some thousands of miles) under a current flowing in the opposite direction and, at the same time, so powerful as the Gulf-stream, it could pass much more easily under comparatively still water, or water flowing in the same direction as itself. And if this be so, then all our difficulties disappear, and we satisfactorily explain the presence of cold polar water at the bottom of inter-tropical seas without having recourse to the hypothesis advanced by Dr. Carpenter.

But we have an example of an under current more inexplicable on the gravitation hypothesis than even that of the polar current, viz., the warm under current of Davis Strait.

There is a strong current flowing north from the Atlantic through Davis Strait into the Arctic Ocean underneath a surface current passing southwards in an opposite direction. Large icebergs have been seen to be carried northwards by this under current at the rate of four knots an hour against both the wind and the surface current, ripping and tearing their way with terrific force through surface ice of great thickness.* A current so powerful and rapid as this cannot, as Dr. Carpenter admits, be referred to difference of specific gravity. But even supposing that it could, still difference of temperature between the equatorial and polar seas would not account for it; for the current in question flows in the *wrong direction*. Nor will it help the matter the least to adopt Maury's explanation, viz., that the warm under current from the south, in consequence of its greater saltness, is denser than the cold one from the polar regions. For if the water of the Atlantic, notwithstanding its higher temperature, is in consequence of its greater saltness so much denser than the polar water on the west of Greenland as to produce an under current of four knots an hour in the direction of the pole, then surely the same thing to a certain extent will hold true in reference to the ocean on the east side of Greenland. Thus instead of there being, as Dr. Carpenter supposes, an underflow of polar water south into the Atlantic in virtue of its *greater* density, there ought, on the contrary, to be a surface flow in consequence of its lesser density.

The true explanation no doubt is, that the warm under current from the south and the cold upper-current from the north are both parts of one grand system of circulation produced by the winds, difference of specific gravity having no share whatever either in impelling the currents, or in determining which shall be the upper and which the lower.

The wind in Baffin's Bay and Davis Strait blows nearly always in one direction, viz. from the north. The tendency of this is to produce a surface or upper current from the north down into the Atlantic, and to prevent or retard any surface current from the south. The warm current from the Atlantic, taking the path of least resistance, dips under the polar current and pursues its course as an under current.

Mr. Clements Markham, in his "Threshold of the Unknown Region," is inclined to attribute the motion of the icebergs to

* See "Physical Geography of the Sea," chap. ix., new edition, and Dr. A. Mühry "On Ocean-currents in the Circumpolar Basin of the North Hemisphere.'

tidal action or to counter under currents. That the motion of the icebergs cannot reasonably be attributed to the tides is, I think, evident from the descriptions given both by Midshipman Griffin and by Captain Duncan, who distinctly saw the icebergs moving at the rate of about four knots an hour against a surface current flowing southwards. And Captain Duncan states that the bergs continued their course northwards for several days, till they ultimately disappeared. The probability is that this northward current is composed partly of Gulf-stream water and partly of that portion of polar water which is supposed to flow round Cape Farewell from the east coast of Greenland. This stream, composed of both warm and cold water, on reaching to about latitude 65° N., where it encounters the strong northerly winds, dips down under the polar current and continues its northward course as an under current.

We have on the west of Spitzbergen, as has already been noticed, a similar example of a warm current from the south passing under a polar current. A portion of the Gulf-stream which passes round the west coast of Spitzbergen flows under an arctic current coming down from the north; and it does so no doubt because it is here in the region of prevailing northerly winds, which favour the polar current but oppose the Gulf-stream. Again, we have a cold and rapid current sweeping round the east and south of Spitzbergen, a current of which Mr. Lamont asserts that he is positive he has seen it running at the rate of seven or eight miles an hour. This current, on meeting the Gulf-stream about the northern entrance to the German Ocean, dips down under that stream and pursues its course southwards as an under current.

Several other cases of under currents might be adduced which cannot be explained on the gravitation theory, and which must be referred to a system of oceanic circulation produced by the impulse of the wind; but these will suffice to show that the assumption that the winds can produce only a mere surface drift is directly opposed to facts. And it will not do to affirm that a current which forms part of a general system of circulation produced by the impulse of the winds cannot possibly be an under current; for in the case referred to we have proof that the thing is not only possible but actually exists. This point, however, will be better understood after we have considered the evidence in favour of a general system of oceanic currents.

Much of the difficulty experienced in comprehending how under currents can be produced by the wind, or how an impulse imparted to the surface of the ocean can ever be transmitted to the bottom, appears to me to result, to a considerable extent at least, from a slight deception of the imagination. The thing which impresses us most forcibly in regard to the ocean is its profound depth. A mean depth of, say, three miles produces a striking impression; but if we could represent to the mind the vast area of the ocean as correctly as we can its depth, *shallowness* rather than *depth* would be the impression produced. If in crossing a meadow we found a sheet of water one hundred yards in diameter and only an inch in depth, we should not call that a *deep*, but a very *shallow* pool. The probability is that we should speak of it as simply a piece of ground covered with a thin layer of water. Yet such a thin layer of water would be a correct representation in miniature of the ocean; for the ocean in relation to its superficial area is as shallow as the pool of our illustration. In reference to such a pool or thin film of water, we have no difficulty in conceiving how a disturbance on its surface would be transmitted to its bottom. In fact our difficulty is in conceiving how any disturbance extending over its entire surface should not extend to the bottom. Now if we could form as accurate a sensuous impression of the vast area of the ocean as we do of such a pool, all our difficulty in understanding how the impulses of the wind acting on the vast area of the ocean should communicate motion down to its bottom would disappear. It is certainly true that sudden commotions caused by storms do not generally extend to great depths. Neither will winds of short continuance produce a current extending far below the surface. But prevailing winds which can produce such immense surface flow as that of the great equatorial currents of the globe and the Gulf-stream, which follow definite directions, must communicate their motion to great depths, unless water be frictionless, a thing which it is not. Suppose the upper layer of the ocean to be forced on by the direct action of the winds with a constant velocity of, say, four miles an hour, the layer immediately below will be dragged along with a constant velocity somewhat less than four miles an hour. The layer immediately below this second layer will in turn be also dragged along with a constant velocity somewhat less than the one above it. The same will take place in regard to each succeeding layer, the constant velocity of each layer being somewhat less than the one immediately above it, and greater than the one below it. The question to be determined is, at what depth will all motion cease? I presume that at present we have not sufficient data for properly determining this point. The depth will depend, other things being equal, upon the amount of molecular resistance offered by the water to motion—in other words, on the amount of the shearing-force of the one layer over the other. The fact, however, that motion imparted to the surface will extend to great depths can be easily shown by direct experiment. If a constant motion be imparted to the surface of water, say, in a vessel, motion will ultimately be communicated to the bottom, no matter how wide or how deep the vessel may be. The same effect will take place whether the vessel be 5 feet deep or 500 feet deep.

The known Condition of the Ocean inconsistent with Dr. Carpenter's Hypothesis.—Dr. Carpenter says that he looks forward with great satisfaction to the results of the inquiries which are being prosecuted by the Circumnavigation Expedition, in the hope that the facts brought to light may establish his theory of a general oceanic circulation; and he specifies certain of these facts which, if found to be correct, will establish his theory. It seems to me, however, that the facts to which he refers are just as explicable on the theory of under currents as on the theory of a general oceanic circulation. He begins by saying, "If the views I have propounded be correct, it may be expected that near the border of the great antarctic ice-barrier a temperature below 30° will be met with (as it has been by Parry, Martens, and Weyprecht near Spitzbergen) at no great depth beneath the surface, and that instead of rising at still greater depths, the thermometer will fall to near the freezing-point of salt water" (§ 39).

Dr. Carpenter can hardly claim this as evidence in favour of his theory; for near the borders of the ice-barrier the water, as a matter of course, could not be expected to have a much higher temperature than the ice itself. And if the observations be made during summer months, the temperature of the water at the surface will no doubt be found to be higher than that of the bottom; but if they be carried on during winter, the surface-temperature will doubtless be found to be as low as the bottom-temperature. These are results which do not depend upon any particular theory of oceanic circulation.

"The bottom temperature of the North Pacific," he continues, "will afford a crucial test of the truth of the doctrine. For since the sole communication of this vast oceanic area with the arctic basin is a strait so shallow as only to permit an inflow of warm surface water, its deep cold stratum must be entirely derived from the antarctic area; and if its bottom temperature is not actually higher than that of the South Pacific, the glacial stratum ought to be found at a greater depth north of the equator than south of it" (§ 39).

This may probably show that the water came from the antarctic regions, but cannot possibly prove that it came in the manner which he supposes.

"In the North Atlantic, again, the comparative limitation of communication with the arctic area may be expected to prevent its bottom temperature from being reduced as low as that of the Southern Atlantic" (§ 39). Supposing the bottom temperature of the South Atlantic should be found to be lower than the bottom temperature of the North Atlantic, this fact will be just as consistent with the theory of under currents as with his theory of a general movement of the ocean.

I am also wholly unable to comprehend how he should imagine, because the bottom temperature of the South Atlantic happens to be lower, and the polar water to lie nearer to the surface in this ocean than in the North Atlantic, that therefore this proves the truth of his theory. This condition of matters is just as consistent, and even more so, as will be shown in Chapter XIII., with my theory as with his. When we consider the immense quantity of warm surface water which, as has been shown (Chapter V.), is being constantly transferred from the South into the North Atlantic, we readily understand how the polar water comes nearer to the surface in the former ocean than in the latter. Every pound of water, of course, passing from the southern to the northern hemisphere must be compensated by an equal amount passing from the northern to the southern hemisphere. But nevertheless the warm water drained off the South Atlantic is not replaced directly by water from the north, but by that cold antarctic current, the existence of which is, unfortunately, too well known to navigators from the immense masses of icebergs which it brings along with it. In fact, the whole of the phenomena are just as easily explained upon the principle of under currents as upon Dr. Carpenter's theory. But we shall have to return to this point in Chapter XIII., when we come to discuss a class of facts which appear to be wholly irreconcilable with the gravitation theory.

Indeed I fear that even although Dr. Carpenter's expectations should eventually be realised in the results of the Circumnavigation Expedition, yet the advocates of the wind theory will still remain unconverted. In fact the Director of this Expedition has already, on the wind theory, offered an explanation of nearly all the phenomena on which Dr. Carpenter relies;* and the same has also been done by Dr. Petermann,† who, as is well known, is equally opposed to

* "Depths of the Sea," *Nature* for July 28, 1870.

† "Memoir on the Gulf-stream," *Geographische Mittheilungen*, vol. xvi. (1870).

Dr. Carpenter's theory. Dr. Carpenter directs attention to the necessity of examining the broad and deep channel separating Iceland from Greenland. The observations which have already been made, however, show that nearly the entire channel is occupied, on the surface at least, by water flowing southward from the polar area—a direction the opposite of what it ought to be according to the gravitation theory. In fact the surface of one half of the entire area of the ocean, extending from Greenland to the North Cape, is moving in a direction the opposite of that which it ought to take according to the theory under review. The western half of this area is occupied by water which at the surface is flowing southwards; while the eastern half, which has hitherto been regarded by almost everybody but Dr. Carpenter himself and Mr. Findlay as an extension of the Gulf-stream, is moving polewards. The motion of the western half must be attributed to the winds and not to gravity; for it is moving in the wrong direction to be accounted for by the latter cause; but had it been moving in the opposite direction, no doubt its motion would have been referred to gravitation. To this cause the motion of the eastern half, which is in the proper direction, is attributed;* but why not assign this motion also to the impulse of the winds, more especially since the direction of the prevailing winds blowing over that area coincides with that of the water? If the wind can produce the motion of the water in the western half, why may not it do the same in the eastern half?

If there be such a difference of density between equatorial and polar waters as to produce a general flow of the upper portion of the ocean poleward, how does it happen that one half of the water in the above area is moving in opposition to gravity? How is it that in a wide open sea gravitation should act so powerfully in the one half of it and with so little effect in the other half? There is probably little doubt that the ice-cold water of the western half extends from the surface down to

* Dr. Carpenter "On the Gulf-stream," Proceedings of Royal Geographical Society for January 9, 1871, § 29.

the bottom. And it is also probable that the bottom water is moving southwards in the same direction as the surface water. The bottom water in such a case would be moving in harmony with the gravitation theory; but would Dr. Carpenter on this account attribute its motion to gravity? Would he attribute the motion of the lower half to gravity and the upper half to the wind? He could not in consistency with his theory attribute the motion of the upper half to gravity: for although the ice-cold water extended to the surface, this could not explain how gravity should move it southward instead of polewards, as according to theory it ought to move. He might affirm, if he chose, that the surface water moves southwards because it is dragged forward by the bottom water; but if this view be held, he is not entitled to affirm, as he does, that the winds can only produce a mere surface drift. If the viscosity and molecular resistance of water be such that, when the lower strata of the ocean are impelled forward by gravity or by any other cause, the superincumbent strata extending to the surface are perforce dragged after them, then, for the same reason, when the upper strata are impelled forward by the wind or any other cause, the underlying strata must also be dragged along after them.

If the condition of the ocean between Greenland and the north-western shore of Europe is irreconcilable with the gravitation theory, we find the case even worse for that theory when we direct our attention to the condition of the ocean on the southern hemisphere; for according to the researches of Captain Duperrey and others on the currents of the Southern Ocean, a very large portion of the area of that ocean is occupied by water moving on the surface more in a northward than a poleward direction. Referring to the deep trough between the Shetland and the Faroe Islands, called by him the "Lightning Channel," Dr. Carpenter says, "If my view be correct, a current-drag suspended in the *upper* stratum ought to have a perceptible movement in the N.E. direction; whilst another, suspended in the *lower* stratum, should move S.W." (§ 40).

Any one believing in the north-eastern extension of the Gulf-stream and in the Spitzbergen polar under current, to which I have already referred, would not feel surprised to learn that the surface-strata have a perceptible north-eastward motion, and the bottom strata a perceptible south-westward motion. North-east and east of Iceland there is a general flow of cold polar water in a south-east direction towards the left edge of the Gulf-stream. This water, as Professor Mohn concludes, "descends beneath the Gulf-stream and partially finds an outlet in the lower half of the Faroe-Shetland channel."*

An Objection Considered.—In *Nature*, vol. ix. p. 423, Dr. Carpenter has advanced the following objection to the foregoing theory of under currents:—"According to Mr. Croll's doctrine, the whole of that vast mass of water in the North Atlantic, averaging, say, 1,500 fathoms in thickness and 3,600 miles in breadth, the temperature of which (from 40° downwards), as ascertained by the *Challenger* soundings, clearly shows it to be mainly derived from a polar source, is nothing else than *the reflux of the Gulf-stream.* Now, even if we suppose that the whole of this stream, as it passes Sandy Hook, were to go on into the closed arctic basin, it would only force out an equivalent body of water. And as, on comparing the sectional areas of the two, I find that of the Gulf-stream to be about 1-900th that of the North Atlantic underflow; and as it is admitted that a large part of the Gulf-stream returns into the Mid-Atlantic circulation, only a branch of it going on to the north-east, the extreme improbability (may I not say impossibility?) that so vast a mass of water can be put in motion by what is by comparison a mere rivulet (the north-east motion of which, as a distinct current, has not been traced eastward of 30° W. long.) seems still more obvious."

In this objection three things are assumed: (1) that the mass of cold water 1,500 fathoms deep and 3,600 miles in breadth is in a state of motion towards the equator; (2) that it cannot be the reflux of the Gulf-stream, because its sectional

* Dr. Petermann's *Mittheilungen* for 1872, p. 315.

area is 900 times as great as that of the Gulf-stream; (3) that the immense mass of water is, according to my views, set in motion by the Gulf-stream.

As this objection has an important bearing on the question under consideration, I shall consider these three assumptions separately and in their order: (1) That this immense mass of cold water came originally from the polar regions I, of course, admit, but that the whole is in a state of motion I certainly do not admit. There is no warrant whatever for any such assumption. According to Dr. Carpenter himself, the heating-power of the sun does not extend to any great depth below the surface; consequently there is nothing whatever to heat this mass but the heat coming through the earth's crust. But the amount of heat derived from this source is so trifling, that an under current from the arctic regions far less in volume than that of the Gulf-stream would be quite sufficient to keep the mass at an ice-cold temperature. Taking the area of the North Atlantic between the equator and the Tropic of Cancer, including also the Caribbean Sea and the Gulf of Mexico, to be 7,700,000 square miles, and the rate at which internal heat passes through the earth's surface to be that assigned by Sir William Thomson, we find that the total quantity of heat derived from the earth's crust by the above area is equal to about 88×10^{15} foot-pounds per day. But this amount is equal to only 1-894th that conveyed by the Gulf-stream, on the supposition that each pound of water carries 19,300 foot-pounds of heat. Consequently an under current from the polar regions of not more than 1-35th the volume of the Gulf-stream would suffice to keep the entire mass of water of that area within 1° of what it would be were there no heat derived from the crust of the earth; that is to say, were the water conveyed by the under current at 32°, internal heat would not maintain the mass of the ocean in the above area at more than 33°. The entire area of the North Atlantic from the equator to the arctic circle is somewhere about 16,000,000 square miles. An under current of less than 1-17th that of the Gulf-stream coming from the arctic regions would therefore suffice to keep the entire North Atlantic basin filled with ice-cold water. In short, whatever theory we adopt regarding oceanic circulation, it follows equally as a necessary consequence that the entire mass of the ocean below the stratum heated by the sun's rays must consist of cold water. For if cold water be continually coming from the polar regions either in the form of under currents, or in the form of a general underflow as Dr. Carpenter supposes, the entire under portion of the ocean must ultimately become occupied by cold water; for there is no source from which this influx of water can derive heat, save from the earth's crust. But the amount thus derived is so trifling as to produce no sensible effect. For example, a polar under current one half the size of the Gulf-stream would be sufficient to keep the entire water of the globe (below the stratum heated by the sun's rays) at an ice-cold temperature. Internal heat would not be sufficient under such circumstances to maintain the mass 1° Fahr. above the temperature it possessed when it left the polar regions.

It follows therefore that the presence of the immense mass of ice-cold water in the great depths of the ocean is completely accounted for by under currents, and there is no necessity for supposing it to be all in a state of motion towards the equator. In fact, this very state of things, which the general oceanic circulation hypothesis was devised to explain, results as a necessary consequence of polar under currents. Unless these were entirely stopped it is physically impossible that the ocean could be in any other condition.

But suppose that this immense mass of cold water occupying the great depths of the ocean were, as Dr. Carpenter assumes it to be, in a state of constant motion towards the equator, and that its sectional area were 900 times that of the Gulf-stream, it would not therefore follow that the quantity of water passing through this large sectional area must be greater than that flowing through a sectional area of the Gulf-stream; for the quantity of water flowing through this large sectional area depends entirely on the rate of motion.

I am wholly unable to understand how it could be supposed that this underflow, according to my view, is set in motion by the Gulf-stream, seeing that I have shown that the return under current is as much due to the impulse of the wind as the Gulf-stream itself.

Dr. Carpenter lays considerable stress on the important fact established by the *Challenger* expedition, that the great depths of the sea in equatorial regions are occupied by ice-cold water, while the portion heated by the sun's rays is simply a thin stratum at the surface. It seems to me that it would be difficult to find a fact more hostile to his theory than this. Were it not for this upper stratum of heated water there would be no difference between the equatorial and polar columns, and consequently nothing to produce motion. But the thinner this stratum is the less is the difference, and the less there is to produce motion.

CHAPTER IX.

EXAMINATION OF THE GRAVITATION THEORY OF OCEANIC CIRCULATION.—THE MECHANICS OF DR. CARPENTER'S THEORY.

Experimental Illustration of the Theory.—The Force exerted by Gravity.—Work performed by Gravity.—Circulation not by Convection.—Circulation depends on Difference in Density of the Equatorial and Polar Columns.—Absolute Amount of Work which can be performed by Gravity.—How Underflow is produced.—How Vertical Descent at the Poles and Ascent at the Equator is produced.—The Gibraltar Current.—Mistake in Mechanics concerning it.—The Baltic Current.

Experiment to illustrate Theory.—In support of the theory of a general movement of water between equatorial and polar regions, Dr. Carpenter adduces the authority of Humboldt and of Prof. Buff.* I have been unable to find anything in the writings of either from which it can be inferred that they have given this matter special consideration. Humboldt merely alludes to the theory, and that in the most casual manner; and that Prof. Buff has not carefully investigated the subject is apparent from the very illustration quoted by Dr. Carpenter from the "Physics of the Earth." "The water of the ocean at great depths," says Prof. Buff, "has a temperature, even under the equator, nearly approaching to the freezing-point. This low temperature cannot depend on any influence of the sea-bottom. . . . The fact, however, is explained by a continual current of cold water flowing from the polar regions towards the equator. The following well-known experiment clearly illustrates the manner of this movement. A glass vessel is to be filled with water with which some powder has been mixed, and is then to be heated *at bottom.* It will soon be seen, from

* Proceedings of the Royal Society, vol. xvii., p. 187, xviii., p. 463.

the motion of the particles of powder, that currents are set up in opposite directions through the water. Warm water rises from the bottom up through the middle of the vessel, and spreads over the surface, while the colder and therefore heavier liquid falls down at the sides of the glass."

This illustration is evidently intended to show not merely the form and direction of the great system of oceanic circulation, but also the mode in which the circulation is induced by heat. It is no doubt true that if we apply heat (say that of a spirit-lamp) to the bottom of a vessel filled with water, the water at the bottom of the vessel will become heated and rise to the surface; and if the heat be continued an ascending current of warm water will be generated; and this, of course, will give rise to a compensating under current of colder water from all sides. In like manner it is also true that, if heat were applied to the bottom of the ocean in equatorial regions, an ascending current of hot water would be also generated, giving rise to an under current of cold water from the polar regions. But all this is the diametrically opposite of what actually takes place in nature. The heat is not applied to the bottom of the ocean, so as to make the water there lighter than the water at the surface, and thus to generate an ascending current; but the heat is applied to the surface of the ocean, and the effect of this is to prevent an ascending current rather than to produce one, for it tends to keep the water at the surface lighter than the water at the bottom. In order to show how the heat of the sun produces currents in the ocean, Prof. Buff should have applied the heat, not to the bottom of his vessel, but to the upper surface of the water. But this is not all, the form of the vessel has something to do with the matter. The wider we make the vessel in proportion to its depth, the more difficult it is to produce currents by means of heat. But in order to represent what takes place in nature, we ought to have the same proportion between the depth and the superficial area of the water in our vessel as there is between the depth and the superficial area of the sea. The mean depth of the sea may be taken roughly to be about three miles.* The distance between pole and pole we shall take in round numbers to be 12,000 miles. The sun may therefore be regarded as shining upon a circular sea 12,000 miles in diameter and three miles deep. The depth of the sea to its diameter is therefore as 1 to 4,000. Suppose, now, that in our experiment we make the depth of our vessel one inch, we shall require to make its diameter 4,000 inches, or 333 feet, say, in round numbers, 100 yards in diameter. Let us, then, take a pool of water 100 yards in diameter, and one inch deep. Suppose the water to be at 32°. Apply heat to the upper surface of the pool, so as to raise the temperature of the surface of the water to 80° at the centre of the pool, the temperature diminishing towards the edge, where it is at 32°. It is found that at a depth of two miles the temperature of the water at the equator is about as low as that of the poles. We must therefore suppose the water at the centre of our pool to diminish in temperature from the surface downwards, so that at a depth of half an inch the water is at 32°. We have in this case a thin layer of warm water half an inch thick at the centre, and gradually thinning off to nothing at the edge of the pool. The lightest water, be it observed, is at the surface, so that an ascending or a descending current is impossible. The only way whereby the heat applied can have any tendency to produce motion is this:—The heating of the water expands it, consequently the surface of the pool must stand at a little higher level at its centre than at its edge, where no expansion takes place; and therefore, in order to restore the level of the pool, the water at the centre will tend to flow towards the sides. But what is the amount of this tendency? Its amount will depend upon the amount of slope, but the slope in the case under consideration amounts to only 1 in 7,340,000.

Dr. Carpenter's Experiment.—In order to obviate the objection to Professor Buff's experiment Dr. Carpenter has devised

* The average depth of the Pacific Ocean, as found by the soundings of Captain Belknap, of the U.S. steamer *Tuscarora*, made during January and February, 1874, is about 2,400 fathoms. The depth of the Atlantic is somewhat less.

another mode. But I presume his experiment was intended rather to illustrate the way in which the circulation of the ocean, according to his theory, takes place, than to prove that it actually does take place. At any rate, all that can be claimed for the experiment is the proof that water will circulate in consequence of difference of specific gravity resulting from difference of temperature. But this does not require proof, for no physicist denies it. The point which requires to be proved is this. Is the difference of specific gravity which exists in the ocean sufficient to produce the supposed circulation? Now his mode of experimenting will not prove this, unless he makes his experiment agree with the conditions already stated.

But I decidedly object to the water being heated in the way in which it has been done by him in his experiment before the Royal Geographical Society; for I feel somewhat confident that in this experiment the circulation resulted not from difference of specific gravity, as was supposed, but rather from the way in which the heat was applied. In that experiment the one half of a thick metallic plate was placed in contact with the upper surface of the water at one end of the trough; the other half, projecting over the end of the trough, was heated by means of a spirit-lamp. It is perfectly obvious that though the temperature of the great mass of the water under the plate might not be raised over 80° or so, yet the molecules in contact with the metal would have a very high temperature. These molecules, in consequence of their expansion, would be unable to sink into the cooler and denser water underneath, and thus escape the heat which was being constantly communicated to them from the heated plate. But escape they must, or their temperature would continue to rise until they would ultimately burst into vapour. They cannot ascend, neither can they descend: they therefore must be expelled by the heat from the plate in a horizontal direction. The next layer of molecules from beneath would take their place and would be expelled in a similar manner, and this process would continue so long as the heat was applied to the plate. A circulation would thus be established by the direct expansive force of vapour, and not in any way due to difference of specific gravity, as Dr. Carpenter supposes.

But supposing the heated bar to be replaced by a piece of ice, circulation would no doubt take place; but this proves nothing more than that difference of density will produce circulation, which is what no one calls in question.

The case referred to by Dr. Carpenter of the heating apparatus in London University is also unsatisfactory. The water leaves the boiler at 120° and returns to it at 80°. The difference of specific gravity between the water leaving the boiler and the water returning to it is supposed to produce the circulation. It seems to me that this difference of specific gravity has nothing whatever to do with the matter. The cause of the circulation must be sought for in the boiler itself, and not in the pipes. The heat is applied to the bottom of the boiler, not to the top. What is the temperature of the molecules in contact with the bottom of the boiler directly over the fire, is a question which must be considered before we can arrive at a just determination of the causes which produce circulation in the pipes of a heating apparatus such as that to which Dr. Carpenter refers. But, in addition to this, as the heat is applied to the bottom of the boiler and not to the top, convection comes into play, a cause which, as we shall find, does not come into play in the theory of oceanic circulation at present under our consideration.

The Force exerted by Gravity.—Dr. Carpenter speaks of his doctrine of a general oceanic circulation sustained by difference of temperature alone, "as one of which physical geographers could not recognise the importance, so long as they remained under the dominant idea that the temperature of the deep sea is everywhere 39°." And he affirms that "until it is clearly apprehended that sea-water becomes more and more dense as its temperature is reduced, the immense motive power of polar cold cannot be understood." But in chap. vii. and also in the Phil. Mag. for October, 1870 and 1871, I proved that if we take 39° as the temperature of maximum density the force exerted by gravity tending to produce circulation is just as great as when we take 32°. The reason for this is that when we take 32° as the temperature of maximum density, although we have, it is true, a greater elevation of the ocean above the place of maximum density, yet this latter occurs at the poles; while on the other hand, when we take 39°, the difference of level is less—the place not being at the poles but in about lat. 56°. Now the shorter slope from the equator to lat. 56° is as steep as the larger one from the equator to the poles, and consequently gravity exerts as much force in the production of motion in the one case as in the other. Sir John Herschel, taking 39° as the temperature of maximum density, estimated the slope at 1-32nd of an inch per mile, whereas we, taking 32° as the actual temperature of maximum density of the polar seas and calculating from modern data, find that the slope is not one-half that amount, and that the force of gravity tending to produce circulation is much less than Herschel concluded it to be. The reason, therefore, why physical geographers did not adopt the theory that oceanic circulation is the result of difference of temperature could not possibly be the one assigned by Dr. Carpenter, viz., that they had under-estimated the force of gravity by taking 39° instead of 32° as the temperature of maximum density.

The Work performed by Gravity.—But in order clearly to understand this point, it will be better to treat the matter according to the third method, and consider not the mere *force* of gravity impelling the waters, but the amount of *work* which gravitation is capable of performing.

Let us then assume the correctness of my estimate, that the height of the surface of the ocean at the equator above that at the poles is 4 feet 6 inches, for in representing the mode in which difference of specific gravity produces circulation it is of no importance what we may fix upon as the amount of the slope. In order, therefore, to avoid fractions of a foot, I shall take the slope at 4 feet instead of 4½ feet, which it actually is. A pound of water in flowing down this slope from the equator to either of the poles will perform 4 foot-pounds of work; or, more properly speaking, gravitation will. Now it is evident that when this pound of water has reached the pole, it is at the bottom of the slope, and consequently cannot descend further. Gravity, therefore, cannot perform any more work upon it; as it can only do so while the thing acted upon continues to descend —that is, moves under the force exerted. But the water will not move under the influence of gravity unless it move downward; it being in this direction only that gravity acts on the water. "But," says Dr. Carpenter, "the effect of surface-cold upon the water of the polar basin will be to reduce the temperature of its whole mass below the freezing-point of fresh water, the surface stratum *sinking* as it is cooled in virtue of its diminished bulk and increased density, and being replaced by water not yet cooled to the same degree."* By the cooling of the whole mass of polar water by cold and the heating of the water at the equator by the sun's rays the polar column of water, as we have seen, is rendered denser than the equatorial one, and in order that the two may balance each other, the polar column is necessarily shorter than the equatorial by 4 feet; and thus it is that the slope of 4 feet is formed. It is perfectly true that the water which leaves the equator warm and light, becomes by the time it reaches the pole cold and dense. But unless it be denser than the underlying polar water it will not sink down *through* it.† We are not told, however, why it should be colder than the whole mass underneath, which, according to Dr. Carpenter, is cooled by polar cold. But that he does suppose it to sink to the bottom in consequence of its contraction by cold would appear from the following quotation:—

"Until it is clearly apprehended that sea-water becomes

* Proceedings of Royal Geographical Society, vol. xv., § 22.

† It is a well-established fact that in polar regions the temperature of the sea decreases from the surface downwards; and the German Polar Expedition found that the water in very high latitudes is actually less dense at the surface than at considerable depths, thus proving that the surface-water could not sink in consequence of its greater density.

more and more dense as its temperature is reduced, and that it consequently continues to sink until it freezes, the immense motor power of polar cold cannot be apprehended. But when this has been clearly recognised, it is seen that the application of *cold at the surface* is precisely equivalent as a moving power to that application of *heat at the bottom* by which the circulation of water is sustained in every heating apparatus that makes use of it" (§ 25).

The application of cold at the surface is thus held to be equivalent as a motor power to the application of heat at the bottom. But heat applied to the bottom of a vessel produces circulation by *convection.* It makes the molecules at the bottom expand, and they, in consequence of buoyancy, rise *through* the water in the vessel. Consequently if the action of cold at the surface in polar regions is equivalent to that of heat, the cold must contract the molecules at the surface and make them sink *through* the mass of polar water beneath. But assuming this to be the meaning in the passage just quoted, how much colder is the surface water than the water beneath? Let us suppose the difference to be one degree. How much work, then, will gravity perform upon this one pound of water which is one degree colder than the mass beneath supposed to be at 32°? The force with which the pound of water will sink will not be proportional to its weight, but to the difference of weight between it and a similar bulk of the water through which it sinks. The difference between the weight of a pound of water at 31° and an equal volume of water at 32° is 1-29,000th of a pound. Now this pound of water in sinking to a depth of 10,000 feet, which is about the depth at which a polar temperature is found at the equator, would perform only one-third of a foot-pound of work. And supposing it were three degrees colder than the water beneath, it would in sinking perform only one foot-pound. This would give us only $4 + 1 = 5$ foot-pounds as the total amount that could be performed by gravitation on the pound of water from the time that it left the equator till it returned to the point from which it started. The amount of work performed in descending the slope from the equator to the pole and in sinking to a depth of 10,000 feet or so through the polar water assumed to be warmer than the surface water, comprehends the total amount of work that gravitation can possibly perform; so that the amount of force gained by such a supposition over and above that derived from the slope is trifling.

It would appear, however, that this is not what is meant after all. What Dr. Carpenter apparently means is this: when a quantity of water, say a layer one foot thick, flows down from the equator to the pole, the polar column becomes then heavier than the equatorial by the weight of this additional layer. A layer of water equal in quantity is therefore pressed away from the bottom of the column and flows off in the direction of the equator as an under current, the polar column at the same time sinking down one foot until equilibrium of the polar and equatorial columns is restored. Another foot of water now flows down upon the polar column and another foot of water is displaced from below, causing, of course, the column to descend an additional foot. The same process being continually repeated, a constant downward motion of the polar column is the result. Or, perhaps, to express the matter more accurately, owing to the constant flow of water from the equatorial regions down the slope, the weight of the polar column is kept always in excess of that of the equatorial; therefore the polar column in the effort to restore equilibrium is kept in a constant state of descent. Hence he terms it a "vertical" circulation. The following will show Dr. Carpenter's theory in his own words:—

"The action of cold on the surface water of each polar area will be exerted as follows:—

"(*a*) In diminishing the height of the polar column as compared with that of the equatorial, so that a lowering of its *level* is produced, which can only be made good by a surface-flow from the latter towards the former.

"(*b*) In producing an excess in the downward *pressure* of the column when this inflow has restored its level, in virtue of the increase of specific gravity it has gained by its reduction in volume; whereby a portion of its heavy bottom-water is displaced laterally, causing a further reduction of level, which draws in a further supply of the warmer and lighter water flowing towards its surface.

"(*c*) In imparting a downward *movement* to each new surface-stratum as its temperature undergoes reduction; so that the *entire column* may be said to be in a state of constant descent, like that which exists in the water of a tall jar when an opening is made at its bottom, and the water which flows away through it is replaced by an equivalent supply poured into the top of the jar" (§ 23).

But if this be his theory, as it evidently is, then the 4 foot-pounds (the amount of work performed by the descent of the water down the slope) comprehends all the work that gravitation can perform on a pound of water in making a complete circuit from the equator to the pole and from the pole back to the equator.

This, I trust, will be evident from the following considerations. When a pound of water has flowed down from the equator to the pole, it has descended 4 feet, and is then at the foot of the slope. Gravity has therefore no more power to pull it down to a lower level. It will not sink through the polar water, for it is not denser than the water beneath on which it rests. But it may be replied that although it will not sink through the polar water, it has nevertheless made the polar column heavier than the equatorial, and this excess of pressure forces a pound of water out from beneath and allows the column to descend. Suppose it may be argued that a quantity of water flows down from the equator, so as to raise the level of the polar water by, say, one foot. The polar column will now be rendered heavier than the equatorial by the weight of one foot of water. The pressure of the one foot will thus force a quantity of water laterally from the bottom and cause the entire column to descend till the level of equilibrium is restored. In other words, the polar column will sink one foot. Now in the sinking of this column work is performed by gravity. A certain amount of work is performed by gravity in causing the water to flow down the slope from the equator to the pole, and, in addition to this, a certain amount is performed by gravity in the vertical descent of the column.

I freely admit this to be sound reasoning, and admit that so much is due to the slope and so much to the vertical descent of the water. But here we come to the most important point, viz., is there the full slope of 4 feet and an additional vertical movement? Dr. Carpenter seems to conclude that there is, and that this vertical force is something in addition to the force which I derive from the slope. And here, I venture to think, is a radical error into which he has fallen in regard to the whole matter. Let it be observed that, when water circulates from difference of specific gravity, this vertical movement is just as real a part of the process as the flow down the slope; but the point which I maintain is that *there is no additional power derived from this vertical movement over and above what is derived from the full slope*—or, in other words, that this *primum mobile,* which he says I have overlooked, has in reality no existence.

Perhaps the following diagram will help to make the point still clearer:—

Fig. 1.

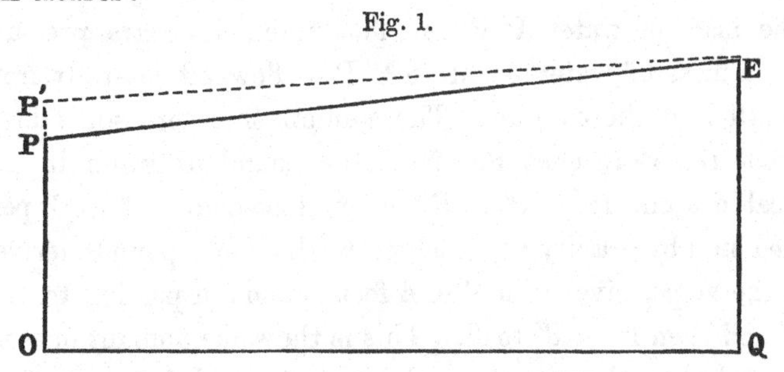

Let P (fig. 1) be the surface of the ocean at the pole, and E the surface at the equator; P O a column of water at the pole, and E Q a column at the equator. The two columns are of equal weight, and balance each other; but as the polar water is colder, and consequently denser than the equatorial, the polar column is shorter than the equatorial, the difference in the length of the two columns being 4 feet. The surface of the ocean at the equator E is 4 feet higher than the surface of the ocean at the pole P; there is therefore a slope of 4 feet from E to P. The molecules of water at E tend to flow down this slope towards P. The amount of work performed by gravity in the descent of a pound of water down this slope from E to P is therefore 4 foot-pounds.

But of course there can be no permanent circulation while the full slope remains. In order to circulation the polar column must be heavier than the equatorial. But any addition to the weight of the polar column is at the expense of the slope. In proportion as the weight of the polar column increases the less becomes the slope. This, however, makes no difference in the amount of work performed by gravity.

Suppose now that water has flowed down till an addition of one foot of water is made to the polar column, and the difference of level, of course, diminished by one foot. The surface of the ocean in this case will now be represented by the dotted line P′ E, and the slope reduced from 4 feet to 3 feet. Let us then suppose a pound of water to leave E and flow down to P′; 3 foot-pounds will be the amount of work performed. The polar column being now too heavy by the extent of the mass of water P′ P one foot thick, its extra pressure causes a mass of water equal to P′ P to flow off laterally from the bottom of the column. The column therefore sinks down one foot till P′ reaches P. Now the pound of water in this vertical descent from P′ to P has one foot-pound of work performed on it by gravity; this added to the 3 foot-pounds derived from the slope, gives a total of 4 foot-pounds in passing from E to P′ and then from P′ to P. This is the same amount of work that would have been performed had it descended directly from E to P. In like manner it can be proved that 4 foot-pounds is the amount of work performed in the descent of every pound of water of the mass P′ P. The first pound which left E flowed down the slope directly to P, and performed 4 foot-pounds of work. The last pound flowed down the slope E P′, and performed only 3 foot-pounds; but in descending from P′ to P it performed the other one foot-pound. A pound leaving at a period exactly intermediate between the two flowed down $3\frac{1}{2}$ feet of slope and descended vertically half a foot. Whatever path a pound of water might take, by the time that it reached P, 4 foot-pounds of work would be performed. But no further work can be performed after it reaches P.

But some will ask, in regard to the vertical movement, is it only in the descent of the water from P′ to P that work is performed? Water cannot descend from P′ to P, it will be urged, unless the entire column P O underneath descend also. But the column P O descends by means of gravity. Why, then, it will be asked, is not the descent of the column a motive power as real as the descent of the mass of water P′ P?

That neither force nor energy can be derived from the mere descent of the polar column P O is demonstrable thus:—The reason why the column P O descends is because, in consequence of the mass of water P′ P resting on it, its weight is in excess of the equatorial column E Q. But the force with which the column descends is equal, not to the weight of the column, but to the weight of the mass P′ P; consequently as much work would be performed by gravity in the descent of the mass P′ P (the one foot of water) alone as in the descent of the entire column P′ O, 10,000 feet in height. Suppose a ton weight is placed in each scale of a balance: the two scales balance each other. Place a pound weight in one of the scales along with the ton weight and the scale will descend. But it descends, not with the pressure of a ton and a pound, but with the pressure of the pound weight only. In the descent of the scale, say, one foot, gravity can perform only one foot-pound of work. In like manner, in the descent of the polar column, the only work available is the work of the mass P′ P laid on the top of the column. But it must be observed that in the descent of the column from P′ to P, a distance of one foot, each pound of water of the mass P′ P does not perform one foot-pound of work; for the moment that a molecule of water reaches P, it then ceases to perform further work. The molecules at the surface P′ descend one foot before reaching P; the molecules midway between P′ and P descend only half a foot before reaching P, and the molecules at the bottom of the mass are already at P, and therefore cannot perform any work. The mean distance through which the entire mass performs work is therefore half a foot. One foot-pound per pound of water represents in this case the amount of work derived from the vertical movement.

That such is the case is further evident from the following considerations. Before the polar column begins to descend, it is heavier than the equatorial by the weight of one foot of water; but when the column has descended half a foot, the polar column is heavier than the equatorial by the weight of only half a foot of water; and, as the column continues to descend, the force with which it descends continues to diminish, and when it has sunk to P the force is zero. Consequently the mean pressure or weight with which the one foot of water P′ P descended was equal to that of a layer of half a foot of water; in other words, each pound of water, taking the mass as a whole, descended with the pressure or weight of half a pound. But a half pound descending one foot performs half a foot-pound; so that whether we consider the *full pressure acting through the mean distance, or the mean pressure acting through the full distance, we get the same result,* viz. a half foot-pound as the work of vertical descent.

Now it will be found, as we shall presently see, that if we calculate the mean amount of work performed in descending the slope from the equator to the pole, 3½ foot-pounds per pound of water is the amount. The water at the bottom of the mass P P′ moved, of course, down the full slope E P 4 feet. The water at the top of the mass which descended from E to P′ descended a slope of only 3 feet. The mean descent of the whole mass is therefore 3½ feet. And this gives 3½ foot-pounds as the mean amount of work per pound of water in descending the slope; this, added to the half foot-pound derived from vertical descent, gives 4 foot-pounds as the total amount of work per pound of the mass.

I have in the above reasoning supposed one foot of water accumulated on the polar column before any vertical descent takes place. It is needless to remark that the same conclusion would have been arrived at, viz., that the total amount of work performed is 4 foot-pounds per pound of water, supposing we had considered 2 feet, or 3 feet, or even 4 feet of water to have accumulated on the polar column before vertical motion took place.

I have also, in agreement with Dr. Carpenter's mode of representing the operation, been considering the two effects, viz., the flowing of the water down the slope and the vertical descent of the polar column as taking place alternately. In nature, however, the two effects take place simultaneously; but it is needless to add that the amount of work performed would be the same whether the effects took place alternately or simultaneously.

I have also represented the level of the ocean at the equator as remaining permanent while the alterations of level were taking place at the pole. But in representing the operation as it would actually take place in nature, we should consider the equatorial column to be lowered as the polar one is being raised. We should, for example, consider the one foot of water P′ P put upon the polar column as so much taken off the equatorial column. But in viewing the problem thus we arrive at exactly the same results as before.

Let P (Fig. 2), as in Fig. 1, be the surface of the ocean at the pole, and E the surface at the equator, there being a slope of 4 feet from E to P. Suppose now a quantity of water, E E′, say, one foot thick, to flow from off the equatorial regions down upon the polar. It will thus lower the level of the equatorial column by one foot, and raise the level of the polar column by the same amount. I may, however, observe that the one foot of water in passing from E to P would have its temperature reduced from 80° to 32°, and this would produce a slight con-

Fig. 2.

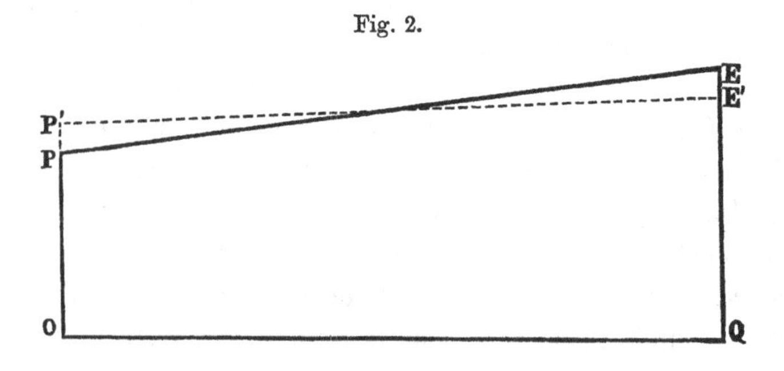

traction. But as the weight of the mass would not be affected, in order to simplify our reasoning we may leave this contraction out of consideration. Any one can easily satisfy himself that the assumption that E E′ is equal to P′ P does not in any way affect the question at issue—the only effect of the contraction being to *increase* by an infinitesimal amount the work done in descending the slope, and to *diminish* by an equally infinitesimal amount the work done in the vertical descent. If, for example, 3 foot-pounds represent the amount of work performed in descending the slope, and one foot-pound the amount performed in the vertical descent, on the supposition that E′ E does not contract in passing to the pole, then 3·0024 foot-pounds will represent the work of the slope, and 0·9976 foot-pounds the work of vertical descent when allowance is made for the contraction. But the total amount of work performed is the same in both cases. Consequently, to simplify our reasoning, we may be allowed to assume P′ P to be equal to E E′.

The slope E P being 4 feet, the slope E′ P′ is consequently 2 feet; the mean slope for the entire mass is therefore 3 feet. The mean amount of work performed by the descent of the mass will of course be 3 foot-pounds per pound of water. The amount of work performed by the vertical descent of P′ P ought therefore to be one foot-pound per pound. That this is the amount will be evident thus:—The transference of the one foot of water from the equatorial column to the polar disturbs the equilibrium by making the equatorial column too light by one foot of water and the polar column too heavy by the same amount of water. The polar column will therefore tend to sink, and the equatorial to rise till equilibrium is restored. The difference of weight of the two columns being equal to 2 feet of water, the polar column will begin to descend with a pressure of 2 feet of water; and the equatorial column will begin to rise with an equal amount of pressure. When the polar column has descended half a foot the equatorial column will have risen half a foot. The pressure of the descending polar column will now be reduced to one foot of water. And when the polar column has descended another foot, P′ will have reached P, and E′ will have reached E; the two columns will then be in equilibrium. It therefore follows that the mean pressure with which the polar column descended the one foot was equal to the pressure of one foot of water. Consequently the mean amount of work performed by the descent of the mass was equal to one foot-pound per pound of water; this, added to the 3 foot-pounds derived from the slope, gives a total of 4 foot-pounds.

In whatever way we view the question, we are led to the conclusion that if 4 feet represent the amount of slope between the equatorial and polar columns when the two are in equilibrium, then 4 foot-pounds is the total amount of work that gravity can perform upon a pound of water in overcoming the resistance to motion in its passage from the equator to the pole down the slope, and then in its vertical descent to the bottom of the ocean.

But it will be replied, not only does the one foot of water P′P descend, but the entire column P O, 10,000 feet in length, descends also. What, then, it will be asked, becomes of the force which gravity exerts in the descent of this column? We shall shortly see that this force is entirely applied in work against gravity in other parts of the circuit; so that not a single foot-pound of this force goes to overcome cohesion, friction, and other resistances; it is all spent in counteracting the efforts which gravity exerts to stop the current in another part of the circuit.

I shall now consider the next part of the movement, viz., the under or return current from the bottom of the polar to the bottom of the equatorial column. What produces this current? It is needless to say that it cannot be caused directly by gravity. Gravitation cannot directly draw any body horizontally along the earth's surface. The water that forms this current is pressed out laterally by the weight of the polar column, and flows, or rather is pushed, towards the equator to supply the vacancy caused by the ascent of the equatorial column. There is a constant flow of water from the equator to the poles along the surface, and this draining of the water from the equator is supplied by the under or return current from the poles. But the only power which can impel the water from the bottom of the polar column to the bottom of the equatorial column is the pressure of the polar column. But whence does the polar column derive its pressure? It can only press to the extent that its weight exceeds that of the equatorial column. That which exerts the pressure is therefore the mass of water which has flowed down the slope from the equator upon the polar column. It is in this case the vertical movement that causes this under current. The energy which produces this current must consequently be derived from the 4 foot-pounds resulting from the slope; for the energy of the vertical movement, as has already been proved, is derived from this source; or, in other words, whatever power this vertical movement may exert is so much deducted from the 4 foot-pounds derived from the full slope.

Let us now consider the fourth and last movement, viz., the ascent of the under current to the surface of the ocean at the equator. When this cold under current reaches the equatorial regions, it ascends to the surface to the point whence it originally started on its circuit. What, then, lifts the water from the bottom of the equatorial column to its top? This cannot be done directly, either by heat or by gravity. When heat, for example, is applied to the bottom of a vessel, the heated water at the bottom expands and, becoming lighter than the water above, rises through it to the surface; but if the heat be applied to the surface of the water instead of to the bottom, the heat will not produce an ascending current. It will tend rather to prevent such a current than to produce one—the reason being that each successive layer of water will, on account of the heat applied, become hotter and consequently lighter than the layer below it, and colder and consequently heavier than the layer above it. It therefore cannot ascend, because it is too heavy; nor can it descend, because it is too light. But the sea in equatorial regions is heated from above, and not from below; consequently the water at the bottom does not rise to the surface at the equator in virtue of any heat which it receives. A layer of water can never raise the temperature of a layer below it to a higher temperature than itself; and since it cannot do this, it cannot make the layer under it lighter than itself. That which raises the water at the equator, according to Dr. Carpenter's theory, must be the downward pressure of the polar column. When water flows down the slope from the equator to the pole, the polar column, as we have seen, becomes too heavy and the equatorial column too light; the former then sinks and the latter rises. It is the sinking of the polar column which raises the equatorial one. When the polar column descends, as much water is pressed in underneath the equatorial column as is pressed from underneath the polar column. If one foot of water is pressed from under the polar column, a foot of water is pressed in under the equatorial column. Thus, when the polar column sinks a foot, the equatorial column rises to the same extent. The equatorial water continuing to flow down the slope, the polar column descends: a foot of water is again pressed from underneath the polar column and a foot pressed in under the equatorial. As foot after foot is thus removed from the bottom of the polar column while it sinks, foot after foot is pushed in under the equatorial column while it rises; so by this means the water at the surface of the ocean in polar regions descends to the bottom, and the water at the bottom in equatorial regions ascends to the surface—the effect of solar heat and polar cold continuing, of course, to maintain the surface of the ocean in equatorial regions at a higher level than at the poles, and thus keeping up a constant state of disturbed equilibrium. Or, to state the matter in Dr. Carpenter's own words, "The cold and dense polar water, as it flows in at the bottom of the equatorial column, will not directly take the place of that which has been drafted off from the surface; but this place will be filled by the rising of the whole superincumbent column, which, being warmer, is also lighter than the cold stratum beneath. Every new arrival from the poles will take its place below that which precedes it, since its temperature will have been less affected by contact with the warmer water above it. In this way an ascending movement will be imparted to the whole equatorial column, and in due course every portion of it will come under the influence of the surface-heat of the sun."*

But the agency which raises up the water of the under current to the surface is the pressure of the polar column. The equatorial column cannot rise directly by means of gravity. Gravity, instead of raising the column, exerts all its powers to prevent its rising. Gravity here is a force acting against the current. It is the descent of the polar column, as has been stated, that raises the equatorial column. Consequently the entire amount of work performed by gravity in pulling down the polar column is spent in raising the equatorial column. Gravity performs exactly as much work in preventing motion in the equatorial column as it performs in producing motion in the polar column; so that, so far as the vertical parts of Dr. Carpenter's circulation are concerned, gravity may be said neither to produce motion nor to prevent it. And this remark, be it observed, applies not only to P O and E Q, but also to the parts P′ P and E E′ of the two columns. When a mass of

* Proceedings of the Royal Society, vol. xix., p. 215.

water E E′, say one foot deep, is removed off the equatorial column and placed upon the polar column, the latter column is then heavier than the former by the weight of two feet of water. Gravity then exerts more force in pulling the polar column down than it does in preventing the equatorial column from rising; and the consequence is that the polar column begins to descend and the equatorial column to rise. But as the polar column continues to descend and the equatorial to rise, the power of gravity to produce motion in the polar column diminishes, and the power of gravity to prevent motion in the equatorial column increases; and when P′ descends to P and E′ rises to E, the power of gravity to prevent motion in the equatorial column is exactly equal to the power of gravity to produce motion in the polar column, and consequently motion ceases. It therefore follows that the entire amount of work performed by the descent of P′ P is spent in raising E′ E against gravity.

It follows also that inequalities in the sea-bottom cannot in any way aid the circulation; for although the cold under current should in its progress come to a deep trough filled with water less dense than itself, it would no doubt sink to the bottom of the hollow; yet before it could get out again as much work would have to be performed against gravity as was performed by gravity in sinking it. But whilst inequalities in the bed of the ocean would not aid the current, they would nevertheless very considerably retard it by the obstructions which they would offer to the motion of the water.

We have been assuming that the weight of P′ P is equal to that of E E′; but the mass P′ P must be greater than E E′ because P′ P has not only to raise E E′, but to impel the under current—to push the water along the sea-bottom from the pole to the equator. So we must have a mass of water, in addition to P′ P, placed on the polar column to enable it to produce the under current in addition to the raising of the equatorial column.

It follows also that the amount of work which can be performed by gravity depends entirely on the *difference* of temperature between the equatorial and the polar waters, and is wholly independent of the way in which the temperature may decrease from the equator to the poles. Suppose, in agreement with Dr. Carpenter's idea,* that the equatorial heat and polar cold should be confined to limited areas, and that through the intermediate space no great difference of temperature should prevail. Such an arrangement as this would not increase the amount of work which gravity could perform; it would simply make the slope steeper at the two extremes and flatter in the intervening space. It would no doubt aid the surface-flow of the water near the equator and the poles, but it would retard in a corresponding degree the flow of the water in the intermediate regions. In short, it would merely destroy the uniformity of the slope without aiding in the least degree the general motion of the water.

It is therefore demonstrable that *the energy derived from the full slope, whatever that slope may be, comprehends all that can possibly be obtained from gravity.*

It cannot be urged as an objection to what has been advanced that I have determined simply the amount of the force acting on the water at the surface of the ocean and not that on the water at all depths—that I have estimated the amount of work which gravity can perform on a given quantity of water at the surface, but not the total amount of work which gravity can perform on the entire ocean. This objection will not stand, because it is at the surface of the ocean where the greatest difference of temperature, and consequently of density, exists between the equatorial and polar waters, and therefore there that gravity exerts its greatest force. And if gravity be unable to move the water at the surface, it is much less able to do so under the surface. So far as the question at issue is concerned, any calculations as to the amount of force exerted by gravity at various depths are needless.

It is maintained also that the winds cannot produce a vertical current except under some very peculiar conditions. We have

* *Nature* for July 6, 1871.

already seen that, according to Dr. Carpenter's theory, the vertical motion is caused by the water flowing off the equatorial column, down the slope, upon the polar column, thus destroying the equilibrium between the two by diminishing the weight of the equatorial column and increasing that of the polar column. In order that equilibrium may be restored, the polar column sinks and the equatorial one rises. Now must not the same effect occur, supposing the water to be transferred from the one column to the other, by the influence of the winds instead of by the influence of gravity? The vertical descent and ascent of these columns depend entirely upon the difference in their weights, and not upon the nature of the agency which makes this difference. So far as difference of weight is concerned, 2 feet of water, propelled down the slope from the equatorial column to the polar by the winds, will produce just the same effect as though it had been propelled by gravity. If vertical motion follows as a necessary consequence from a transference of water from the equator to the poles by gravity, it follows equally as a necessary consequence from the same transference by the winds; so that one is not at liberty to advocate a vertical circulation in the one case and to deny it in the other.

Gravitation Theory of the Gibraltar Current.—If difference of specific gravity fails to account for the currents of the ocean in general, it certainly fails in a still more decided manner to account for the Gibraltar current. The existence of the submarine ridge between Capes Trafalgar and Spartel, as was shown in the Phil. Mag. for October, 1871, p. 269, affects currents resulting from difference of specific gravity in a manner which does not seem to have suggested itself to Dr. Carpenter. The pressure of water and other fluids is not like that of a solid —not like that of the weight in the scale of a balance, simply a downward pressure. Fluids press downwards like the solids, but they also press laterally. The pressure of water is hydrostatic. If we fill a basin with water or any other fluid, the fluid remains in perfect equilibrium, provided the sides of the basin be sufficiently strong to resist the pressure. The Mediterranean and Atlantic, up to the level of the submarine ridge referred to, may be regarded as huge basins, the sides of which are sufficiently strong to resist all pressure. It follows that, however much denser the water of the Mediterranean may be than that of the Atlantic, it is only the water above the level of the ridge that can possibly exercise any influence in the way of disturbing equilibrium, so as to cause the level of the Mediterranean to stand lower than that of the Atlantic. The water of the Atlantic below the level of this ridge might be as light as air, and that of the Mediterranean as heavy as molten lead, but this could produce no disturbance of equilibrium; and if there be no difference of density between the Atlantic and the Mediterranean waters from the surface down to the level of the top of the ridge, then there can be nothing to produce the circulation which Dr. Carpenter infers. Suppose both basins empty, and dense water to be poured into the Mediterranean, and water less dense into the Atlantic, until they are both filled up to the level of the ridge, it is evident that the heavier water in the one basin can exercise no influence in raising the level of the lighter water in the other basin, the entire pressure being borne by the sides of the basins. But if we continue to pour in water till the surface is raised, say one foot, above the level of the ridge, then there is nothing to resist the lateral pressure of this one foot of water in the Mediterranean but the counter pressure of the one foot in the Atlantic. But as the Mediterranean water is denser than the Atlantic, this one foot of water will consequently exert more pressure than the one foot of water of the Atlantic. We must therefore continue to pour more water into the Atlantic until its lateral pressure equals that of the Mediterranean. The two seas will then be in equilibrium, but the surface of the Atlantic will of course be at a higher level than the surface of the Mediterranean. The difference of level will be proportionate to the difference in density of the waters of the two seas. But here we come to the point of importance. In determining the difference of level between the two seas, or, which is the same thing, the difference of level between a column of the Atlantic and a column of the Mediterranean, we must take into consideration *only the water which lies above the level of the ridge.* If there be one foot of water above the ridge, then there is a difference of level proportionate to the difference of pressure between the one foot of water of the two seas. If there be 2 feet, 3 feet, or any number of feet of water above the level of the ridge, the difference of level is proportionate to the 2 feet, 3 feet, or whatever number of feet there may be of water above the ridge. If, for example, 13 should represent the density of the Mediterranean water and 12 the density of the Atlantic water, then if there were one foot of water in the Mediterranean above the level of the ridge, there would require to be one foot one inch of water in the Atlantic above the ridge in order that the two might be in equilibrium. The difference of level would therefore be one inch. If there were 2 feet of water, the difference of level would be 2 inches; if 3 feet, the difference would be 3 inches, and so on. And this would follow, no matter what the actual depth of the two basins might be; the water below the level of the ridge exercising no influence whatever on the level of the surface.

Taking Dr. Carpenter's own data as to the density of the Mediterranean and Atlantic waters, what, then, is the difference of density? The submarine ridge comes to within 167 fathoms of the surface; say, in round numbers, to within 1,000 feet. What are the densities of the two basins down to the depth of 1,000 feet? According to Dr. Carpenter there is little, if any, difference. His own words on this point are these:—"A comparison of these results leaves no doubt that there is an excess of salinity in the water of the Mediterranean above that of the Atlantic; but that this excess *is* slight in the surface-water, whilst somewhat greater in the deeper water" (§ 7). "Again, it was found by examining samples of water taken from the surface, from 100 fathoms, from 250 fathoms, and from 400 fathoms respectively, that whilst the *first two* had the *characteristic temperature and density of Atlantic water*, the last two had the characteristics and density of Mediterranean water" (§ 13). Here, at least to the depth of 100 fathoms or 600 feet, there is little difference of density between the waters of the two basins. Consequently down to the depth of 600 feet, there is nothing to produce any sensible disturbance of equilibrium. If there be any sensible disturbance of equilibrium, it must be in consequence of difference of density which may exist between the depths of 600 feet and the surface of the ridge. We have nothing to do with any difference which may exist between the water of the Mediterranean and the Atlantic below the ridge; the water in the Mediterranean basin may be as heavy as mercury below 1,000 feet: but this can have no effect in disturbing equilibrium. The water to the depth of 600 feet being of the same density in both seas, the length of the two columns acting on each other is therefore reduced to 400 feet—that is, to that stratum of water lying at a depth of from 600 to the surface of the ridge 1,000 feet below the surface. But, to give the theory full justice, we shall take the Mediterranean stratum at the density of the deep water of the Mediterranean, which he found to be about 1·029, and the density of the Atlantic stratum at 1·026. The difference of density between the two columns is therefore ·003. Consequently, if the height of the Mediterranean column be 400 feet, it will be balanced by the Atlantic column of 401·2 feet; the difference of level between the Mediterranean and the Atlantic cannot therefore be more than 1·2 foot. The amount of work that can be performed by gravity in the case of the Gibraltar current is little more than one foot-pound per pound of water, an amount of energy evidently inadequate to produce the current.

It is true that in his last expedition Dr. Carpenter found the bottom-water on the ridge somewhat denser than Atlantic water at the same depth, the former being 1·0292 and the latter 1·0265; but it also proved to be denser than Mediterranean water at the same depth. He found, for example, that "the dense Mediterranean water lies about 100 fathoms nearer the surface over a 300-fathoms bottom, than it does where the bottom sinks to more than 500 fathoms" (§ 51). But any excess of density which might exist at the ridge could have no tendency whatever to make the Mediterranean column preponderate over the Atlantic column, any more than could a weight placed over the fulcrum of a balance have a tendency to make the one scale weigh down the other.

If the objection referred to be sound, it shows the mechanical impossibility of the theory. It proves that whether there be an under current or not, or whether the dense water lying in the deep trough of the Mediterranean be carried over the submarine ridge into the Atlantic or not, the explanation offered by Dr. Carpenter is one which cannot be admitted. It is incumbent on him to explain either (1) how the almost infinitesimal difference of density which exists between the Atlantic and Mediterranean columns down to the level of the ridge can produce the upper and under currents carrying the deep and dense water of the Mediterranean over the ridge, or (2) how all this can be done by means of the difference of density which exists below the level of the ridge.* What the true cause of the Gibraltar current really is will be considered in Chap. XIII.

The Baltic Current.—The entrance to the Baltic Sea is in some places not over 50 or 60 feet deep. It follows, therefore, from what has already been proved in regard to the Gibraltar current, that the influence of gravity must be even still less in causing a current in the Baltic strait than in the Gibraltar strait.

* Since the above objection to the Gravitation Theory of the Gibraltar Current was advanced three years ago, Dr. Carpenter appears to have abandoned the theory to a great extent. He now admits (Proceedings of Royal Geographical Society, vol. xviii., pp. 319—334, 1874) that the current is almost wholly due not to difference of specific gravity, but to an excess of evaporation in the Mediterranean over the return by rain and rivers.

CHAPTER X.

EXAMINATION OF THE GRAVITATION THEORY OF OCEANIC CIRCULATION.—DR. CARPENTER'S THEORY.—OBJECTIONS CONSIDERED.

Modus Operandi of the Matter.—Polar Cold considered by Dr. Carpenter the *Primum Mobile*.—Supposed Influence of Heat derived from the Earth's Crust.—Circulation without Difference of Level.—A Confusion of Ideas in Reference to the supposed Agency of Polar Cold.—M. Dubuat's Experiments.—A Begging of the Question at Issue.—Pressure as a Cause of Circulation.

In the foregoing chapter, the substance of which appeared in the Phil. Mag. for October, 1871, I have represented the manner in which difference of specific gravity produces circulation. But Dr. Carpenter appears to think that there are some important points which I have overlooked. These I shall now proceed to consider in detail.

"Mr. Croll's whole manner of treating the subject," he says, "is so different from that which it appears to me to require, and he has so completely misapprehended my own view of the question, that I feel it requisite to present this in fuller detail in order that physicists and mathematicians, having both sides fully before them, may judge between us" (§ 26).*

He then refers to a point so obvious as hardly to require consideration, viz., the effect which results when the surface of the entire area of a lake or pond of water is cooled. The whole of the surface-film, being chilled at the same time, sinks through the subjacent water, and a new film from the warmer layer immediately beneath the surface rises into its place. This being cooled in its turn, sinks, and so on. He next considers

* Proceedings of Royal Society, No. 138, § 26.

what takes place when only a portion of the surface of the pond is cooled, and shows that in this case the surface-film which descends is replaced not from beneath, but by an inflow from the neighbouring area.

"That such must be the case," says Dr. Carpenter, "appears to me so self-evident that I am surprised that any person conversant with the principles of physical science should hesitate in admitting it, still more that he should explicitly deny it. But since others may feel the same difficulty as Mr. Croll, it may be worth while for me to present the case in a form of yet more elementary simplicity" (§ 29).

Then, in order to show the mode in which the general oceanic circulation takes place, he supposes two cylindrical vessels, W and C, of equal size, to be filled with sea-water. Cylinder W represents the equatorial column, and the water contained in it has its temperature maintained at 60°; whilst the water in the other cylinder C, representing the polar column, has its temperature maintained at 30° by means of the constant application of cold at the top. Free communication is maintained between the two cylinders at top and bottom; and the water in the cold cylinder being, in virtue of its low temperature, denser than the water in the warm cylinder, the two columns are therefore not in static equilibrium. The cold, and hence heavier column tends to produce an outflow of water from its bottom to the bottom of the warm column, which outflow is replaced by an inflow from the top of the warm column to the top of the cold column. In fact, we have just a simple repetition of what he has given over and over again in his various memoirs on the subject. But why so repeatedly enter into the *modus operandi* of the matter? Who feels any difficulty in understanding how the circulation is produced?

Polar Cold considered by Dr. Carpenter the Primum Mobile.— It is evident that Dr. Carpenter believes that he has found in polar *cold* an agency the potency of which, in producing a general oceanic circulation, has been overlooked by physicists; and it is with the view of developing his ideas on this subject that he has entered so fully and so frequently into the exposition of his theory. "If I have myself done anything," he says, "to strengthen the doctrine, it has been by showing that polar cold, rather than equatorial heat, is the *primum mobile* of this circulation."*

The influence of the sun in heating the waters of the inter-tropical seas is, in Dr. Carpenter's manner of viewing the problem, of no great importance. The efficient cause of motion he considers resides in *cold* rather than in *heat*. In fact, he even goes the length of maintaining that, as a power in the production of the general interchange of equatorial and polar water, the effect of polar cold is so much superior to that of inter-tropical heat, that the influence of the latter may be *practically disregarded.*

"Suppose two basins of ocean-water," he says, "connected by a strait to be placed under such different climatic conditions that the surface of one is exposed to the heating influence of tropical sunshine, whilst the surface of the other is subjected to the extreme cold of the sunless polar winter. The effect of the surface-heat upon the water of the tropical basin will be for the most part limited (as I shall presently show) to its uppermost stratum, and may here be *practically disregarded.*" †

Dr. Carpenter's idea regarding the efficiency of cold in producing motion seems to me to be not only opposed to the generally received views on the subject, but wholly irreconcileable with the ordinary principles of mechanics. In fact, there are so many points on which Dr. Carpenter's theory of a "General *Vertical* Oceanic Circulation" differs from the generally received views on the subject of circulation by means of difference of specific gravity, that I have thought it advisable to enter somewhat minutely into the consideration of the mechanics of that theory, the more so as he has so repeatedly asserted that eminent physicists agree with what he has advanced on the subject.

According to the generally received theory, the circulation

* Proceedings of Royal Geographical Society, January 9, 1871. † Ibid.

is due to the *difference of density* between the sea in equatorial and polar regions. The real efficient cause is gravity; but gravity cannot act when there is no difference of specific gravity. If the sea were of equal density from the poles to the equator, gravity could exercise no influence in the production of circulation; and the influence which it does possess is in proportion to the difference of density. But the difference of density between equatorial and polar waters is in turn due not absolutely either to polar cold or to tropical heat, but to both—or, in other words, to the *difference* of temperature between the polar and equatorial seas. This difference, in the very nature of things, must be as much the result of equatorial heat as of polar cold. If the sea in equatorial regions were not being heated by the sun as rapidly as the sea in polar regions is being cooled, the difference of temperature between them, and consequently the difference of density, would be diminishing, and in course of time would disappear altogether. As has already been shown, it is a necessary consequence that the water flowing from equatorial to polar regions must be compensated by an equal amount flowing from polar to equatorial regions. Now, if the water flowing from polar to equatorial regions were not being heated as rapidly as the water flowing from equatorial to polar regions is being cooled, the equatorial seas would gradually become colder and colder until no sensible difference of temperature existed between them and the polar oceans. In fact, *equality of the two rates* is necessary to the very existence of such a general circulation as that advocated by Dr. Carpenter. If he admits that the general interchange of equatorial and polar water advocated by him is caused by the difference of density between the water at the equator and the poles, resulting from difference of temperature, then he must admit also that this difference of density is just as much due to the heating of the equatorial water by the sun as it is to the cooling of the polar water by radiation and other means—or, in other words, that it is as much due to equatorial heat as to polar cold. And if so, it cannot be true that polar cold rather than equatorial heat is the "*primum mobile*" of this circulation; and far less can it be true that the heating of the equatorial water by the sun is of so little importance that it may be "practically disregarded."

Supposed Influence of Heat derived from the Earth's Crust.—There is, according to Dr. Carpenter, another agent concerned in the production of the general oceanic circulation, viz., the heat derived by the bottom of the ocean from the crust of the earth.* We have no reason to believe that the quantity of internal heat coming through the earth's crust is greater in one part of the globe than in another; nor have we any grounds for concluding that the bottom of inter-tropical seas receives more heat from the earth's crust than the bottom of those in polar regions. But if the polar seas receive as much heat from this source as the seas within the tropics, then the difference of density between the two cannot possibly be due to heat received from the earth's crust; and this being so, it is mechanically impossible that internal heat can be a cause in the production of the general oceanic circulation.

Circulation without Difference of Level.—There is another part of the theory which appears to me irreconcilable with mechanics. It is maintained that this general circulation takes place without any difference of level between the equator and the poles. Referring to the case of the two cylinders W and C, which represent the equatorial and polar columns respectively, Dr. Carpenter says:—

"The force which will thus lift up the entire column of water in W is that which causes the descent of the entire column in C, namely, the excess of gravity constantly acting in C,—the levels of the two columns, and consequently their heights, being maintained at a *constant equality* by the free passage of surface-water from W to C."

"The whole of Mr. Croll's discussion of this question, however," he continues, "proceeds upon the assumption that the levels of the polar and equatorial columns are *not kept at an*

* See §§ 20, 34; also Brit. Assoc. Report for 1872, p. 49, and other places.

equality, &c." (§ 30.) And again, "Now, so far from asserting (as Captain Maury has done) that the trifling difference of level arising from inequality of temperature is adequate to the production of ocean-currents, I simply affirm that as fast as the level is disturbed by change of temperature it will be restored by gravity." (§ 23.)*

In order to understand more clearly how the circulation under consideration cannot take place without a difference of level, let W E (Fig. 3) represent the equatorial column, and C P the polar column. The equatorial column is warmer than the polar column because it receives *more* heat from the sun than the latter; and the polar is colder than the equatorial column

Fig. 3.

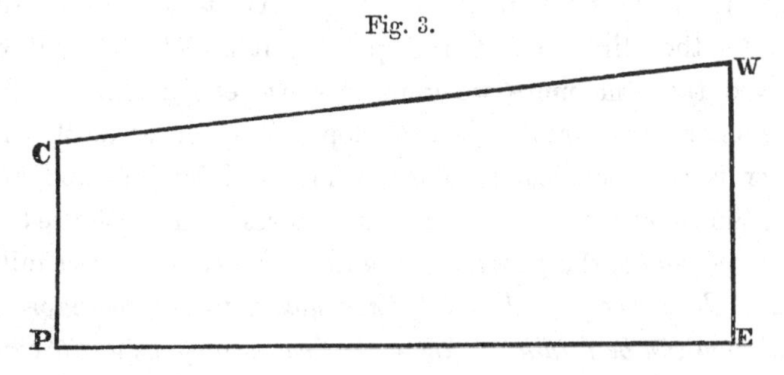

because it receives *less*. The difference in the density of the two columns results from their difference of temperature; and the difference of temperature results in turn from the difference in the quantity of heat received from the sun by each. Or, to express the matter in other words, the difference of density (and consequently the circulation under consideration) is due to the excess of heat received from the sun by the equatorial over that received by the polar column; so that to leave out of account the superheating of the inter-tropical waters by the sun is to leave out of account the very thing of all others that is absolutely essential to the existence of the circulation. The water being assumed to be the same in both columns and differing only as regards temperature, and the equatorial column possessing more heat than the polar, and being therefore less

* See also to the same effect Brit. Assoc. Report, 1872, p. 50.

dense than the latter, it follows, in order that the two columns may be in static equilibrium, that the surface of the equatorial column must stand at a higher level than that of the polar. This produces the slope W C from the equator to the pole. The extent of the slope will of course depend upon the extent of the difference of their temperatures. But, as was shown on a former occasion,* it is impossible that static equilibrium can ever be fully obtained, because the slope occasioned by the elevation of the equatorial column above the polar produces what we may be allowed to call a *molecular* disturbance of equilibrium. The surface of the ocean, or the molecules of water lying on the slope, are not in a position of equilibrium, but tend, in virtue of gravity, to roll down the slope in the direction of the polar column C. It will be observed that the more we gain of static equilibrium of the entire ocean the greater is the slope, and consequently the greater is the disturbance of molecular equilibrium; and, *vice versâ*, the more molecular equilibrium is restored by the reduction of the slope, the greater is the disturbance of static equilibrium. *It is therefore absolutely impossible that both conditions of equilibrium can be fulfilled at the same time so long as a difference of temperature exists between the two columns.* And this conclusion holds true even though we should assume water to be a perfect fluid absolutely devoid of viscosity. It follows, therefore, that a general oceanic circulation without a difference of level is a *mechanical impossibility.*

In a case of actual circulation due to difference of gravity, there is always a constant disturbance of both *static* and molecular equilibrium. Column C is always higher and column W always lower than it ought to be were the two in equilibrium; but they never can be at the same level.

It is quite conceivable, of course, that the two conditions of equilibrium may be fulfilled alternately. We can conceive column C remaining stationary till the water flowing from column W has restored the level. And after the level is re-

* Phil. Mag. for Oct. 1871.

stored we can conceive the polar column C sinking and the equatorial column W rising till the two perfectly balance each other. Such a mode of circulation, consisting of an alternate surface-flow and vertical descent and ascent of the columns, though conceivable, is in reality impossible in nature; for there are no means by which the polar column C could be supported from sinking till the level had been restored. But Dr. Carpenter does not assume that the general oceanic circulation takes place in this intermitting manner; according to him, the circulation is *constant*. He asserts that there is a "*continual* transference of water from the bottom of C to the bottom of W, and from the top of W to the top of C, with a *constant* descending movement in C and a *constant* ascending movement in W" (§ 29). But such a condition of things is irreconcilable with the idea of "the levels of the two columns, and consequently their heights, being maintained at a *constant* equality" (§ 29).

Although Dr. Carpenter does not admit the existence of a permanent difference of level between the equator and the pole, he nevertheless speaks of a depression of level in the polar basin resulting from the contraction by cooling of the water flowing into it. This reduction of level induces an inflow of water from the surrounding area; "and since what is drawn away," to quote his own words, "is supplied from a yet greater distance, the continued cooling of the surface-stratum in the polar basin will cause a 'set' of waters towards it, to be propagated backwards through the whole intervening ocean in communication with it until it reaches the tropical area." The slope produced between the polar basin and the surrounding area, if sufficiently great, will enable the water in the surrounding area to flow polewards; but unless this slope extend to the equator, it will not enable the tropical waters also to flow polewards. One of two things necessarily follows: either the slope extends from the equator to the pole, or water can flow from the equator to the pole without a slope. If Dr. Carpenter maintains the former, he contradicts himself; and if he adopts the latter, he contradicts an obvious principle of mechanics.

A Confusion of Ideas in Reference to the supposed Agency of Polar Cold.—It seems to me that Dr. Carpenter has been somewhat misled by a slight confusion of ideas in reference to the supposed agency of polar cold. This is brought out forcibly in the following passage from his memoir in the Proceedings of the Royal Geographical Society, vol. xv.

"Mr. Croll, in arguing against the doctrine of a general oceanic circulation sustained by difference of temperature, and *justly maintaining* that such a circulation cannot be produced by the application of heat at the surface, has entirely ignored the agency of cold."

It is here supposed that there are two agents at work in the production of the general oceanic circulation. The one agent is *heat*, acting at the equatorial regions; and the other agent is *cold*, acting at the polar regions. It is supposed that the agency of cold is far more powerful than that of heat. In fact so trifling is the agency of equatorial heat in comparison with that of polar cold that it may be "practically disregarded"—left out of account altogether,—polar cold being the *primum mobile* of the circulation. It is supposed also that I have considered the efficiency of one of the agents, viz., heat, and found it totally inadequate to produce the circulation in question; and it is admitted also that my conclusions are perfectly correct. But then I am supposed to have left out of account the other agent, viz., polar cold, the only agent possessing real potency. Had I taken into account polar cold, it is supposed that I should have found at once a cause perfectly adequate to produce the required effect.

This is a fair statement of Dr. Carpenter's views on the subject; I am unable, at least, to attach any other meaning to his words. And I have no doubt they are also the views which have been adopted by those who have accepted his theory.

It must be sufficiently evident from what has already been stated, that the notion of there being two separate agents at work producing circulation, namely heat and cold, the one of which is assumed to have much more potency than the other, is not only opposed to the views entertained by physicists, but is also wholly irreconcilable with the ordinary principles of mechanics. But more than this, if we analyze the subject a little so as to remove some of the confusion of ideas which besets it, we shall find that these views are irreconcilable with even Dr. Carpenter's own explanation of the cause of the general oceanic circulation.

Cold is not a something positive imparted to the polar waters giving them motion, and of which the tropical waters are deprived. If, dipping one hand into a basin filled with tropical water at 80° and the other into one filled with polar water at 32°, we refer to our *sensations*, we call the water in the one *hot* and that in the other *cold;* but so far as the water itself is concerned heat and cold simply mean difference in the amounts of heat possessed. Both the polar and the tropical water possess a certain amount of energy in the form of heat, only the polar water does not possess so much of it as the tropical.

How, then, according to Dr. Carpenter, does polar cold impart motion to the water? The warm water flowing in upon the polar column becomes chilled by cold, but it is not cooled below that of the water underneath; for, according to Dr. Carpenter, the ocean in polar regions is as cold and as dense underneath as at the surface. The cooled surface-water does not sink through the water underneath, like the surface-water of a pond chilled during a frosty night. "The descending motion in column C will not consist," he says, "in a successional descent of surface-films from above downwards, but it will be a downward movement of the *entire mass,* as if water in a tall jar were being drawn off through an orifice at the bottom" (§ 29). There is a downward motion of the entire column, producing an outflow of water at the bottom towards the equatorial column W, which outflow is compensated by an inflow from the top of the equatorial column to the top of the polar column C. But what causes column C to descend? The cause of the descent is its excess of weight over that of column W. Column C descends and column W ascends, for the same reason that in a balance the heavy scale descends and the light scale rises. Column C descends not simply because it is cold, but because it is *colder* than column W. Column C descends not simply because in consequence of being cold it is dense and therefore heavy, but because in consequence of being cold it is *denser* and therefore *heavier* than column W. It might be as cold as frozen mercury and as heavy as lead; but it would not on that account descend unless it were heavier than column W. The descent of column C and ascent of column W, and consequently the general oceanic circulation, results, therefore, according to Dr. Carpenter's explanation, from the *difference* in the weights of the two columns; and the difference in the weights of the two columns results from their *difference* of density; and the difference of density of the two columns in turn results from their *difference* of temperature. But it has already been proved that the difference of temperature between the polar and equatorial columns depends wholly on the difference in the amount of heat received by each from the sun. The equatorial column W possesses more heat than the polar column C, solely because it receives more heat from the sun than column C. Consequently Dr. Carpenter's statement that the circulation is produced by polar cold rather than by equatorial heat, is just as much in contradiction to his own theory as it is to the principles of mechanics. Again, his admission that the general oceanic circulation "cannot be produced by the application of heat to the surface," is virtually a giving up the whole point in debate; for according to his gravitation theory, and every form of that theory, the circulation results from *difference* of temperature between equatorial and polar seas; but this difference, as we have seen, is entirely owing to the difference in the amount of heat received from the sun at these two places. The heat received, however, is "surface-heat;" for it is at the surface that the ocean receives all its heat from the sun; and consequently if surface-heat cannot produce the effect required, nothing else can.

M. Dubuat's Experiments.—Referring to the experiments of

M. Dubuat adduced by me to show that water would not run down a slope of 1 in 1,820,000,* he says, "Now the experiments of M. Dubuat had reference, not to the slow restoration of level produced by the motion of water on itself, but to the sensible movement of water flowing over solid surfaces and retarded by its friction against them" (§ 22). Dr. Carpenter's meaning, I presume, is that if the incline consist of any solid substance, water will not flow down it; but if it be made of *water* itself, *water* will flow down it. But in M. Dubuat's experiments it was only the molecules in actual *contact* with the solid incline that could possibly be retarded by friction against it. The molecules not in contact with the solid incline evidently rested upon an *incline of water*, and were at perfect liberty to roll down that incline if they chose; but they did not do so; and consequently M. Dubuat's experiment proved that water will not flow over itself on an incline of 1 in 1,000,000.

A Begging of the Question at Issue.—"It is to be remembered," says Dr. Carpenter, "that, however small the original amount of movement may be, a *momentum* tending to its continuance *must* be generated from the instant of its commencement; so that if the initiating force be in constant action, there will be a *progressive acceleration* of its rate, until the increase of resistance equalises the tendency to further acceleration. Now, if it be admitted that the propagation of the disturbance of equilibrium from one column to another is simply *retarded, not* prevented, by the viscosity of the liquid, I cannot see how the conclusion can be resisted, that the constantly maintained difference of gravity between the polar and equatorial columns really acts as a *vis viva* in maintaining a circulation between them" (§ 35).

If it be true, as Dr. Carpenter asserts, that in the case of the general oceanic circulation advocated by him "viscosity" simply *retards* motion, but does not *prevent* it, I certainly agree with him "that the constantly maintained difference of gravity between the polar and equatorial columns really acts as a *vis*

* The actual slope, however, does not amount to more than 1 in 7,000,000.

viva in maintaining a circulation between them." But to assert that it merely retards, but does not prevent, motion, is simply *begging the question* at *issue*. It is an established principle that if the *force* resisting motion be greater than the force tending to produce it, then no motion can take place and no work can be performed. The experiments of M. Dubuat prove that the *force* of the molecular resistance of water to motion is *greater* than the *force* derived from a slope of 1 in 1,000,000; and therefore it is simply begging the question at issue to assert that it is *less*. The experiments of MM. Barlow, Rainey, and others, to which he alludes, are scarcely worthy of consideration in relation to the present question, because we know nothing whatever regarding the actual amount of force producing motion of the water in these experiments, further than that it must have been enormously greater than that derived from a slope of 1 in 1,000,000.

Supposed Argument from the Tides.—Dr. Carpenter advances Mr. Ferrel's argument in regard to the tides. The power of the moon to disturb the earth's water, he asserts, is, according to Herschel, only 1-11,400,000th part of gravity, and that of the sun not over 1-25,736,400th part of gravity; yet the moon's attractive force, even when counteracted by the sun, will produce a rise of the ocean. But as the disturbance of gravity produced by difference of temperature is far greater than the above, it ought to produce circulation.

It is here supposed that the force exerted by gravity on the ocean, resulting from difference of temperature, tending to produce the general oceanic circulation, is much greater than the force exerted on the ocean by the moon in the production of the tides. But if we examine the subject we shall find that the opposite is the case. The attraction of the moon tending to lift the waters of the ocean acts directly on every molecule from the surface to the bottom; but the force of gravity tending to produce the circulation in question acts directly on only a portion of the ocean. Gravity can exercise no direct force in impelling the underflow from the polar to the equatorial regions, nor in raising the water to the surface when it reaches the equatorial regions. Gravity can exercise no direct influence in pulling the water horizontally along the earth's surface, nor in raising it up to the surface. The pull of gravity is always *downwards,* never *horizontally* nor upwards. Gravity will tend to pull the surface-water from the equator to the poles because here we have *descent.* Gravity will tend to sink the polar column because here also we have *descent.* But these are the only parts of the circuit where gravity has any tendency to produce motion. Motion in the other parts of the circuit, viz., along the bottom of the ocean from the poles to the equator and in raising the equatorial column, is produced by the *pressure* of the polar column; and consequently it is only *indirectly* that gravity may be said to produce motion in those parts. It is true that on certain portions of the ocean the force of gravity tending to produce motion is greater than the force of the moon's attraction, tending to produce the tides; but this portion of the ocean is of inconsiderable extent. The total force of gravity acting on the entire ocean tending to produce circulation is in reality prodigiously less than the total force of the moon tending to produce the tides.

It is no doubt a somewhat difficult problem to determine accurately the total amount of force exercised by gravity on the ocean; but for our present purpose this is not necessary. All that we require at present is a very rough estimate indeed. And this can be attained by very simple considerations. Suppose we assume the mean depth of the sea to be, say, three miles. The mean depth may yet be found to be somewhat less than this, or it may be found to be somewhat greater; a slight mistake, however, in regard to the mass of the ocean will not materially affect our conclusions. Taking the depth at 3 miles, the force or direct pull of gravity on the entire waters of the ocean tending to the production of the general circulation will not amount to more than 1-24,000,000,000th that of gravity, or only about 1-2,100th that of the attraction of the moon in the production of the tides. Let it be observed that I am referring to the force or pull of gravity, and not to hydrostatic pressure.

The moon, by raising the waters of the ocean, will produce a slope of 2 feet in a quadrant; and because the raised water sinks and the level is restored, Mr. Ferrel concludes that a similar slope of 2 feet produced by difference of temperature will therefore be sufficient to produce motion and restore level. But it is overlooked that the restoration of level in the case of the tides is as truly the work of the moon as the disturbance of that level is. For the water raised by the attraction of the moon at one time is again, six hours afterwards, pulled down by the moon when the earth has turned round a quadrant.

No doubt the earth's gravity alone would in course of time restore the level; but this does not follow as a logical consequence from Mr. Ferrel's premises. If we suppose a slope to be produced in the ocean by the moon and the moon's attraction withdrawn so as to allow the water to sink to its original level, the raised side will be the heaviest and the depressed side the lightest; consequently the raised side will tend to sink and the depressed side will tend to rise, in order that the ocean may regain its static equilibrium. But when a difference of level is produced by difference of temperature, the raised side is always the lightest and the depressed side is always the heaviest; consequently the very effort which the ocean makes to maintain its equilibrium tends to prevent the level being restored. The moon produces the tides chiefly by means of a simple yielding of the entire ocean considered as a mass; whereas in the case of a general oceanic circulation the level is restored by a *flow* of water at or near the surface. Consequently the amount of friction and molecular resistance to be overcome in the restoration of level in the latter case is much greater than in the former. The moon, as the researches of Sir William Thomson show, will produce a tide in a globe composed of a substance where no currents or general flow of the materials could possibly take place.

Pressure as a Cause of Circulation.—We shall now briefly refer to the influence of pressure (the indirect effects of gravity) in the production of the circulation under consideration. That which causes the polar column C to descend and the equatorial column W to ascend, as has repeatedly been remarked, is the difference in the weight of the two columns. The efficient cause in the production of the movement is, properly speaking, gravity; *cold* at the poles and *heat* at the equator, or, what is the same thing, the *excess* of heat received by the equator over that received by the poles is what maintains the difference of temperature between the two columns, and consequently is that also which maintains the difference of weight between them. In other words, difference of temperature is the cause which maintains the *state of disturbed equilibrium.* But the efficient cause of the circulation in question is gravity. Gravity, however, could not act without this state of disturbed equilibrium; and difference of temperature may therefore be called, in relation to the circulation, a necessary *condition*, while gravity may be termed the *cause.* Gravity sinks column C *directly*, but it raises column W *indirectly* by means of pressure. The same holds true in regard to the motion of the bottom-waters from C to W, which is likewise due to pressure. The pressure of the excess of the weight of column C over that of column W impels the bottom-water equatorwards and lifts the equatorial column. But on this point I need not dwell, as I have in the preceding chapter entered into a full discussion as to how this takes place.

We come now to the most important part of the inquiry, viz., how is the surface-water impelled from the equator to the poles? Is pressure from behind the impelling force here as in the case of the bottom-water of the ocean? It seems to me that, in attempting to account for the surface-flow from the equator to the poles, Dr. Carpenter's theory signally fails. The force to which he appeals appears to be wholly inadequate to produce the required effect.

The experiments of M. Dubuat, as already noticed, prove that any slope which can possibly result from the difference of temperature between the equator and the poles is wholly insufficient to enable gravity to move the waters; but it does not necessarily prove that the *pressure* resulting from the raised water at the equator may not be sufficient to produce motion. This point will be better understood from the following figure, where, as before, P C represents the polar column and E W the equatorial column.

Fig. 4.

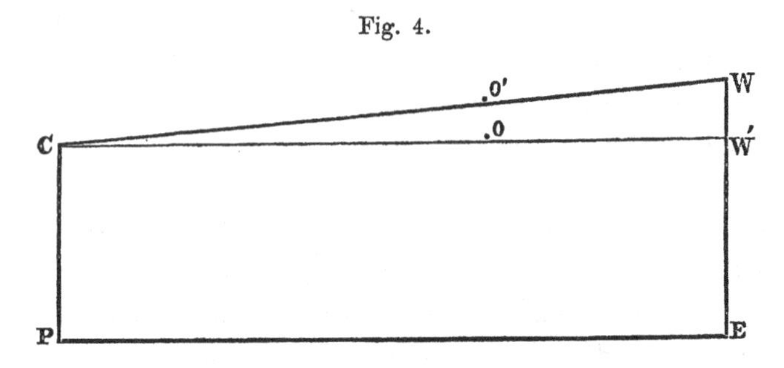

It will be observed that the water in that wedge-shaped portion W C W′ forming the incline cannot be in a state of static equilibrium. A molecule of water at O, for example, will be pressed more in the direction of C than in the direction of W′, and the amount of this excess of pressure towards C will depend upon the height of W above the line C W′. It is evident that the pressure tending to move the molecule at O towards C will be far greater than the direct pull of gravity tending to draw a molecule at O′ lying on the surface of the incline towards C. The experiments of M. Dubuat prove that the direct force of gravity will not move the molecule at O′—that is, cause it to roll down the incline W C; but they do not prove that it may not yield to pressure from above, or that the pressure of the column W W′ will not move the molecule at O. The pressure is caused by gravity, and cannot, of course, enable gravity to perform more work than what is derived from the energy of gravity; it will enable gravity, however, to overcome resistance, which it could not do by direct action. But whether the pressure resulting from the greater height of the water at the equator due to its higher temperature be actually sufficient to produce displacement of the water is a question which I am wholly unable to answer.

If we suppose 4 feet 6 inches to be the height of the equatorial surface above the polar required to make the two columns balance each other, the actual difference of level between the two columns will certainly not be more than one-half that amount, because, if a circulation exist, the weight of the polar column must always be in excess of that of the equatorial. But this excess can only be obtained at the expense of the surface-slope, as has already been shown at length. The surface-slope probably will not be more than 2 feet or 2 feet 6 inches. Suppose the ocean to be of equal density from the poles to the equator, and that by some means or other the surface of the ocean at the equator is raised, say, 2 feet above that of the poles, then there can be little doubt that in such a case the water would soon regain its level; for the ocean at the equator being heavier than at the poles by the weight of a layer 2 feet in thickness, it would sink at the former place and rise at the latter until equilibrium was restored, producing, of course, a very slight displacement of the bottom-waters towards the poles. It will be observed, however, that restoration of level in this case takes place by a simple yielding, as it were, of the entire mass of the ocean without displacement of the molecules of the water over each other to any great extent. In the case of a slope produced by difference of temperature, however, the raised portion of the ocean is not heavier but lighter than the depressed portion, and consequently has no tendency to sink. Any movement which the ocean as a mass makes in order to regain equilibrium tends, as we have seen, rather to increase the difference of level than to reduce it. Restoration of level can only be produced by the forces which are in operation in the wedge-shaped mass W C W′, constituting the slope itself. But it will be observed by a glance at the Figure that, in order to the restoration of level, a large portion of the water W W′ at the equator will require to flow to C, the pole.

According to the general *vertical* oceanic circulation theory, pressure from behind is not one of the forces employed in the production of the flow from the equator to the poles. This is evident; for there can be no pressure from behind acting on the water if there be no slope existing between the equator and the poles. Dr. Carpenter not only denies the actual existence of a slope, but denies the necessity for its existence. But to deny the existence of a slope is to deny the existence of pressure, and to deny the necessity for a slope is to deny the necessity for pressure. That in Dr. Carpenter's theory the surface-water is supposed to be *drawn* from the equator to the poles, and not *pressed* forward by a force from behind, is further evident from the fact that he maintains that the force employed is not *vis a tergo* but *vis a fronte*.*

* Proc. of Roy. Geog. Soc., January 9, 1871, § 29.

CHAPTER XI.

THE INADEQUACY OF THE GRAVITATION THEORY PROVED BY ANOTHER METHOD.

Quantity of Heat which can be conveyed by the General Oceanic Circulation trifling.—Tendency in the Advocates of the Gravitation Theory to underestimate the Volume of the Gulf-stream.—Volume of the Stream as determined by the *Challenger*.—Immense Volume of Warm Water discovered by Captain Nares.—Condition of North Atlantic inconsistent with the Gravitation Theory.—Dr. Carpenter's Estimate of the Thermal Work of the Gulf-stream.

I SHALL now proceed by another method to prove the inadequacy of such a general oceanic circulation as that which Dr. Carpenter advocates. By contrasting the quantity of heat carried by the Gulf-stream from inter-tropical to temperate and polar regions with such amount as can possibly be conveyed in the same direction by means of a general oceanic circulation, it will become evident that the latter sinks into utter insignificance before the former.

In my earlier papers on the amount of heat conveyed by the Gulf-stream,* I estimated the volume of that stream as *equal to that* of a current 50 miles broad and 1,000 feet deep, flowing (from the surface to the bottom) at 4 miles an hour. Of course I did not mean, as Dr. Carpenter seems to suppose, that the stream at any particular place is 50 miles broad and 1,000 feet deep, or that it actually flows at the uniform rate of 4 miles an hour at surface and bottom. All I meant was, that the Gulf-stream is *equal to that* of a current of the above size and velocity. But in my recent papers on Ocean-currents, the substance of which appears in the present volume, to obviate any objections

* Trans. of Geol. Soc. of Glasgow for April, 1867; Phil. Mag. for June, 1867.

on the grounds of having over-estimated the volume, I have taken that at one half this estimate, viz., equal to a current 50 miles broad and 1,000 feet deep flowing at the rate of 2 miles an hour. I have estimated the mean temperature of the stream as it passes the Straits of Florida to be 65°, and have supposed that the water in its course becomes ultimately cooled down on an average to 40°. In this case each pound of water conveys 19,300 foot-pounds of heat from the Gulf of Mexico, to be employed in warming temperate and polar regions. Assuming these data to be correct, it follows that the amount of heat transferred from the Gulf of Mexico by this stream per day amounts to 77,479,650,000,000,000,000 foot-pounds. This enormous quantity of heat is equal to one-fourth of all that is received from the sun by the whole of the Atlantic Ocean from the Tropic of Cancer up to the Arctic Circle.

This is the amount of heat conveyed from inter-tropical to temperate and polar regions by the Gulf-stream. What now is the amount conveyed by means of the General Oceanic Circulation?

According to this theory there ought to be as much warm water flowing from inter-tropical regions towards the Antarctic as towards the Arctic Circle. We may, therefore, in our calculations, consider that the heat which is received in tropical regions to the south of the equator goes to warm the southern hemisphere, and that received on the north side of the equator to warm the northern hemisphere. The warm currents found in the North Atlantic in temperate regions we may conclude came from the regions lying to the north of the equator,—or, in other words, from that part of the Atlantic lying between the equator and the Tropic of Cancer. At least, according to the gravitation theory, we have no reason to believe that the quantity of warm water flowing from tropical to temperate and polar regions in the Atlantic is greater than the area between the equator and the Tropic of Cancer can supply—because it is affirmed that a very large proportion of the cold water found in the North Atlantic comes, not from the arctic, but from the antarctic regions. But if the North Atlantic is cooled by a cold stream from the southern hemisphere, the southern hemisphere in turn must be heated by a warm current from the North Atlantic—unless we assume that the compensating current flowing from the Atlantic into the southern hemisphere is as cold as the antarctic current, which is very improbable. But Dr. Carpenter admits that the quantity of warm water flowing from the Atlantic in equatorial regions towards the south is even greater than that flowing northwards. "The unrestricted communication," he says, "which exists between the antarctic area and the great Southern Ocean-basins would involve, if the doctrine of a general oceanic circulation be admitted, a much more considerable interchange of waters between the antarctic and the equatorial areas than is possible in the northern hemisphere." *

We have already seen that, were it not for the great mass of warm water which finds its way to the polar regions, the temperature of these regions would be enormously lower than they really are. It has been shown likewise that the comparatively high temperature of north-western Europe is due to the same cause. But if it be doubtful whether the Gulf-stream reaches our shores, and if it be true that, even supposing it did, it "could only affect the *most superficial* stratum," and that the great mass of warm water found by Dr. Carpenter in his dredging-expeditions came directly from the equatorial regions, and not from the Gulf-stream, then the principal part of the heating-effect must be attributed, not to the Gulf-stream, but to the general flow of water from the equatorial regions. It surely would not, then, be too much to assume that the quantity of heat conveyed from equatorial regions by this general flow of water into the North Atlantic is at least equal to that conveyed by the Gulf-stream. If we assume this to be the amount of heat conveyed by the two agencies into the Atlantic from inter-tropical regions, it will, of course, be equal to twice that conveyed by the Gulf-stream alone.

* *Nature*, vol. i., p. 541. Proc. Roy. Soc., vol. xviii., p. 473.

We shall now consider whether the area of the Atlantic to the north of the equator is sufficient to supply the amount of heat demanded by Dr. Carpenter's theory.

The entire area of the Atlantic, extending from the equator to the Tropic of Cancer, including the Caribbean Sea and the Gulf of Mexico, is about 7,700,000 square miles.

The quantity of heat conveyed by the Gulf-stream through the Straits of Florida is, as we have already endeavoured to show, equal to all the heat received from the sun by 1,560,935 square miles at the equator. The annual quantity of heat received from the sun by the torrid zone per unit surface, taking the mean of the whole zone, is to that received by the equator as 39 to 40, consequently the quantity of heat conveyed by the Gulf-stream is equal to all the heat received by 1,600,960 square miles of the Atlantic in the torrid zone.

But if, according to Dr. Carpenter's views, the quantity of heat conveyed from the tropical regions is double that conveyed by the Gulf-stream, the amount of heat in this case conveyed into the Atlantic in temperate regions will be equal to all the heat received from the sun by 3,201,920 square miles of the Atlantic between the equator and the Tropic of Cancer. This is 32-77ths of all the heat received from the sun by that area.

Taking the annual quantity received per unit surface at the equator at 1,000, the quantities received by the three zones would be respectively as follows:—

Equator	1000
Torrid zone	975
Temperate zone	757
Frigid zone	454

Now, if we remove from the Atlantic in tropical regions 32-77ths of the heat received from the sun, we remove 405 parts from every 975 received from the sun, and consequently only 570 parts per unit surface remain.

It has been shown* that the quantity of heat conveyed by

* Chapter II.

the Gulf-stream from the equatorial regions into the temperate regions is equal to 100-412ths of all the heat received by the Atlantic in temperate regions. But according to the theory under consideration the quantity removed is double this, or equal to 100-206ths of all the heat received from the sun. But the amount received from the sun is equal to 757 parts per unit surface; add then to this 100-206ths of 757, or 367, and we have 1,124 parts of heat per unit surface as the amount possessed by the Atlantic in temperate regions. The Atlantic should in this case be much warmer in temperate than in tropical regions; for in temperate regions it would possess 1,124 parts of heat per unit surface, whereas in tropical regions it would possess only 570 parts per unit surface. Of course the heat conveyed from tropical regions does not all remain in temperate regions; a very considerable portion of it must pass into the arctic regions. Let us, then, assume that one half goes to warm the Arctic Ocean, and the other half remains in the temperate regions. In this case 183·5 parts would remain, and consequently 757+183·5=940·5 parts would be the quantity possessed by the Atlantic in temperate regions, a quantity which still exceeds by no less than 370·5 parts the heat possessed by the Atlantic in tropical regions.

As one half of the amount of heat conveyed from the tropical regions is assumed to go into the Arctic Ocean, the quantity passing into that ocean would therefore be equal to that which passes through the Straits of Florida, an amount which, as we have found, is equal to all the heat received from the sun by 3,436,900 square miles of the Arctic Ocean.* The entire area covered by sea beyond the Arctic Circle is under 5,000,000 square miles; but taking the Arctic Ocean in round numbers at 5,000,000 square miles, the quantity of heat conveyed into it by currents to that received from the sun would therefore be as 3,436,900 to 5,000,000.

The amount received on the unit surface of the arctic regions we have seen to be 454 parts. The amount received from the

* Chapter II.

currents would therefore be 312 parts. This gives 766 parts of heat per unit surface as the quantity possessed by the Arctic Ocean. Thus the Arctic Ocean also would contain more heat than the Atlantic in tropical regions; for the Atlantic in these regions would, in the case under consideration, possess only 570 parts, while the Arctic Ocean would possess 766 parts. It is true that more rays are cut off in arctic regions than in tropical; but still, after making due allowance for this, the Arctic Ocean, if the theory we are considering were true, ought to be as warm as, if not warmer than, the Atlantic in tropical regions. The relative quantities of heat possessed by the three zones would therefore be as follows:—

Atlantic, in torrid zone	570
" in temperate zone	940
" in frigid zone	766

It is here assumed, however, that none of the heat possessed by the Gulf-stream is derived from the southern hemisphere, which, we know, is not the case. But supposing that as much as one half of the heat possessed by the stream came from the southern hemisphere, and that the other half was obtained from the seas lying between the equator and the Tropic of Cancer, the relative proportions of heat possessed by the three zones per given area would be as follows:—

Atlantic, in torrid zone	671
" in temperate zone	940
" in frigid zone	766

This proves incontestably that, supposing there is such a general oceanic circulation as is maintained, the quantity of heat conveyed by means of it into the North Atlantic and Arctic Oceans must be trifling in comparison with that conveyed by the Gulf-stream; for if it nearly equalled that conveyed by the Gulf-stream, then not only the North Atlantic in temperate regions, but even the Arctic Ocean itself would be much warmer than the inter-tropical seas. In fact, so far as the distribution of heat over the globe is concerned, it is a matter of indifference whether there really is or is not such a thing as this general oceanic circulation. The enormous amount of heat conveyed by the Gulf-stream alone puts it beyond all doubt that ocean-currents are the great agents employed in distributing over the globe the excess of heat received by the sea in inter-tropical regions.

It is therefore, so far as concerns the theory of a General Oceanic Circulation, of the utmost importance that the advocates of that theory should prove that I have over-estimated the thermal power of the Gulf-stream. This, however, can only be done by detecting some error either in my computation or in the data on which it is based; yet neither Dr. Carpenter nor any one else, as far as I know, has challenged the accuracy of my figures. The question at issue is the correctness of the data; but the only part of the data which can possibly admit of being questioned is my estimate of the *volume* and *temperature* of the stream. Dr. Carpenter, however, does not maintain that I have over-estimated the temperature of the stream; on the contrary, he affirms that I have really under-estimated it. "If we assume," he remarks, "the limit of the stratum above 60° as that of the real Gulf-stream current, we shall find its average temperature to be somewhat higher than it has been stated by Mr. Croll, who seems to have taken 65° as the average of the water flowing through the entire channel. The average surface temperature of the Florida channel for the whole year is 80°; and we may fairly set the average of the entire outgoing stream, down to the plane of 60°, at 70°, instead of 65° as estimated by Mr. Croll" (§ 141). It follows, then, that every pound of water of the Gulf-stream actually conveys 5 units of heat more than I have estimated it to do—the amount conveyed being 30 units instead of 25 units as estimated by me. Consequently, if the Gulf-stream be equal to that of a current of merely 41½ miles broad and 1,000 feet deep, flowing at the rate of 2 miles an hour, it will still convey the estimated quantity of heat. But this estimate of the volume of the stream, let it be observed, barely exceeds *one-third* of that given by Herschel, Maury, and Colding,* and is little more than one-half that assigned to it by Mr. Laughton, while it very little exceeds that given by Mr. Findlay,† an author whom few will consider likely to overrate either the volume or heating-power of the stream.

The important results obtained during the *Challenger* expedition have clearly proved that I have neither over-estimated the temperature nor the volume of the Gulf-stream. Between Bermuda and Sandy Hook the stream is 60 miles broad and 600 feet deep, with a maximum velocity of from 3½ to 4 miles an hour. If the mean velocity of the entire section amounts to 2¼ miles an hour, which it probably does, the volume of the stream must equal that given in my estimate. But we have no evidence that all the water flowing through the Straits of Florida passes through the section examined by the officers of the *Challenger*. Be this, however, as it may, the observations made between St. Thomas and Sandy Hook reveal the existence of an immense flow of warm water, 2,300 feet deep, entirely distinct from the water included in the above section of the Gulf-stream proper. As the thickest portion of this immense body of water joins the warm water of the Gulf-stream, Captain Nares considers that "it is evidently connected with it, and probably as an offshoot." At Sandy Hook, according to him, it extends 1,200 feet deeper than the Gulf-stream itself, but off Charleston, 600 miles nearer the source, the same temperature is found at the same depth. But whether it be an offshoot of the Gulf-stream or not, one thing is certain, it can only come from the Gulf of Mexico or from the Caribbean Sea. This mass of water, after flowing northwards for about 1,000 miles, turns to the right and crosses the Atlantic in the direction of the Azores, where it appears to thin out.

If, therefore, we take into account the combined heat con-

* Chapter II.

† Mr. Findlay considers that the daily discharge does not exceed 333 cubic miles (Brit. Assoc. Rep., 1869, p. 160). My estimate makes it 378 cubic miles. Mr. Laughton's estimate is 630 cubic miles (Paper "On Ocean-currents," Journal of Royal United-Service Institution, vol. xv.).

veyed by both streams, my estimate of the heat transferred from inter-tropical regions into the North Atlantic will be found rather under than above the truth.

Dr. Carpenter's Estimate of the Thermal Work of the Gulf-stream.—In the appendix to an elaborate memoir on Oceanic Circulation lately read before the Geographical Society, Dr. Carpenter endeavours to show that I have over-estimated the thermal work of the Gulf-stream. In that memoir* he has also favoured us with his own estimate of the sectional area, rate of flow, and temperature of the stream. Even adopting his data, however, I find myself unable to arrive at his conclusions.

Let us consider first his estimate of the sectional area of the stream. He admits that "it is impossible, in the present state of our knowledge, to arrive at any exact estimate of the sectional area of the stream; since it is for the most part only from the temperatures of its different strata that we can judge whether they are, or are not, in movement, and what is the direction of their movement." Now it is perfectly evident that our estimate of the sectional area of the stream will depend upon what we assume to be its bottom temperature. If, for example, we assume 70° to be the bottom temperature, we shall have a small sectional area. Taking the temperature at 60°, the sectional area will be larger, and if 50° be assumed to be the temperature, the sectional area will be larger still, and so on. Now the small sectional area obtained by Dr. Carpenter arises from the fact of his having assumed the high temperature of 60° to be that of the bottom of the stream. He concludes that all the water below 60° has an inward flow, and that it is only that portion from 60° and upwards which constitutes the Gulf-stream. I have been unable to find any satisfactory evidence for assuming so high a temperature for the bottom of the stream. It must be observed that the water underlying the Gulf-stream is not the ordinary water of the Atlantic, but the cold current from the arctic regions. In fact, it is the same

* Proceedings of the Royal Geographical Society, vol. xviii., p. 393.

water which reaches the equator at almost every point with a temperature not much above the freezing-point. It is therefore highly improbable that the under surface of the Gulf-stream has a temperature so high as 60°.

Dr. Carpenter's method of measuring the mean velocity of the Gulf-stream is equally objectionable. He takes the mean annual rate at the surface in the "Narrows" to be two miles an hour and the rate at the bottom to be zero, and he concludes from this that the average rate of the whole is one mile an hour—the arithmetical mean between these two extremes. Now it will be observed that this conclusion only holds true on the supposition that the breadth of the stream is as great at the bottom as at the surface, which of course it is not. All admit that the sides of the Gulf-stream are not perpendicular, but slope somewhat in the manner of the banks of a river. The stream is broad at the surface and narrows towards the bottom. It is therefore evident that the upper half of the section has a much larger area than the lower; the quantity of water flowing through the upper half with a greater velocity than one mile an hour must be much larger than the quantity flowing through the lower half with a less velocity than one mile an hour.

His method of estimating the mean temperature of the stream is even more objectionable. He says, "The average surface temperature of the Florida Channel for the whole year is 80°, and we may set the average of the entire outgoing stream down to the plane of 60° at 70°, instead of 65°, as estimated by Mr. Croll." If 80° be the surface and 60° be the bottom temperature, temperature and rate of velocity being assumed of course to decrease uniformly from the surface downwards, how is it possible that 70° can be the average temperature? The amount of water flowing through the upper half of the section, with a temperature above 70°, is far more than the amount flowing through the under half of the section, with a temperature below 70°. Supposing the lower half of the section to be as large as the upper half, which it is not, still the quantity of water flowing through it would only equal one-third of that flowing through the upper half, because the mean velocity of the water in the lower half would be only half a mile per hour, whereas the mean velocity of that in the upper half would be a mile and a half an hour. But the area of the lower half is much less than that of the upper half, consequently the amount of water whose temperature is under 70° must be even much under one-third of that, the temperature of which is above 70°.

Had Dr. Carpenter taken the proper method of estimating the mean temperature, he would have found that 75°, even according to his own data, was much nearer the truth than 70°. I pointed out, several years ago,* the fallacy of estimating the mean temperature of a stream in this way.

So high a mean temperature as 75° for the Gulf-stream, even in the Florida Channel, is manifestly absurd, but if 60° be the bottom temperature of the stream, the mean temperature cannot possibly be much under that amount. It is, of course, by under-estimating the sectional area of the stream that its mean temperature is over-estimated. We cannot reduce the mean temperature without increasing the sectional area. If my estimate of 65° be taken as the mean temperature, which I have little doubt will yet be found to be not far from the truth, Dr. Carpenter's estimate of the sectional area must be abandoned. For if 65° be the mean temperature of the stream, its bottom temperature must be far under 60°, and if the bottom temperature be much under 60°, then the sectional area must be greater than he estimates it to be.

Be this, however, as it may; even if we suppose that 60° will eventually be found to be the actual bottom temperature of the Gulf-stream, nevertheless, if the total quantity of heat conveyed by the stream from inter-tropical regions be estimated in the proper way, we shall still find that amount to be so enormous, that there is not sufficient heat remaining in those

* Phil. Mag. for October, 1871, p. 274.

regions to supply Dr. Carpenter's oceanic circulation with a quantity as great for distribution in the North Atlantic.

It therefore follows (and so far as regards the theory of Secular changes of climate, this is all that is worth contending for) that Ocean-currents and not a General Oceanic Circulation resulting from gravity, are the great agents employed in the distribution of heat over the globe.

CHAPTER XII.

MR. A. G. FINDLAY'S OBJECTIONS CONSIDERED.

Mr. Findlay's Estimate of the Volume of the Gulf-stream.—Mean Temperature of a Cross Section less than Mean Temperature of Stream.—Reason of such Diversity of Opinion regarding Ocean-currents.—More rigid Method of Investigation necessary.

At the conclusion of the reading of Dr. Carpenter's paper before the Royal Geographical Society, on January 9th, 1871, Mr. Findlay made the following remarks:—

"When, by the direction of the United States Government, ten or eleven years ago, the narrowest part of the Gulf-stream was examined, figures were obtained which shut out all idea of its ever reaching our shores as a heat-bearing current. In the narrowest part, certainly not more than from 250 to 300 cubic miles of water pass per diem. Six months afterwards that water reaches the banks of Newfoundland, and nine or twelve months afterwards the coast of England, by which time it is popularly supposed to cover an area of 1,500,000 square miles. The proportion of the water that passes through the Gulf of Florida will not make a layer of water more than 6 inches thick per diem over such a space. Every one knows how soon a cup of tea cools; and yet it is commonly imagined that a film of only a few inches in depth, after the lapse of so long a time, has an effect upon our climate. There is no need for calculations; the thing is self-evident."*

About five years ago, Mr. Findlay objected to the conclusions which I had arrived at regarding the enormous heating-power of the Gulf-stream on the ground that I had over-estimated the

* Proceedings of the Royal Geographical Society, vol. xv.

volume of the stream. He stated that its volume was only about the half of what I had estimated it to be. To obviate this objection, I subsequently reduced the volume to one-half of my former estimate.* But taking the volume at this low estimate, it was nevertheless found that the quantity of heat conveyed into the Atlantic through the Straits of Florida by means of the stream was equal to about *one-fourth* of all the heat received from the sun by the Atlantic from the latitude of the Strait of Florida up to the Arctic Circle.

Mr. Findlay, in his paper read before the British Association, affirmed that the volume of the stream is somewhere from 294 to 333 cubic miles per day; but in his remarks at the close of Dr. Carpenter's address, he stated it to be not greater than from 250 to 300 cubic miles per day. I am unable to reconcile any of those figures with the data from which he appears to have derived them. In his paper to the British Association, he remarks that "the Gulf-stream at its outset is not more than 39½ miles wide, and 1,200 feet deep." "From all attainable data, he computes the mean annual rate of motion to be 65·4 miles per day; but as the rate decreases with the depth, the mean velocity of the whole mass does not exceed 49·4 miles per day. When he speaks of the mean velocity of the Gulf-stream being so and so, he must refer to the mean velocity at some particular place. This is evident; for the mean velocity entirely depends upon the sectional area of the stream. The place where the mean velocity is 49·4 miles per day must be the place where it is 39½ miles broad and 1,200 feet deep; for he is here endeavouring to show us how small the volume of the stream actually is. Now, unless the mean velocity refers to the place where he gives us the breadth and depth of the stream, his figures have no bearing on the point in question. But a stream 39½ miles broad and 1,200 feet deep has a sectional area of 8·97 square miles, and this, with a mean velocity of 49·4 miles per day, will give 443 cubic miles of water. The amount, according to my estimate, is 459 cubic

* Phil. Mag., February, 1870.

miles per day; it therefore exceeds Mr. Findlay's estimate by only 16 cubic miles.

Mr. Findlay does not, as far as I know, consider that I have over-estimated the mean temperature of the stream. He states* that between Sand Key and Havanna the Gulf-stream is about 1,200 feet deep, and that it does not reach the summit of a submarine ridge, which he states has a temperature of 60°. It is evident, then, that the bottom of the stream has a temperature of at least 60°, which is within 5° of what I regard as the mean temperature of the mass. But the surface of the stream is at least 17° above this mean. Now, when we consider that it is at the upper parts of the stream, the place where the temperature is so much above 65°, that the motion is greatest, it is evident that the mean temperature of the entire moving mass must, according to Mr. Findlay, be considerably over 65°. It therefore follows, according to his own data, that the Gulf-stream conveys into the Atlantic an amount of heat equal to one-fourth of all the heat which the Atlantic, from the latitude of the Straits of Florida up to the arctic regions, derives from the sun.

But it must be borne in mind that although the mean temperature of the cross section should be below 65°, it does not therefore follow that the mean temperature of the *water flowing through this cross section* must be below that temperature, for it is perfectly obvious that the mean temperature of the mass of water flowing through the cross section in a given time must be much higher than that of the cross section itself. The reason is very simple. It is in the upper half of the section where the high temperature exists; but as the velocity of the stream is much greater in its upper than in its lower half, the greater portion of the water passing through this cross section is water of high temperature.

But even supposing we were to halve Mr. Findlay's own estimate, and assume that the volume of the stream is equal to only 222 cubic miles of water per day instead of 443, still the

* Brit. Assoc. Report, 1869, Sections, p. 160.

amount of heat conveyed would be equal to one-eighth part of the heat received from the sun by the Atlantic. But would not the withdrawal of an amount of heat equal to one-eighth of that received from the sun greatly affect the climate of the Atlantic? Supposing we take the mean temperature of the Atlantic at, say, 56°; this will make its temperature 295° above that of space. Extinguish the sun and stop the Gulf-stream, and the temperature ought to sink 295°. How far, then, ought the temperature to sink, supposing the sun to remain and the Gulf-stream to stop? Would not the withdrawal of the stream cause the temperature to sink some 30°? Of course, if the Gulf-stream were withdrawn and everything else were to remain the same, the temperature of the Atlantic would not actually remain 30° lower than at present; for heat would flow in from all sides and partly make up for the loss of the stream. But nevertheless 30° represents the amount of temperature maintained by means of the heat from the stream. And this, be it observed, is taking the volume of the stream at a lower estimate than even Mr. Findlay himself would be willing to admit. Mr. Findlay says that, by the time the Gulf-stream reaches the shores of England, it is supposed to cover a space of 1,500,000 square miles. "The proportion of water that passes through the Straits of Florida will not make," according to him, "a layer of water more than 6 inches thick per diem over such a space." But a layer of water 6 inches thick cooling 25° will give out 579,000 foot-pounds of heat per square foot. If, therefore, the Gulf-stream, as he asserts, supplies 6 inches per day to that area, then every square foot of the area gives off per day 579,000 foot-pounds of heat. The amount of heat received from the sun per square foot in latitude 55°, which is not much above the mean latitude of Great Britain, is 1,047,730 foot-pounds per day, taking, of course, the mean of the whole year; *consequently this layer of water gives out an amount of heat equal to more than* one-half *of all that is received from the sun.* But assuming that the stream should leave the half of its heat on the

American shores and carry to the shores of Britain only $12\frac{1}{2}°$ of heat, still we should have 289,500 foot-pounds per square foot, which notwithstanding *is more than equal to* one-fourth *of that received from the sun.* If an amount of heat so enormous cannot affect climate, what can?

I shall just allude to one other erroneous notion which prevails in regard to the Gulf-stream; but it is an error which I by no means attribute either to Mr. Findlay or to Dr. Carpenter. The error to which I refer is that of supposing that when the Gulf-stream widens out to hundreds of miles, as it does before it reaches our shores, its depth must on this account be much less than when it issues from the Gulf of Mexico. Although the stream may be hundreds of miles in breadth, there is no necessity why it should be only 6 inches, or 6 feet, or 60 feet, or even 600 feet in depth. It may just as likely be 6,000 feet deep as 6 inches.

The Reason why such Diversity of Opinion prevails in Regard to Ocean-currents.—In conclusion I venture to remark that more than nine-tenths of all the error and uncertainty which prevail, both in regard to the cause of ocean-currents and to their influence on climate, is due, not, as is generally supposed, to the intrinsic difficulties of the subject, but rather to the defective methods which have hitherto been employed in its investigation—that is, in not treating the subject according to the rigid methods adopted in other departments of physics. What I most particularly allude to is the disregard paid to the modern method of determining the amount of effects in *absolute measure.*

But let me not be misunderstood on this point. I by no means suppose that the *absolute quantity* is the thing always required for its own sake. It is in most cases required simply as a means to an end; and very often that end is the knowledge of the *relative* quantity. Take, for example, the Gulf-stream. Suppose the question is asked, to what extent does the heat conveyed by that stream influence the climate of the North Atlantic? In order to the proper answering of this question, the principal thing required is to know what proportion the amount of heat conveyed by the stream into the Atlantic bears to that received from the sun by that area. We want the *relative proportions* of these two quantities. But how are we to obtain them? We can only do so by determining first the *absolute* quantity of each. We must first measure each before we can know how much the one is greater than the other, or, in other words, before we can know their relative proportions. We have the means of determining the absolute amount of heat received from the sun by a given area at any latitude with tolerable accuracy; but the same cannot be done with equal accuracy in regard to the amount of heat conveyed by the Gulf-stream, because the volume and mean temperature of the stream are not known with certainty. Nevertheless we have sufficient data to enable us to fix upon such a maximum and minimum value to these quantities as will induce us to admit that the truth must lie somewhere between them. In order to give full justice to those who maintain that the Gulf-stream exercises but little influence on climate, and to put an end to all further objections as to the uncertainty of my data, I shall take a minimum to which none of them surely can reasonably object, viz. that the volume of the stream is not over 230 cubic miles per day, and the heat conveyed per pound of water not over 12½ units. Calculating from these data, we find that the amount of heat carried into the North Atlantic is equal to one-sixteenth of all the heat received from the sun by that area. There are, I presume, few who will not admit that the actual proportion is much higher than this, probably as high as 1 to 3, or 1 to 4. But, who, without adopting the method I have pursued, could ever have come to the conclusion that the proportion was even 1 to 16? He might have guessed it to be 1 to 100 or 1 to 1000, but he never would have guessed it to be 1 to 16. Hence the reason why the great influence of the Gulf-stream as a heating agent has been so much under-estimated.

The same remarks apply to the gravitation theory of the cause of currents. Viewed simply as a theory it looks very reasonable. There is no one acquainted with physics but will admit that the tendency of the difference of temperature between the equator and the poles is to cause a surface-current from the equator towards the poles, and an under-current from the poles to the equator. But before we can prove that this tendency does actually produce such currents, another question must be settled, viz. is this force sufficiently great to produce the required motion? Now when we apply the method to which I refer, and determine the absolute amount of the force resulting from the difference of specific gravity, we discover that not to be the powerful agent which the advocates of the gravitation theory suppose, but a force so infinitesimal as not to be worthy of being taken into account when considering the causes by which currents are produced.

CHAPTER XIII.

THE WIND THEORY OF OCEANIC CIRCULATION.

Ocean-currents not due alone to the Trade-winds.—An Objection by Maury.—Trade-winds do not explain the Great Antarctic Current.—Ocean-currents due to the System of Winds.—The System of Currents agrees with the System of the Winds.—Chart showing the Agreement between the System of Currents and System of Winds.—Cause of the Gibraltar Current.—North Atlantic an immense Whirlpool.—Theory of Under Currents.—Difficulty regarding Under Currents obviated.—Work performed by the Wind in impelling the Water forward.—The *Challenger's* crucial Test of the Wind and Gravitation Theories.—North Atlantic above the Level of Equator.—Thermal Condition of the Southern Ocean irreconcilable with the Gravitation Theory.

Ocean-currents not due alone to the Trade-winds.—The generally received opinion amongst the advocates of the wind theory of oceanic circulation is that the Gulf-stream and other currents of the ocean are due to the impulse of the trade-winds. The tendency of the trade-winds is to impel the inter-tropical waters along the line of the equator from east to west; and were those regions not occupied in some places by land, this equatorial current would flow directly round the globe. Its westward progress, however, is arrested by the two great continents, the old and the new. On approaching the land the current bifurcates, one portion trending northwards and the other southwards. The northern branch of the equatorial current of the Atlantic passes into the Caribbean Sea, and after making a circuit of the Gulf of Mexico, flows northward and continues its course into the Arctic Ocean. The southern branch, on the other hand, is deflected along the South-American coast, constituting what is known as the Brazilian current. In the Pacific a similar deflection occurs against the Asiatic coast, forming a current somewhat resembling the Gulf-stream, a portion of which (Kamtschatka current) in like manner passes into the arctic regions. In reference to all these various currents, the impelling cause is supposed to be the force of the trade-winds.

It is, however, urged as an objection by Maury and other advocates of the gravitation theory, that a current like the Gulf-stream, extending as far as the arctic regions, could not possibly be impelled and maintained by a force acting at the equatorial regions. But this is a somewhat weak objection. It seems to be based upon a misconception of the magnitude of the force in operation. It does not take into account that this force acts on nearly the whole area of the ocean in inter-tropical regions. If, in a basin of water, say three feet in diameter, a force is applied sufficient to produce a surface-flow one foot broad across the centre of the basin, the water impelled against the side will be deflected to the extremes of the vessel. And this result does not in any way depend upon the size of the basin. The same effect which occurs in a small basin will occur in a large one, provided the proportion between the breadth of the belt of water put in motion and the size of the vessel be the same in both cases. It does not matter, therefore, whether the diameter of the basin be supposed to be three feet, or three thousand miles, or ten thousand miles.

There is a more formidable objection, however, to the theory. The trade-winds will account for the Gulf-stream, Brazil, Japan, Mozambique, and many other currents; but there are currents, such as some of the polar currents, which cannot be so accounted for. Take, for example, the great antarctic current flowing northward into the Pacific. This current does not bend to the left under the influence of the earth's rotation and continue its course in a north-westerly direction, but actually bends round to the right and flows eastward against the South-American coast, in direct opposition both to the influence of rotation and to the trade-winds. The trade-wind theory, therefore, is insufficient to account for all the facts. But there is yet another explanation, which satisfactorily solves our difficulties. The currents of the ocean owe their origin, not to the trade-winds alone, but to the *prevailing* winds of the globe (including, of course, the trade-winds).

Ocean-currents due to the System of Winds.—If we leave out of account a few small inland sheets of water, the globe may be said to have but one sea, just as it possesses only one atmosphere. We have accustomed ourselves, however, to speak of parts or geographical divisions of the one great ocean, such as the Atlantic and the Pacific, as if they were so many separate oceans. And we have likewise come to regard the currents of the ocean as separate and independent of one another. This notion has no doubt to a considerable extent militated against the acceptance of the theory that the currents are caused by the winds, and not by difference of specific gravity; for it leads to the conclusion that currents in a sea must flow in the direction of the prevailing winds blowing over that particular sea. The proper view of the matter, as I hope to be able to show, is that which regards the various currents merely as members of one grand system of circulation produced, not by the trade-winds alone, nor by the prevailing winds proper alone, but by the combined action of all the prevailing winds of the globe, regarded as one system of circulation.

If the winds be the impelling cause of currents, the *direction* of the currents will depend upon two circumstances, viz.:—(1) the direction of the prevailing winds of the globe, including, of course, under this term the prevailing winds proper and the trade-winds; and (2) the conformation of land and sea. It follows, therefore, that as a current in any given sea is but a member of a general system of circulation, its direction is determined, not alone by the prevailing winds blowing over the sea in question, but by the general system of prevailing winds. It may consequently sometimes happen that the general system of winds may produce a current directly opposite to the prevailing wind blowing over the current. The accompanying Chart (Plate I.) shows how exactly the system of ocean-currents agrees with the system of the prevailing winds. The fine

PLATE I.

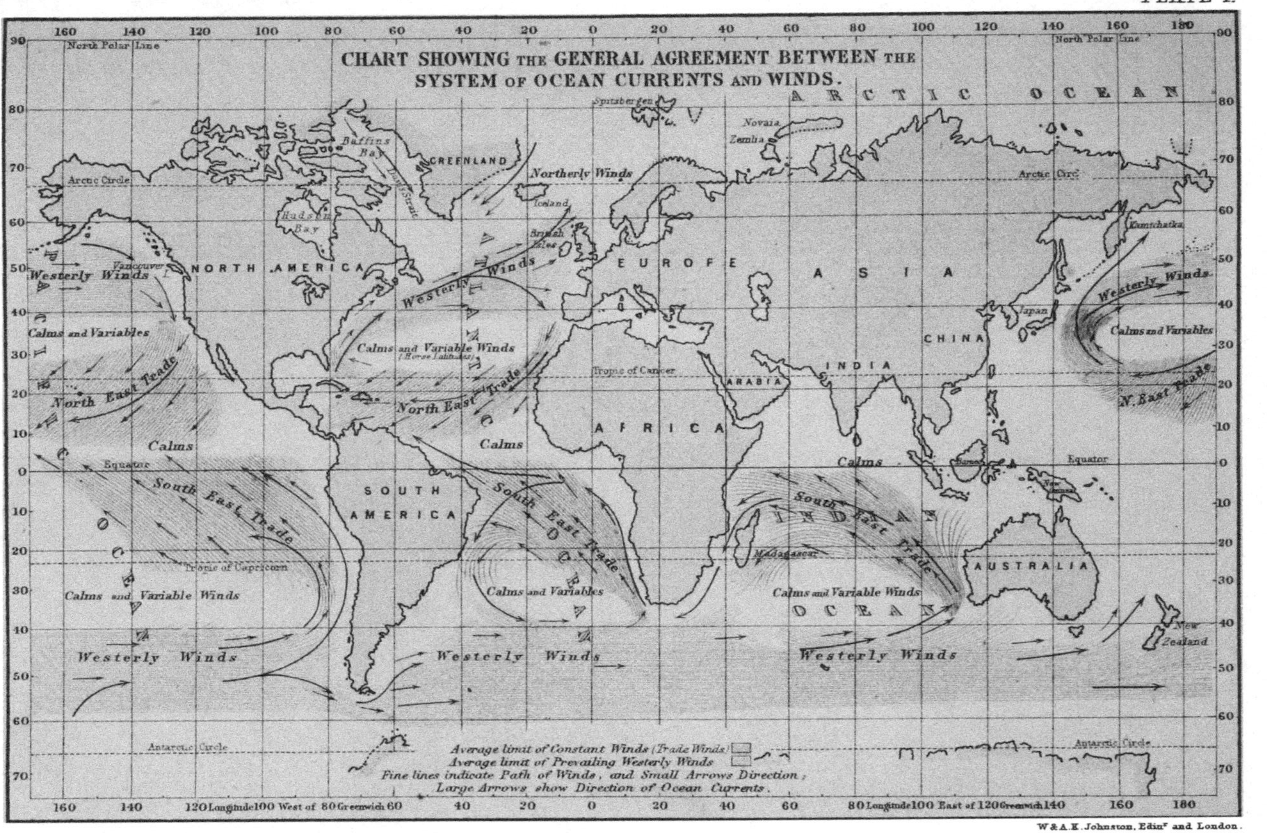

W & A. K. Johnston, Edin[r] and London.

lines indicate the paths of the prevailing winds, and the fine arrows the direction in which the wind blows along those paths. The large arrows show the direction of the principal ocean-currents.

The directions and paths of the prevailing winds have been taken from Messrs. Johnston's small physical Atlas, which, I find, agrees exactly with the direction of the prevailing winds as deduced from the four quarterly wind charts lately published by the Hydrographic Department of the Admiralty. The direction of the ocean-currents has been taken from the Current-chart published by the Admiralty.

In every case, without exception, the direction of the main currents of the globe agrees exactly with the direction of the prevailing winds. There could not possibly be a more convincing proof that those winds are the cause of the ocean-currents than this general agreement of the two systems as indicated by the chart. Take, for example, the North Atlantic. The Gulf-stream follows exactly the path of the prevailing winds. The Gulf-stream bifurcates in mid-Atlantic; so does the wind. The left branch of the stream passes north-eastwards into the arctic regions, and the right branch south-eastwards by the Azores; so does the wind. The south-eastern branch of the stream, after passing the Canaries, re-enters the equatorial current and flows into the Gulf of Mexico; the same, it will be observed, holds true of the wind. A like remarkable agreement exists in reference to all the other leading currents of the ocean. This is particularly seen in the case of the great antarctic current between long. 140° W. and 160° W. This current, flowing northwards from the antarctic regions, instead of bending to the left under the influence of rotation, turns to the right when it enters the regions of the westerly winds, and flows eastwards towards the South-American shores. In fact, all the currents in this region of strong westerly winds flow in an easterly or north-easterly direction.

Taking into account the effects resulting from the conformation of sea and land, the system of ocean-currents agrees precisely with the system of the winds. All the principal currents of the globe are in fact moving in the exact direction in which they ought to move, assuming the winds to be the sole impelling cause. In short, so perfect is the agreement between the two systems, that, given the system of winds and the conformation of sea and land, and the direction of all the currents of the ocean, or more properly the system of oceanic circulation, might be determined *à priori*. Or given the system of the ocean-currents together with the conformation of sea and land, and the direction of the prevailing winds could also be determined *à priori*. Or, thirdly, given the system of winds and the system of currents, and the conformation of sea and land might be roughly determined. For example, it can be shown by this means that the antarctic regions are probably occcupied by a continent and not by a number of separate islands, nor by sea.

While holding that the currents of the ocean form one system of circulation, we must not be supposed to mean that the various currents are connected end to end, having the same water flowing through them all in succession like that in a heating-apparatus. All that is maintained is simply this, that the currents are so mutually related that any great change in one would modify the conditions of all the others. For example, a great increase or decrease in the easterly flow of antarctic water in the Southern Ocean would decrease or increase, as the case might be, the strength of the West Australian current; and this change would modify the equatorial current of the Indian Ocean, a modification which in like manner would affect the Agulhas current and the Southern Atlantic current—this last leading in turn to a modification of the equatorial current of the Atlantic, and consequently of the Brazilian current and the Gulf-stream. Furthermore, since a current impelled by the winds, as Mr. Laughton in his excellent paper on Ocean-currents justly remarks, tends to leave a vacancy behind, it follows that a decrease or increase in the Gulf-stream would affect the equatorial current, the

Agulhas current, and all the other currents back to the antarctic currents. Again, a large modification in the great antarctic drift-current would in like manner affect all the currents of the Pacific. On the other hand, any great change in the currents of the Pacific would ultimately affect the currents of the Atlantic and Indian Oceans, through its influence on the Cape Horn current, the South Australian current, and the current passing through the Asiatic archipelago; and *vice versâ*, any changes in the currents of the Atlantic or Indian Oceans would modify the currents of the Pacific.

Cause of Gibraltar Current.—I may now consider the cause of the Gibraltar current. There can be little doubt that this current owes its origin (as Mr. Laughton points out) to the Gulf-stream. "I conceive," that author remarks, "that the Gibraltar current is distinctly a stream formed by easterly drift of the North Atlantic, which, although it forms a southerly current on the coast of Portugal, is still strongly pressed to the eastward and seeks the first escape it can find. So great indeed does this pressure seem to be, that more water is forced through the Straits than the Mediterranean can receive, and a part of it is ejected in reverse currents, some as lateral currents on the surface, some, it appears, as an under current at a considerable depth."* The funnel-shaped nature of the strait through which the water is impelled helps to explain the existence of the under current. The water being pressed into the narrow neck of the channel tends to produce a slight banking up; and as the pressure urging the water forward is greatest at the surface and diminishes rapidly downwards, the tendency to the restoration of level will cause an underflow towards the Atlantic, because below the surface the water will find the path of least resistance. It is evident indeed that this underflow will not take place toward the Mediterranean, from the fact that that sea is already filled to overflowing by the current received from the outside ocean.

* Journal of Royal United-Service Institute, vol. xv.

If we examine the Current-chart published by the Hydrographic Department of the Admiralty, we shall find the Gibraltar current represented as merely a continuation of the S.E. flow of Gulf-stream water. Now, if the arrows shown upon this chart indicate correctly the direction of the flow, we must become convinced that the Gulf-stream water cannot possibly avoid passing through the Gibraltar Strait. Of course the excess of evaporation over that of precipitation within the Mediterranean area would alone suffice to produce a considerable current through the Strait; but this of itself would not fill that inland sea to overflowing.*

The Atlantic may, in fact, be regarded as an immense whirlpool with the Saragossa Sea as its vortex; and although it is true, as will be seen from an inspection of the Chart, that the wind blows round the Atlantic along the very path taken by the water, impelling the water forward along every inch of its course, yet nevertheless it must hold equally true that the water has a tendency to flow off in a straight line at a tangent to the circular course in which it is moving. But the water is so hemmed in on all sides that it cannot leave this circular path except only at two points; and at these two points it actually does flow outwards. On the east and west sides the land prevents any such outflow. Similarly, in the south the escape of the water is frustrated by the pressure of the opposing currents flowing from that quarter; while in the north it is prevented by the pressure exerted by polar currents from Davis Strait and the Arctic Ocean. But in the Strait of Gibraltar and in the north-eastern portion of the Atlantic between

* Dr. Carpenter (Proc. of Roy. Geog. Soc., vol. xviii., p. 334) misapprehends me in supposing that I attribute the Gibraltar current wholly to the Gulf-stream. In the very page from which he derives or could derive his opinion as to my views on the subject (Phil. Mag. for March, 1874, p. 182), I distinctly state that "the excess of evaporation over that of precipitation within the Mediterranean area would of itself produce a considerable current through the Strait." That the Gibraltar current is due to two causes, (1) the pressure of the Gulf-stream, and (2) excess of evaporation over precipitation in the Mediterranean, has always appeared to me so perfectly obvious, that I never held nor could have held any other opinion on the subject.

Iceland and the north-eastern shores of Europe there is no resistance offered: and at these two points an outflow does actually take place. In both cases, however, especially the latter, the outflow is greatly aided by the impulse of the prevailing winds.

No one, who will glance at the accompanying chart (Plate I.) showing how the north-eastern branch of the Gulf-stream bends round and, of course, necessarily presses against the coast, can fail to understand how the Atlantic water should be impelled into the Gibraltar Strait, even although the loss sustained by the Mediterranean from evaporation did not exceed the gain from rain and rivers.

Theory of Under Currents.—The consideration that ocean-currents are simply parts of a system of circulation produced by the system of prevailing winds, and not by the impulse of the trade-winds alone, helps to remove the difficulty which some have in accounting for the existence of under currents without referring them to difference of specific gravity. Take the case of the Gulf-stream, which passes under the polar stream on the west of Spitzbergen, this latter stream passing in turn under the Gulf-stream a little beyond Bear Island. The polar streams have their origin in the region of prevailing northerly winds, which no doubt extends to the pole. The current flowing past the western shores of Spitzbergen, throughout its entire course up to near the point where it disappears under the warm waters of the Gulf-stream, lies in the region of these same northerly winds. Now why should this current cease to be a surface current as soon as it passes out of the region of northerly into that of south-westerly winds? The explanation seems to be this: when the stream enters the region of prevailing south-westerly winds, its progress southwards along the surface of the ocean is retarded both by the wind and by the surface water moving in opposition to its course; but being continually pressed forward by the impulse of the northerly winds acting along its whole course back almost to the pole, perhaps, or as far north at least as the sea is not wholly covered with ice, the polar current cannot stop when it enters the region of opposing winds and currents; it must move forward. But the water thus pressed from behind will naturally take the *path of least resistance*. Now in the present case this path will necessarily lie at a considerable distance below the surface. Had the polar stream simply to contend with the Gulf-stream flowing in the opposite direction, it would probably keep the surface and continue its course along the side of that stream; but it is opposed by the winds, from which it cannot escape except by dipping down under the surface; and the depth to which it will descend will depend upon the depth of the surface current flowing in the opposite direction. There is no necessity for supposing a heaping up of the water in order to produce by pressure a force sufficient to impel the under current. The pressure of the water from behind is of itself enough. The same explanation, of course, applies to the case of the Gulf-stream passing under the polar stream. And if we reflect that these under currents are but parts of the general system of circulation, and that in most cases they are currents compensating for water drained off at some other quarter, we need not wonder at the distance which they may in some cases flow, as, for example, from the banks of Newfoundland to the Gulf of Mexico. The under currents of the Gulf-stream are necessary to compensate for the water impelled southwards by the northerly winds; and again, the polar under currents are necessary to compensate for the water impelled northward by the south and south-westerly winds.

But it may be asked, how do the opposing currents succeed in crossing each other? It is evident that the Gulf-stream must plunge through the whole thickness of the polar stream before it can become an under current, and so likewise must the cold water of the polar-flow pass through the genial water of the Gulf-stream in order to get underneath it and continue on its course towards the south. The accompanying diagram (Plate II., Fig. 1) will render this sufficiently intelligible.

Fig. 3

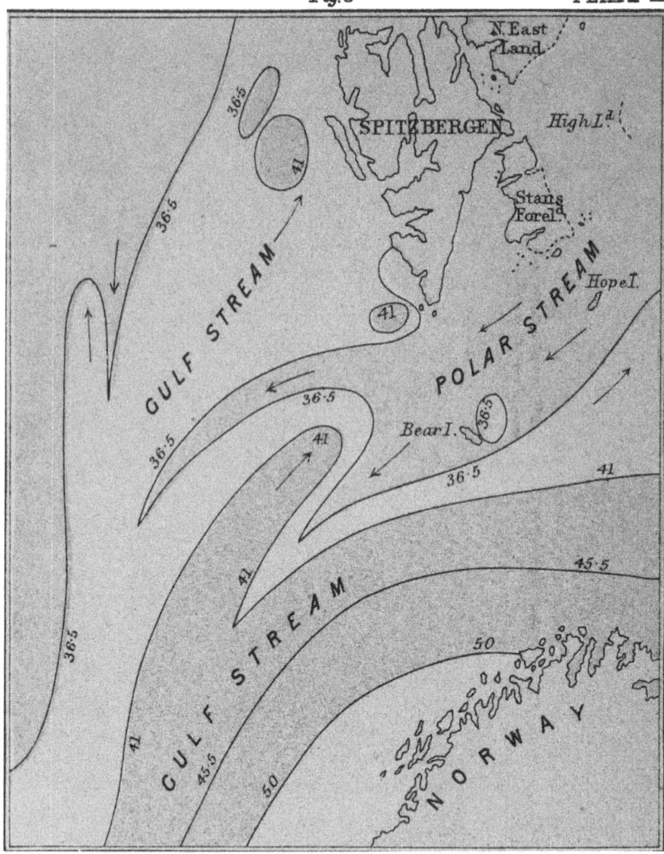

Map shewing meeting of the Gulf-stream and Polar Current from Dr. Petermann's Geographische Mittheilungen. The curved lines are Isotherms; temperatures are in Fahrenheit.

N. Winds — Fig. 1 — S. Winds

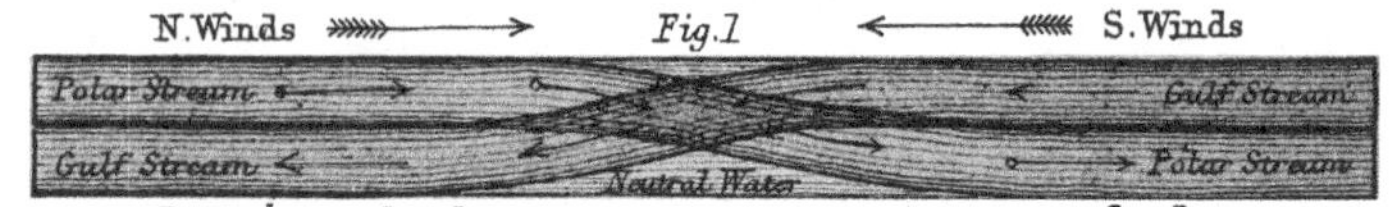

Diagram to shew how two opposing currents intersect each other

Surface Plan to shew how two opposing currents meet each other

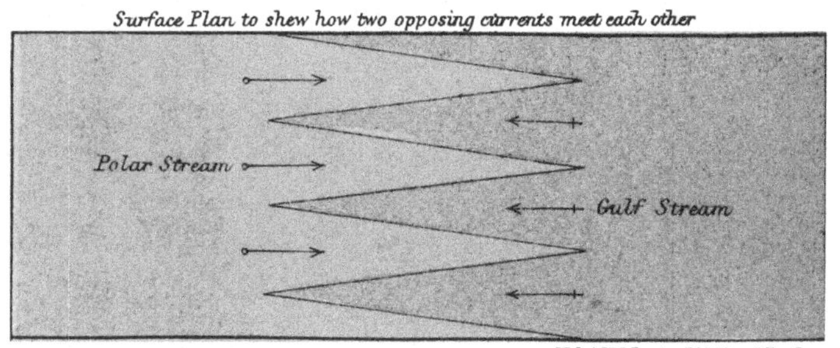

Fig. 2

W. & A. K. Johnston, Edinbr. and London.

Now these two great ocean-currents are so compelled to intersect each other for the simple reason that they cannot turn aside, the one to the left and the other to the right. When two broad streams like those in question are pressed up against each other, they succeed in mutually intersecting each other's path by breaking up into bands or belts—the cold water being invaded and pierced as it were by long tongues of warm water, while at the same time the latter is similarly intersected by corresponding protrusions of cold water. The two streams become in a manner interlocked, and the one passes through the other very much as we pass the fingers of one hand between the fingers of the other. The diagram (Plate II., Fig. 2), representing the surface of the ocean at the place of meeting of two opposing currents, will show this better than description. At the surface the bands necessarily assume the tongue-shaped appearance represented in the diagram, but when they have succeeded in mutually passing down through the whole thickness of the opposing currents, they then unite and form two definite under currents, flowing in opposite directions. The polar bands, after penetrating the Gulf-stream, unite below to form a southward-flowing under current, and in the same way the Gulf-stream bands, uniting underneath the polar current, continue in their northerly course as a broad under current of warm water. That this is a correct representation of what actually occurs in nature becomes evident from an inspection of the current charts. Thus in the chart of the North Atlantic which accompanies Dr. Petermann's Memoir on the Gulf-stream, we observe that south of Spitzbergen the polar current and the Gulf-stream are mutually interpenetrated—long tongues invading and dipping down underneath the Gulf-stream, while in like manner the polar current becomes similarly intersected by well-marked protrusions of warm water flowing from the south. (See Plate II., Fig. 3.)

No accurate observations, as far as I know, have been made regarding the amount of work performed by the wind in impelling the water forward; but when we consider the great retarding effect of objects on the earth's surface, it is quite apparent that the amount of work performed on the surface of the ocean must be far greater than is generally supposed. For example, Mr. Buchan, Secretary to the Scottish Meteorological Society, has shown * that a fence made of slabs of wood three inches in width and three inches apart from each other is a protection even during high winds to objects on the lee side of it, and that a wire screen with meshes about an inch apart affords protection during a gale to flower-pots. The same writer was informed by Mr. Addie that such a screen put up at Rockville was torn to pieces by a storm of wind, the wire screen giving way much in the same way as sails during a hurricane at sea.

The "Challenger's" Crucial Test of the Wind and Gravitation Theories of Oceanic Circulation.—It has been shown in former chapters that all the facts which have been adduced in support of the gravitation theory are equally well explained by the wind theory. We may now consider a class of facts which do not appear to harmonize with either theory. The recent investigations of the *Challenger* Expedition into the thermal state of the ocean reveal a condition of things which appears to me utterly irreconcilable with the gravitation theory.

It is a condition absolutely essential to the gravitation theory that the surface of the ocean should be highest in equatorial regions and slope downwards to either pole. Were water absolutely frictionless, an incline, however small, would be sufficient to produce a surface-flow from the equator to the poles, but to induce such an effect some slope there must be, or gravitation could exercise no power in drawing the surface-water polewards.

The researches of the *Challenger* Expedition bring to light the striking and important fact that the general surface of the North Atlantic in order to produce equilibrium must stand at a higher level than at the equator. In other words the surface of

* Paper read to the Edinburgh Botanical Society on January 8, 1874.

the Atlantic is lowest at the equator, and rises with a gentle slope to well nigh the latitude of England. If this be the case, then it is mechanically impossible that, as far as the North Atlantic is concerned, there can be any such general movement as Dr. Carpenter believes. Gravitation can no more cause the surface-water of the Atlantic to flow towards the arctic regions than it can compel the waters of the Gulf of Mexico up the Mississippi into the Missouri. The impossibility is equally great in both cases.

In order to prove what has been stated, let us take a section of the mid-Atlantic, north and south, across the equator; and, to give the gravitation theory every advantage, let us select that particular section adopted by Dr. Carpenter as the one of all others most favourable to his theory, viz., Section marked No. VIII. in his memoir lately read before the Royal Geographical Society.*

The fact that the polar cold water comes so near the surface at the equator is regarded by Dr. Carpenter as evidence in favour of the gravitation theory. On first looking at Dr. Carpenter's section it forcibly struck me that if it was accurately drawn, the ocean to be in equilibrium would require to stand at a higher level in the North Atlantic than at the equator. In order, therefore, to determine whether this is the case or not I asked the hydrographer of the Admiralty to favour me with the temperature soundings indicated in the section, a favour which was most obligingly granted. The following are the temperature soundings at the three stations A, B, and C. The temperature of C are the mean of six soundings taken along near the equator:—

* Proc. Roy. Geog. Soc., vol. xviii., p. 362. A more advantageous section might have been chosen, but this will suffice. The section referred to is shown in Plate III. The peculiarity of this section, as will be observed, is the thinness of the warm strata at the equator, as compared with that of the heated water in the North Atlantic.

Depth in Fathoms.	A	B	C	
	Lat. 37° 54′ N. Long. 41° 44′ W.	Lat. 23° 10′ N. Long. 38° 42′ W.	Mean of six temperature soundings near equator.	
	Temperature.	Temperature.	Depth in Fathoms.	Temperature.
Surface.	70°·0	72°·0	Surface.	77°·9
100	63·5	67·0	10	77·2
200	60·6	57·6	20	77·1
300	60·0	52·5	30	76·9
400	54·8	47·7	40	71·7
500	46·7	43·7	50	64·0
600	41·6	41·7	60	60·4
700	40·6	40·6	70	59·4
800	38·1	39·4	80	58·0
900	37·8	39·2	90	58·0
1000	37·9	38·3	100	55·6
1100	37·1	38·0	150	51·0
1200	37·1	37·6	200	46·6
1300	37·2	36·7	300	42·2
1400	37·1	36·9	400	40·3
1500	..	36·7	500	38·9
2700	35·2	..	600	39·2
2720	..	35·4	700	39·0
			800	39·1
			900	38·2
			1000	36·9
			1100	37·6
			1200	36·7
			1300	35·8
			1400	36·4
			1500	36·1
			Bottom.	34·7

On computing the extent to which the three columns A, B, and C are each expanded by heat according to Muncke's table of the expansion of sea water for every degree Fahrenheit, I found that column B, in order to be in equilibrium with C (the equatorial column), would require to have its surface standing fully 2 feet 6 inches above the level of column C, and column A fully 3 feet 6 inches above that column. In short, it is evident that there must be a gradual rise from the equator to latitude 38° N. of 3½ feet. Any one can verify the accuracy of these results by making the necessary computations for himself.*

* The temperature of column C in Dr. Carpenter's section is somewhat less than that given in the foregoing table; so that, according to that section, the difference of level between column C and columns A and B would be greater than my estimate.

PLATE III

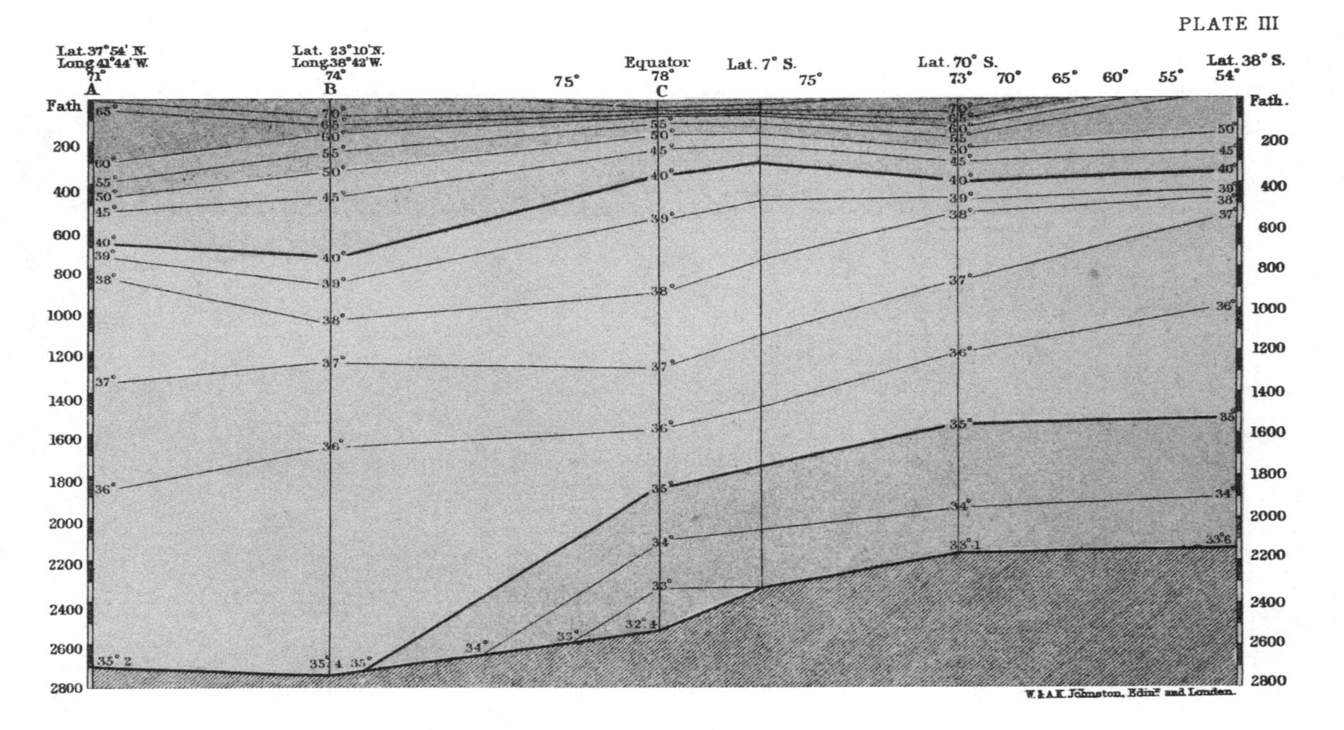

SECTION OF THE ATLANTIC nearly North and South, between LAT. 38° N. & LAT. 38° S.

I may observe that, had column C extended to the same depth as columns A and B, the difference of level would be considerably greater, for column C requires to balance only that portion of columns A and B which lies above the level of its base. Suppose a depth of ocean equal to that of column C to extend to the north pole, and the polar water to have a uniform temperature of 32° from the surface to the bottom, then, in order to produce equilibrium, the surface of the ocean at the equator would require to be 4 feet 6 inches above that at the pole. But the surface of the ocean at B would be 7 feet, and at A 8 feet, above the poles. Gravitation never could have caused the ocean to assume this form. It is impossible that this immense mass of warm water, extending to such a depth in the North Atlantic, could have been brought from equatorial regions by means of gravitation. And, even if we suppose this accumulation of warm water can be accounted for by some other means, still its presence precludes the possibility of any such surface-flow as that advocated by Dr. Carpenter. For so long as the North Atlantic stands 3½ feet above the level of the equator, gravitation can never move the equatorial waters polewards.

There is another feature of this section irreconcilable with the gravitation theory. It will be observed that the accumulation of warm water is all in the North Atlantic, and that there is little or none in the south. But according to the gravitation theory it ought to have been the reverse. For owing to the unrestricted communication between the equatorial and antarctic regions, the general flow of water towards the south pole is, according to that theory, supposed to be greater than towards the north, and consequently the quantity of warm equatorial water in the South Atlantic ought also to be greater. Dr. Carpenter himself seems to be aware of this difficulty besetting the theory, and meets it by stating that "the upper stratum of the North Atlantic is not nearly as much cooled down by its limited polar underflow, as that of the South Atlantic is by the vast movement of antarctic water which is constantly taking place towards the equator." But this "vast movement of antarctic water" necessarily implies a vast counter-movement of warm surface-water. So that if there is more polar water in the South Atlantic to produce the cooling effect, there should likewise be more warm water to be cooled.

According to the wind theory of oceanic circulation the explanation of the whole phenomena is simple and obvious. It has already been shown that owing to the fact that the S. E. trades are stronger than the N. E., and blow constantly over upon the northern hemisphere, the warm surface-water of the South Atlantic is drifted across the equator. It is then carried by the equatorial current into the Gulf of Mexico, and afterwards of course forms a part of the Gulf-stream.

The North Atlantic, on the other hand, not only does not lose its surface heat like the equatorial and South Atlantic, but it receives from the Gulf-stream in the form of warm water an amount of heat, as we have seen, equal to one-fourth of all the heat which it receives from the sun. The reason why the warm surface strata are so much thicker on the North Atlantic than on the equatorial regions is perfectly obvious. The surface-water at the equator is swept into the Gulf of Mexico by the trade winds and the equatorial current, as rapidly as it is heated by the sun, so that it has not time to gather to any great depth. But all this warm water is carried by the Gulf-stream into the North Atlantic, where it accumulates. That this great depth of warm water in the North Atlantic, represented in the section, is derived from the Gulf-stream, and not from a direct flow from the equator due to gravitation, is further evident from the fact that temperature sounding A in latitude 38° N. is made through that immense body of warm water, upwards of 300 fathoms thick, extending from Bermuda to near the Azores, discovered by the *Challenger* Expedition, and justly regarded by Captain Nares as an offshoot of the Gulf-stream. This, in Captain Nares's Report, is No. 8 "temperature sounding," between Bermuda and the Azores; sounding B is No. 6 "temperature curve," between Teneriffe and St. Thomas.

There is an additional reason to the one already stated why the surface temperature of the South Atlantic should be so much below that of the North. It is perfectly true that whatever amount of water is transferred from the southern hemisphere to the northern must be compensated by an equal amount from the northern to the southern hemisphere, nevertheless the warm water which is carried off the South Atlantic by the winds is not directly compensated by water from the north, but by that cold antarctic current whose existence is so well known to mariners from the immense masses of ice which it brings from the Southern Ocean.

Thermal Condition of Southern Ocean.—The thermal condition of the Southern Ocean, as ascertained by the *Challenger* Expedition, appears to me to be also irreconcilable with the gravitation theory. Between the parallels of latitude 65° 42′ S. and 50° 1′ S., the ocean, with the exception of a thin stratum at the surface heated by the sun's rays, was found, down to the depth of about 200 fathoms, to be several degrees colder than the water underneath.* The cold upper stratum is evidently an antarctic current, and the warm underlying water an equatorial under current. But, according to the gravitation theory, the colder water should be underneath.

The very fact of a mass of water, 200 fathoms deep and extending over fifteen degrees of latitude, remaining above water of three or four degrees higher temperature shows how little influence difference of temperature has in producing motion. If it had the potency which some attribute to it, one would suppose that this cold stratum should sink down and displace the warm water underneath. If difference of density is sufficient to move the water horizontally, surely it must be more than sufficient to cause it to sink vertically.

* Captain Nares's Report, July 30, 1874.

CHAPTER XIV

THE WIND THEORY OF OCEANIC CIRCULATION IN RELATION TO CHANGE OF CLIMATE.

Direction of Currents depends on Direction of the Winds.—Causes which affect the Direction of Currents will affect Climate.—How Change of Eccentricity affects the Mode of Distribution of the Winds.—Mutual Reaction of Cause and Effect.—Displacement of the Great Equatorial Current.—Displacement of the Median Line between the Trades, and its Effect on Currents.—Ocean-currents in Relation to the Distribution of Plants and Animals.—Alternate Cold and Warm Periods in North and South.—Mr. Darwin's Views quoted. —How Glaciers at the Equator may be accounted for.—Migration across the Equator.

Ocean-currents in Relation to Change of Climate.—In my attempts to prove that oceanic circulation is produced by the winds and not by difference of specific gravity, and that ocean-currents are the great distributors of heat over the globe, my chief aim has been to show the bearing which these points have on the grand question of secular changes of climate during geological epochs, more particularly in reference to that mystery the cause of the glacial epoch.

In concluding this discussion regarding oceanic circulation, I may therefore be allowed briefly to recapitulate those points connected with the subject which seem to shed most light on the question of changes of climate.

The complete agreement between the systems of ocean-currents and winds not only shows that the winds are the impelling cause of the currents, but it also indicates to what an extent the *directions* of the currents are determined by the winds, or, more properly, to what an extent their directions are determined by the *direction* of the winds.

We have seen in Chapter II. to what an enormous extent

the climatic conditions of the globe are dependent on the distribution of heat effected by means of ocean-currents. It has been there pointed out that, if the heat conveyed from intertropical to temperate and polar regions by oceanic circulation were restored to the former, the equatorial regions would then have a temperature about 55° warmer, and the high polar regions a climate 83° colder than at present. It follows, therefore, that any cause which will greatly affect the currents or greatly change their paths and mode of distribution, will of necessity seriously affect the climatic condition of the globe. But as the existence of these currents depends on the winds, and their direction and form of distribution depend upon the direction and form of distribution of the winds, any cause which will greatly affect the winds will also greatly affect the currents, and consequently will influence the climatic condition of the globe. Again, as the existence of the winds depends mainly on the difference of temperature between equatorial and polar regions, any cause which will greatly affect this difference of temperature will likewise greatly affect the winds; and these will just as surely react on the currents and climatic conditions of the globe. A simple increase or decrease in the difference of temperature between equatorial and polar regions, though it would certainly produce an increase or a decrease, as the case might be, in the strength of the winds, and consequently in the strength of the currents, would not, however, greatly affect the mode of *distribution* of the winds, nor, as a consequence, the mode of *distribution* of the currents. But although a simple change in the difference of temperature between the equator and the poles would not produce a different *distribution* of aërial, and consequently of ocean-currents, nevertheless a *difference in the difference* of temperature between the equator and the two poles would do so; that is to say, any cause that should increase the difference of temperature between the equator and the pole on the one hemisphere, and decrease that difference on the other, would effect a change in the distribution of the aërial currents, which change would in turn produce a corresponding change in the distribution of ocean-currents.

It has been shown * that an increase in the eccentricity of the earth's orbit tends to lower the temperature of the one hemisphere and to raise the temperature of the other. It is true that an increase of eccentricity does not afford more heat to the one hemisphere than to the other; nevertheless it brings about a condition of things which tends to lower the temperature of the one hemisphere and to raise the temperature of the other. Let us imagine the eccentricity to be at its superior limit, 0·07775, and the winter solstice in the aphelion. The midwinter temperature, owing to the increased distance of the sun, would be lowered enormously; and the effect of this would be to cause all the moisture which now falls as rain during winter in temperate regions to fall as snow. Nor is this all; the winters would not merely be colder than now, but they would also be much longer. At present the summer half year exceeds the winter half year by nearly eight days; but at the period in question the winters would be longer than the summers by upwards of thirty-six days. The heat of the sun during the short summer, for reasons which have already been explained, would not be sufficient to melt the snow of winter; so that gradually, year by year, the snow would continue to accumulate on the ground.

On the southern hemisphere the opposite condition of things would obtain. Owing to the nearness of the sun during the winter of that hemisphere, the moisture of the air would be precipitated as rain in regions where at present it falls as snow. This and the shortness of the winter would tend to produce a decrease in the quantity of snow. The difference of temperature between the equatorial and the temperate and polar regions would therefore be greater on the northern than on the southern hemisphere; and, as a consequence, the aërial currents of the former hemisphere would be stronger than those of the latter. This would be more especially the case

* See Chapter IV.

with the trade-winds. The N.E. trades being stronger than the S.E. trades would blow across the equator, and the median line between them would therefore be at some distance to the south of the equator. Thus the equatorial waters would be impelled more to the southern than to the northern hemisphere; and the warm water carried over in this manner to the southern hemisphere would tend to increase the difference of temperature between the two hemispheres. This change, again, would in turn tend to strengthen the N.E. and to weaken the S.E. trades, and would thus induce a still greater flow of equatorial waters into the southern hemisphere—a result which would still more increase the difference of temperature between the northern and southern hemisphere, and so on—the one cause so reacting on the other as to increase its effects, as was shown at length in Chapter IV.

It was this mutual reaction of those physical agents which led, as was pointed out in Chapter IV., to that extraordinary condition of climate which prevailed during the glacial epoch.

There is another circumstance to be considered which perhaps more than any thing else would tend to lower the temperature of the one hemisphere and to raise the temperature of the other; and this is the *displacement of the great equatorial current.* During a glacial period in the northern hemisphere the median line between the trades would be shifted very considerably south of the equator; and the same would necessarily be the case with the great equatorial currents, the only difference being that the equatorial currents, other things being equal, would be deflected farther south than the median line. For the water impelled by the strong N.E. trades would be moving with greater velocity than the waters impelled by the weaker S.E. trades, and, of course, would cross the median line of the trades before its progress southwards could be arrested by the counteracting influence of the S.E. trades. Let us glance briefly at the results which would follow from such a condition of things. In the first place, as was shown on former occasions,* were the equatorial current of the Atlantic (the feeder of the Gulf-stream) shifted considerably south of its present position, it would not bifurcate, as it now does, off Cape St. Roque, owing to the fact that the whole of the waters would strike obliquely against the Brazilian coast and thus be deflected into the Southern Ocean. The effect produced on the climate of the North Atlantic and North-Western Europe by the withdrawal of the water forming the Gulf-stream, may be conceived from what has already been stated concerning the amount of heat conveyed by that stream. The heat thus withdrawn from the North Atlantic would go to raise the temperature of the Southern Ocean and antarctic regions. A similar result would take place in the Pacific Ocean. Were the equatorial current of that ocean removed greatly to the south of its present position, it would not then impinge and be deflected upon the Asiatic coast, but upon the continent of Australia; and the greater portion of its waters would then pass southward into the Southern Ocean, while that portion passing round the north of Australia (owing to the great strength of the N.E. trades) would rather flow into the Indian Ocean than turn round, as now, along the east coast of Asia by the Japan Islands. The stoppage of the Japan current, combined with the displacement of the equatorial current to the south of the equator, would greatly lower the temperature of the whole of the North Pacific and adjoining continents, and raise to a corresponding degree the temperature of the South Pacific and Southern Ocean. Again, the waters of the equatorial current of the Indian Ocean (owing to the opposing N.E. trades), would not, as at present, find their way round the Cape of Good Hope into the North Atlantic, but would be deflected southwards into the Antarctic Sea.

We have in the present state of things a striking example of the extent to which the median line between the two trades may be shifted, and the positionof the great equatorial currents of the ocean may be affected, by a slight difference in the

* Phil. Mag. for August, 1864, February, 1867, March, 1870; see Chap. IV.

relative strength of the two aërial currents. The S.E. trades are at present a little stronger than the N.E.; and the consequence is that they blow across the equator into the northern hemisphere to a distance sometimes of 10 or 15°, so that the mean position of the median line lies at least 6 or 7 degrees north of the equator.

And it is doubtless owing to the superior strength of the S.E. trades that so much warm water crosses the equator from the South to the North Atlantic, and that the main portion of the equatorial current flows into the Caribbean Sea rather than along the Brazilian coast. Were the two trades of equal strength, the transferrencé of heat into the North Atlantic from the southern hemisphere by means of the Southern Atlantic and equatorial currents would be much less than at present. The same would also hold true in regard to the Pacific.

Ocean-currents in Relation to the Distribution of Plants and Animals.—In the fifth and last editions of the "Origin of Species," Mr. Darwin has done me the honour to express his belief that the foregoing view regarding alternate cold and warm periods in north and south during the glacial epoch explains a great many facts in connection with the distribution of plants and animals which have always been regarded as exceedingly puzzling.

There are certain species of plants which occur alike in the temperate regions of the southern and northern hemispheres. At the equator these same temperate forms are found on elevated mountains, but not on the lowlands. How, then, did these temperate forms manage to cross the equator from the northern temperate regions to the southern, and *vice versâ?* Mr. Darwin's solution of the problem is (in his own words) as follows:—

"As the cold became more and more intense, we know that arctic forms invaded the temperate regions; and from the facts just given, there can hardly be a doubt that some of the more vigorous, dominant, and widest-spreading temperate forms invaded the equatorial lowlands. The inhabitants of these hot lowlands would at the same time have migrated to the tropical and subtropical regions of the south; for the southern hemisphere was at this period warmer. On the decline of the glacial period, as both hemispheres gradually recovered their former temperatures, the northern temperate forms living on the lowlands under the equator would have been driven to their former homes or have been destroyed, being replaced by the equatorial forms returning from the south. Some, however, of the northern temperate forms would almost certainly have ascended any adjoining high land, where, if sufficiently lofty, they would have long survived like the arctic forms on the mountains of Europe."

"In the regular course of events the southern hemisphere would in its turn be subjected to a severe glacial period, with the northern hemisphere rendered warmer; and then the southern temperate forms would invade the equatorial lowlands. The northern forms which had before been left on the mountains would now descend and mingle with the southern forms. These latter, when the warmth returned, would return to their former homes, leaving some few species on the mountains, and carrying southward with them some of the northern temperate forms which had descended from their mountain fastnesses. Thus we should have some few species identically the same in the northern and southern temperate zones and on the mountains of the intermediate tropical regions" (p. 339, sixth edition).

Additional light is cast on this subject by the results already stated in regard to the enormous extent to which the temperature of the equator is affected by ocean-currents. Were there no transferrence of heat from equatorial to temperate and polar regions, the temperature of the equator, as has been remarked, would probably be about 55° warmer than at present. In such a case no plant existing on the face of the globe could live at the equator unless on some elevated mountain region. On the other hand, were the quantity of warm water which is being transferred from the equator to be very much increased, the temperature of inter-tropical latitudes might be so lowered as easily to admit of temperate species of plants growing at the equator. A lowering of the temperature at the equator some 20° or 30° is all that would be required; and only a moderate increase in the volume of the currents proceeding from the equator, taken in connection with the effects flowing from the following considerations, might suffice to produce that result. During the glacial epoch, when the one hemisphere was under ice and the other enjoying a warm and equable climate, the median line between the trades may have been shifted to almost the tropical line of the warm hemisphere. Under such a condition of things the warmest part would probably be somewhere about the tropic of the warm hemisphere, and not, as now, at the equator; for since all, or nearly all, the surface-water of the equator would then be impelled over to the warm hemisphere, the tropical regions of that hemisphere would be receiving nearly double their present amount of warm water.

Again, as the equatorial current at this time would be shifted towards the tropic of the warm hemisphere, the surface-water would not, as at present, be flowing in equatorial regions parallel to the equator, but obliquely across it from the cold to the warm hemisphere. This of itself would tend greatly to lower the temperature of the equator.

It follows, therefore, as a necessary consequence, that during the glacial epoch, when the one hemisphere was under snow and ice and the other enjoying a warm and equable climate, the temperature of the equator would be lower than at present. But when the glaciated hemisphere (which we may assume to be the northern) began to grow warmer and the climate of the southern or warm hemisphere to get colder, the median line of the trades and the equatorial currents of the ocean also would begin to move back from the southern tropic towards the equator. This would cause the temperature of the equator to rise and to continue rising until the equatorial currents reached their normal position. When the snow began to accumulate on the southern hemisphere and to disappear on the northern, the median line of the trades and the equatorial currents of the ocean would then begin to move towards the northern tropic as they had formerly towards the southern. The temperature of the equator would then again begin to sink, and continue to do so until the glaciation of the southern hemisphere reached its maximum. This oscillation of the thermal equator to and fro across the geographical equator would continue so long as the alternate glaciation of the two hemispheres continued.

This lowering of the temperature of the equator during the severest part of the glacial epoch will help to explain the former existence of glaciers in inter-tropical regions at no very great elevation above the sea-level, evidence of which appears recently to have been found by Professor Agassiz, Mr. Belt, and others.

The glacial *epoch* may be considered as contemporaneous in both hemispheres. But the epoch consisted of a succession of cold and warm *periods*, the cold periods of one hemisphere coinciding with the warm periods of the other, and *vice versâ*.

Migration across the Equator.—Mr. Belt* and others have felt some difficulty in understanding how, according to theory, the plants and animals of temperate regions could manage to migrate from one hemisphere to the other, seeing that in their passage they would have to cross the thermal equator. The oscillation to and fro of the thermal equator across the geographical, removes every difficulty in regard to how the migration takes place. When, for example, a cold period on the northern hemisphere and the corresponding warm one on the southern were at their maximum, the thermal equator would by this time have probably passed beyond the Tropic of Capricorn. The geographical equator would then be enjoying a subtropical, if not a temperate condition of climate, and the plants and animals of the northern hemisphere would manage then to reach the equator. When the cold began to abate

* Quarterly Journal of Science for October, 1874.

on the northern and to increase on the southern hemisphere, the thermal equator would commence its retreat towards the geographical. The plants and animals from the north, in order to escape the increasing heat as the thermal equator approached them, would begin to ascend the mountain heights; and when that equator had passed to its northern limit, and the geographical equator was again enjoying a subtropical condition of climate, the plants and animals would begin to descend and pursue their journey southwards as the cold abated on the southern hemisphere.

CHAPTER XV.

WARM INTER-GLACIAL PERIODS.

Alternate Cold and Warm Periods.—Warm Inter-glacial Periods a Test of Theories.—Reason why their Occurrence has not been hitherto recognised. —Instances of Warm Inter-glacial Periods.—Dranse, Dürnten, Hoxne, Chapelhall, Craiglockhart, Leith-Walk, Redhall Quarry, Beith, Crofthead, Kilmaurs, Sweden, Ohio, Cromer, Mundesley, &c., &c.—Cave and River Deposits.—Occurrence of Arctic and Warm Animals in some Beds accounted for.—Mr. Boyd Dawkins's Objections.—Occurrence of Southern Shells in Glacial Deposits.—Evidence of Warm Inter-glacial Periods from Mineral Borings.—Striated Pavements.—Reason why Inter-glacial Land-surfaces are so rare.

Alternate Cold and Warm Periods.—If the theory developed in the foregoing chapters in reference to the cause of secular changes of climate be correct, it follows that that long age known as the glacial epoch did not, as has hitherto been generally supposed, consist of one long unbroken period of cold and ice. Neither did it consist, as some have concluded, of two long periods of ice with an intervening mild period, but it must have consisted of a long succession of cold and warm periods; the warm periods of the one hemisphere corresponding in time with the cold periods of the other and *vice versâ.* It follows also from theory that as the cold periods became more and more severe, the warm intervening periods would become more and more warm and equable. As the ice began to accumulate during the cold periods in subarctic and temperate regions in places where it previously did not exist, so in like manner during the corresponding warm periods it would begin to disappear in arctic regions where it had held enduring sway throughout the now closing cycle. As the cold periods in the southern hemisphere became more and more severe, the ice would continue to advance northwards in the temperate regions; but at that very same time the intervening warm periods in the northern hemisphere would become warmer and warmer and more equable, and the ice of the arctic regions would continue to disappear farther and farther to the north, till by the time that the ice had reached a maximum during the cold antarctic periods, Greenland and the arctic regions would, during the warm intervening periods, be probably free of ice and enjoying a mild and equable climate. Or we may say that as the one hemisphere became cold the other became warm, and when the cold reached a maximum in the one hemisphere, the warmth would reach a maximum in the other. The time when the ice had reached its greatest extension on the one hemisphere would be the time when it had disappeared from the other.

Inter-glacial Periods a Test of Theories.—Here we have the grand crucial test of the truth of the foregoing theory of the cause of the glacial epoch. That the glacial epoch should have consisted of a succession of cold and warm periods is utterly inconsistent with all previous theories which have been advanced to account for it. What, then, is the evidence of geology on this subject? If the glacial epoch can be proved from geological evidence to have consisted of such a succession of cold and warm periods, then I have little doubt but the theory will soon be generally accepted. But at the very outset an objection meets us, viz., why call an epoch, which consisted as much of warm periods as of cold, a glacial epoch, or an "Ice Age," as Mr. James Geikie tersely expresses it? Why not as well call it a warm epoch as a cold one, seeing that, according to theory, it was just as much a warm as a cold epoch? The answer to this objection will be fully discussed in the chapter on the Reason of the Imperfection of Geological Records. But in the meantime, I may remark that it will be shown that the epoch known as the glacial has been justly called the glacial epoch or "Ice Age," because the geological evidences of the cold periods remain in a remarkably perfect state, whilst the evidences of the warm periods have to a great extent disappeared. The reason of this difference in the two cases will be discussed in the chapter to which I have referred. Besides, the condition of things during the cold periods was so extraordinary, so exceptional, so totally different from those now prevailing, that even supposing the geological records of the warm periods had been as well preserved as those of the cold, nevertheless we should have termed the epoch in question a glacial epoch. There is yet another reason, however, for our limited knowledge of warm inter-glacial periods. Till very lately, little or no attention was paid by geologists to this part of the subject in the way of keeping records of cases of inter-glacial deposits which, from time to time, have been observed. Few geologists ever dreamt of such a thing as warm periods during the age of ice, so that when intercalated beds of sand and gravel, beds of peat, roots, branches, trunks, leaves, and fruits of trees were found in the boulder clay, no physical importance was attached to them, and consequently no description or record of them ever kept. In fact, all such examples were regarded as purely accidental and exceptional, and were considered not worthy of any special attention. A case which came under my own observation will illustrate my meaning. An intelligent geologist, some years ago, read a paper before one of our local geological societies, giving an account of a fossiliferous bed of clay found intercalated between two distinct beds of till. In this intercalated bed were found rootlets and stems of trees, nuts, and other remains, showing that it had evidently been an old inter-glacial land surface. In the transactions of the society a description of the two beds of till was given, but no mention whatever was made of the intercalated bed containing the organic remains, although this was the only point of any real importance.

Since the theory that the glacial epoch resulted from a high state of eccentricity of the earth's orbit began to receive some little acceptance, geologists have paid a good deal of attention to cases of intercalated beds in the till containing organic remains, and the result is that we have already a great body of evidence of a geological nature in favour of warm inter-glacial periods, and I have little doubt that in the course of a few years the former occurrence of warm inter-glacial periods will be universally admitted.

I shall now proceed to give a very brief outline of the evidence bearing on the subject. But the cases to which I shall have to refer are much too numerous to allow me to enter into details.

Inter-glacial Beds of Switzerland.—The first geologist, so far as I am aware, who directed attention to evidence of a break in the cold of the glacial epoch was M. Morlot. It is now twenty years ago since he announced the existence of a warm period during the glacial epoch from geological evidence connected with the glacial drift of the Alps.*

The rivers of Switzerland, he found, show on their banks three well-marked terraces of regularly stratified and well-rounded shingle, identical with the modern deposits of the rivers. They stand at 50, 100, and 150 feet above the present level of the rivers. These terraces were evidently formed by the present system of rivers when these flowed at a higher level, and extend up the Alps to a height of from 3,000 to 4,000 feet above the level of the sea. There is a terrace bordering the Rhine at Camischollas, above Disentis, 4,400 feet above the level of the sea, proving that during the period of its formation the Alps were free of ice up to the height of 4,400 feet above the sea-level. It is well known that a glacial period must have succeeded the formation of these drifts, for they are in many places covered with erratics. At Geneva, for example, an erratic drift nearly 50 feet thick is seen to rest on the drift of the middle terrace, which rises 100 feet above the level of the lake. But it is also evident that a glacial period must have preceded the formation of the drift beds, for they are found to lie in many places upon the unstratified boulder

* See a paper by M. Morlot, on "The Post-Tertiary and Quaternary Formations of Switzerland." Edin. New Phil. Journal, New Series, vol. ii., 1855.

clay or *till.* M. Morlot observed in the neighbourhood of Clareus, from 7 to 9 feet of drift resting upon a bed of true till 40 feet thick; the latter was composed of a compact blue clay, containing worn and scratched alpine boulders and without any trace of stratification. In the gorge of Dranse, near Thoron, M. Morlot found the whole three formations in a direct superimposed series. At the bottom was a mass of compact till or boulder clay, 12 feet thick, containing boulders of alpine limestone. Over this mass came regularly stratified beds 150 feet thick, made up of rounded pebbles in horizontal beds. Above this again lay a second formation of unstratified boulder clay, with erratic blocks and striated pebbles, which constituted the left lateral moraine of the great glacier of the Rhone, when it advanced for the second time to the Lake of Geneva. A condition of things somewhat similar was observed by M. Ischer in the neighbourhood of Berne.

These facts, M. Morlot justly considers, prove the existence of two glacial periods separated by an intermediate one, during which the ice, which had not only covered Switzerland, but the greater part of Europe, disappeared even in the principal valleys of the Alps to a height of more than 4,400 feet above the present level of the sea. This warm period, after continuing for long ages, was succeeded by a second glacial period, during which the country was again covered with ice as before. M. Morlot even suggests the possibility of these alternations of cold and warm periods depending upon a cosmical cause. "Wild as it may have appeared," he says, "when first started, the idea of general and periodical eras of refrigeration for our planet, connected perhaps with some cosmic agency, may eventually prove correct."*

Shortly afterwards, evidence of a far more remarkable character was found in the glacial drift of Switzerland, namely, the famous lignite beds of Dürnten. In the vicinity of Utznach and Dürnten, on the Lake of Zurich, and near Mörschwyl, on the Lake of Constance, there are beds of coal or

* Edin. New Phil. Journ., New Series, vol. ii., p. 28.

lignite, nearly 12 feet thick, lying directly on the boulder clay. Overlying these beds is another mass of drift and clay 30 feet in thickness, with rounded blocks, and on the top of this upper drift lie long angular erratics, which evidently have been transported on the back of glaciers.* Professor Vogt attributes their transport to floating ice; but he evidently does so to avoid the hypothesis of a warm period during the glacial epoch.

Here we have proof not merely of the disappearance of the ice during the glacial epoch, but of its absence during a period of sufficient length to allow of the growth of 10 or 12 feet of coal. Professor Heer thinks that this coal-bed, when in the condition of peat, must have been 60 feet thick; and assuming that one foot of peat would be formed in a century, he concludes that 6,000 years must have been required for the growth of the coal plants. According to Liebig, 9,600 years would be required. This, as we have already seen, is about the average duration of a warm period.

In these beds have been found the bones of the elephant (*E. Merkii*), stag, cave-bear, and other animals. Numerous insects have also been met with, which further prove the warm, mild condition of climate which must have prevailed at the time of the formation of the lignite.

At Hoxne, near Diss, in Suffolk, a black peaty mass several feet thick, containing fragments of wood of the oak, yew, and fir, was found overlying the boulder clay.† Professor Vogt believes that this peat bed is of the same age as the lignite beds of Switzerland.

In the glacial drift of North America, particularly about Lake Champlain and the valley of the St. Lawrence, there is similar evidence of two glacial periods with an intervening non-glacial or warm period.‡

* Vogt's "Lectures on Man," pp. 318—321.

† See Mr. Prestwich on Flint Implements, Phil. Trans. for 1860 and 1864. Lyell's "Antiquity of Man," Second Edition, p. 168.

‡ Edin. New Phil. Journ., New Series, vol. ii., p. 28. Silliman's Journ., vol. xlvii., p. 259 (1844).

Glacial and Inter-glacial Periods of the Southern Hemisphere—(South Africa).—Mr. G. W. Stow, in a paper on the "Geology of South Africa," * describes a recent glaciation extending over a large portion of Natal, British Kaffraria, the Kaga and Krome mountains, which he attributes to the action of land-ice. He sums up the phenomena as follows:—"The rounding off of the hills in the interiors of the ancient basins; the numerous dome-shaped (*roches moutonnée*) rocks; the enormous erratic boulders in positions where water could not have carried them; the frequency of unstratified clays—clays with imbedded angular boulders; drift and lofty mounds of boulders; large tracts of country thickly spread over with unstratified clays and superimposed fragments of rock; the Oliphant's-Hoek clay, and the vast piles of Enon conglomerate." In addition to these results of ice-action, he records the discovery by himself of distinct ice-scratches or groovings on the surface of the rocks at Reit-Poort in the Tarka, and subsequently † the discovery by Mr. G. Gilfillan of a large boulder at Pniel with *striæ* distinctly marked upon it, and also that the same observer found that almost every boulder in the gravel at "Moonlight Rush" had unmistakable striæ on one or more sides.

In South Africa there is evidence not only of a glacial condition during the Pliocene period, but also of a warmer climate than now prevails in that region. "The evidence," says Mr. Stow, "of the Pliocene shells of the superficial limestone of the Zwartkops heights, and elsewhere, leads us to believe that the climate of South Africa must have been of a far more tropical character than at present.

"Take, for instance, the characteristic *Venericardia* of that limestone. This has migrated along the coast some 29° or 30° and is now found within a few degrees of the equator, near Zanzibar, gradually driven, as I presume it must have been, further and further north by a gradual lowering of the temperature of the more southern parts of this coast since the limestone was deposited."

* Quart. Journ. Geol. Soc., vol. xxvii., p. 534. † Ibid., vol. xxviii., p. 17.

"During the formation of the shell-banks in the Zwartkops estuary, younger than the Pliocene limestone, the immense number of certain species of shells, which have as yet been found living only in latitudes nearer the equator, point to a somewhat similar though a more modified change of temperature."

Inter-glacial Beds of Scotland.—Upwards of a dozen years ago, Professor Geikie arrived, from his own observations of the glacial drift of Scotland, at a similar conclusion to that of M. Morlot regarding the intercalation of warm periods during the glacial epoch; and the facts on which Professor Geikie's conclusions were based are briefly as follows. In a cliff of boulder clay on the banks of the Slitrig Water, near the town of Hawick, he observed a bed of stones or shingle. Over the lower stratum of stones lay a few inches of well-stratified sand, silt, and clay, some of the layers being black and peaty, *with enclosed vegetable fibres* in a crumbling state.* There were some 30 or 40 feet of boulder clay above these stratified beds, and 15 or 20 feet under them. The stones in the shingle band were identical with those of the boulder clay, but they showed no striations, and were more rounded and water-worn, and resembled in every respect the stones now lying in the bed of the Slitrig. The section of the cliff stood as under:—

1. Vegetable soil.
2. Boulder clay, thirty to forty feet.

Stratified beds {
3. Yellowish gravelly sand.
4. Peaty silt and clay.
5. Fine ferruginous sand.
6. Coarse shingle, two to three feet.
}

7. Coarse, stiff boulder clay, fifteen to twenty feet.

A few more cases of intercalation of stratified materials in the true till were also found in the same valley.

In a cliff of stiff brown boulder clay, about 20 feet high, on the banks of the Carmichael Water, Lanarkshire, Professor

* "Glacial Drift of Scotland," p. 54.

Geikie observed a stratified bed of clay about 3 or 4 inches in thickness. About a mile higher up the stream, he found a series of beds of gravel, sand, and clay in the true *till*. "A thin seam of *peaty matter*," he says, "was observed to run for a few inches along the bottom of a bed of clay and then disappear, while in a band of fine laminated clay with thin sandy partings occasional *fragments of mouldering wood* were found."*

At Chapelhall, near Airdrie, a sand-bed has been extensively mined under about 114 feet of till. This bed of finely stratified sand is about 20 feet thick. In it were found lenticular beds of fine pale-coloured clay containing layers of peat and decaying twigs and branches. Professor Geikie found the vegetable fibres, though much decayed, still distinct, and the substance when put into the fire burned with a dull lambent flame. Underlying these stratified beds, and forming the floor of the mine, is a deposit of *the true till* about 24 feet in thickness. In another pit adjoining, the till forming the floor is 30 feet thick, but it is sometimes absent altogether, so as to leave the sand-beds resting directly on the sandstone and shale of the coal-measures. At some distance from this sand-pit an old buried river channel was met with in one of the pit workings. This channel was found to contain a coating of boulder clay, on which the laminated sands and clays reposed, showing, as Professor Geikie has pointed out, that this old channel had been filled with boulder clay, and then re-excavated to allow of the deposition of the stratified deposits. Over all lay a thick mantle of boulder clay which buried the whole.

A case somewhat similar was found by Professor Nicol in a cutting on the Edinburgh and Leith Railway. In many places the till had been worn into hollows as if part of it had been removed by the action of running water.† One of these hollows, about 5 or 6 feet wide by 3 or 4 feet deep, closely resembled the channel of a small stream. It was also filled

* "Glacial Drift of Scotland," p. 58.

† Quart. Journ. Geol. Soc., vol. v., p. 22.

with gravel and sand, in all respects like that found in such a stream at the present day. It was seen to exhibit the same characters on both sides of the cutting, but Professor Nicol was unable to determine how far it may have extended beyond; but he had no doubt whatever that it had been formed by a stream of water. Over this old water-course was a thick deposit of true till.

In reference to the foregoing cases,. Professor Geikie makes the following pertinent remarks:—"Here it is evident that the scooping out of this channel belongs to the era of the boulder clay. It must have been effected during a pause in the deposition of the clay, when a run of water could find its way along the inequalities of the surface of the clay. This pause must have been of sufficient duration to enable the runnel to excavate a capacious channel for itself, and leave in it a quantity of sand and shingle. We can scarcely doubt that when this process was going on the ground must have been a land surface, and could not have been under the sea. And lastly, we see from the upper boulder clay that the old conditions returned, the water-course was choked up, and another mass of chaotic boulder clay was tumbled down upon the face of the country. This indicates that the boulder clay is not the result of one great catastrophe, but of slow and silent, yet mighty, forces acting sometimes with long pauses throughout a vast cycle of time."*

At Craiglockhart Hill, about a mile south of Edinburgh, an extensive bed of fine sand of from one to three feet in thickness was found between two distinct masses of true boulder clay or till. The sand was extensively used for building purposes during the erection of the city poorhouse a few years ago. In this sand-bed I found a great many tree roots in the position in which they had grown. During the time of the excavations I visited the place almost daily, and had every opportunity of satisfying myself that this sand-bed, prior to the time of the formation of the upper boulder clay, must have

* "Glacial Drift of Scotland," p. 64.

been a land surface on which the roots had grown. In no case did I find them penetrating into the upper boulder clay, and in several places I found stones of the upper clay resting directly on the broken ends of the roots. These roots were examined by Professor Balfour, but they were so decayed that he was unable to determine their character.

In digging a foundation for a building in Leith Walk, Edinburgh, a few years ago, two distinct beds of sand were passed through, the upper, about 10 feet in thickness, rested upon what appeared to be a denuded surface of the lower bed. In this lower bed, which evidently had been a land surface, numbers of tree roots were found. I had the pleasure of examining them along with my friend Mr. C. W. Peach, who first directed my attention to them. In no instance were the roots found in the upper bed. That these roots did not belong to trees which had grown on the present surface and penetrated to that depth, was further evident from the fact that in one or two cases we found the roots broken off at the place where they had been joined to the trunk, and there the upper sand-bed over them was more than 10 feet in thickness. If we assume that the roots belonged to trees which had grown on the present surface, then we must also assume, what no one would be willing to admit, that the trunks of the trees had grown downwards into the earth to a depth of upwards of ten feet. I have shown these roots to several botanists, but none of them could determine to what trees they belonged. The surface of the ground at the spot in question is 45 feet above sea-level. Mr. Peach and I have found similar roots in the under sand-bed at several other places in the same neighbourhood. That they belong to an inter-glacial period appears probable for the following reasons:—(1.) This upper sand-bed is overlaid by a tough clay, which in all respects appears to be the same as the Portobello clay, which we know belongs to the glacial series. In company with Mr. Bennie, I found the clay in some places to be contorted in a similar manner to the Portobello clays. (2.) In a sand-pit about one or two hundred yards to the west of where the roots were found, the sand-bed was found contorted in the most extraordinary manner to a depth of about 15 feet. In fact, for a space of more than 30 feet, the bedding had been completely turned up on end without the fine layers being in the least degree broken or disarranged, showing that they had been upturned by some enormous powers acting on a large mass of the sand.

One of the best examples of true till to be met with in the neighbourhood of Edinburgh is at Redhall Quarry, about three miles to the south-west of the city. In recently opening up a new quarry near the old one a bed of peat was found intercalated in the thick mass of till overlying the rock. The clay overlying and underlying the peat-bed was carefully examined by Mr. John Henderson,* and found to be true till.

In a quarry at Overtown, near Beith, Ayrshire, a sedimentary bed of clay, intercalated between two boulder clays, was some years ago observed by Mr. Robert Craig, of the Glasgow Geological Society. This bed filled an elliptical basin about 130 yards long, and about 30 yards broad. Its thickness averaged from one to two feet. This sedimentary bed rested on the till on the north-east end of the basin, and was itself overlaid on the south-west end by the upper bed of till. The clay bed was found to be full of roots and stems of the common hazel. That these roots had grown in the position in which they were found was evident from the fact that they were in many places found to pass into the "cutters" or fissures of the limestone, and were here found in a flattened form, having in growing accommodated themselves to the size and shape of the fissures. Nuts of the hazel were plentifully found.†

At Hillhead, some distance from Overtown, there is a similar intercalated bed full of hazel remains, and a species of fresh-water *Ostracoda* was detected by Mr. David Robertson.

In a railway cutting a short distance from Beith, Mr. Craig pointed out to my colleague, Mr. Jack, and myself, a thin

* Trans. Edin. Geol. Soc., vol. ii., p. 391.

† Trans. of Geol. Soc. of Glasgow, vol. iv., p. 146.

layer of peaty matter, extending for a considerable distance between an upper and lower mass of till; and at one place we found a piece of oak about four feet in length and about seven or eight inches in thickness. This oak boulder was well polished and striated.

Not far from this place is the famous Crofthead inter-glacial bed, so well known from the description given by Mr. James Geikie and others that I need not here describe it. I had the pleasure of visiting the section twice while it was well exposed, once, in company with Mr. James Geikie, and I do not entertain the shadow of a doubt as to its true inter-glacial character.

In the silt, evidently the mud of an inter-glacial lake, were found the upper portion of the skull of the great extinct ox (*Bos primigenius*), horns of the Irish elk or deer, and bones of the horse. In the detailed list of the lesser organic remains found in the intercalated peat-bed by Mr. J. A. Mahony,* are the following, viz., three species of *Desmidaceæ*, thirty-one species of *Diatomaceæ*, eleven species of mosses, nine species of phanerogamous plants, and several species of annelids, crustacea, and insects. This list clearly shows that the inter-glacial period, represented by these remains, was not only mild and warm, but of considerable duration. Mr. David Robertson found in the clay under the peat several species of *Ostracoda*.

The well-known Kilmaurs bed of peaty matter in which the remains of the mammoth and reindeer were found, has now by the researches of the Geological Survey been proved to be of inter-glacial age.†

In Ireland, as shown by Professors Hull and Harkness, the inter-glacial beds, called by them the "manure gravels," contain numerous fragments of shells indicating a more genial climate than prevailed when the boulder clays lying above and below them were formed.‡

* Geol. Mag., vi., p. 391.

† See "Memoirs of Geological Survey of Scotland," Explanation of sheet 22, p. 29. See also Trans. Glasgow Geol. Soc., iv., p. 150.

‡ "Great Ice Age," p. 374.

In Sweden inter-glacial beds of fresh-water origin, containing plants, have been met with by Herr Nathorst and also by Herr Holmström.*

In North America Mr. Whittlesey describes inter-glacial beds of blue clay enclosing pieces of wood, intercalated with beds of hard pan (till). Professor Newberry found at Germantown, Ohio, an immense bed of peat, from 12 to 20 feet in thickness, underlying, in some places 30 feet, and in other places as much as 80 feet, of till, and overlying drift beds. The uppermost layers of the peat contain undecomposed sphagnous mosses, grasses, and sedges, but in the other portions of the bed abundant fragments of coniferous wood, identified as red cedar (*Juniperus virginiana*), have been found. Ash, hickory, sycamore, together with grape-vines and beech-leaves, were also met with, and with these the remains of the mastodon and great extinct beaver.†

Inter-glacial Beds of England.—Scotland has been so much denuded by the ice sheet with which it was covered during the period of maximum glaciation that little can be learned in this part of the island regarding the early history of the glacial epoch. But in England, and more especially in the south-eastern portion of it, matters are somewhat different. We have, in the Norwich Crag and Chillesford beds, a formation pretty well developed, which is now generally regarded as lying at the base of the Glacial Series. That this formation is of a glacial character is evident from the fact of its containing shells of a northern type, such as *Leda lanceolata*, *Cardium Groènlandicum*, *Lucina borealis*, *Cyprina Islandica*, *Panopœa Norvegica*, and *Mya truncata*. But the glacial character of the formation is more strikingly brought out, as Sir Charles Lyell remarks, by the predominance of such species as *Rhynchonella psittacea*, *Tellina calcarea*, *Astarte borealis*, *Scalaria Groènlandica*, and *Fusus carinatus*.

* "Great Ice Age," p. 384.

† "Geological Survey of Ohio, 1869," p. 165. See also "Great Ice Age," chap. xxviii.

The "Forest Beds."—Immediately following this in the order of time comes the famous "Forest Bed" of Cromer. This buried forest has been traced for more than forty miles along the coast from Cromer to near Kessengland, and consists of stumps of trees standing erect, attached to their roots, penetrating the original soil in which they grew. Here and in the overlying fluvio-marine beds we have the first evidence of at least a temperate, if not a warm, inter-glacial period. This is evident from the character of the flora and fauna belonging to these beds. Among the trees we have, for example, the Scotch and spruce fir, the yew, the oak, birch, the alder, and the common sloe. There have also been found the white and yellow water-lilies, the pond-weed, and others. Amongst the mammalia have been met with the *Elephas meridionalis,* also found in the Lower Pliocene beds of the Val d'Arno, near Florence; *Elephas antiquus, Hippopotamus major, Rhinoceros Etruscus,* the two latter Val d'Àrno species, the roebuck, the horse, the stag, the Irish elk, the *Cervus Polignacus,* found also at Mont Perrier, France, *C. verticornis,* and *C. carnutorum,* the latter also found in Pliocene strata of St. Prest, France. In the fluvio-marine series have been found the *Cyclas omnica* and the *Paludina marginata,* a species of mollusc still found in the South of France, but no longer inhabiting the British Isles.

Above the forest bed and fluvio-marine series comes the well-known unstratified Norwich boulder till, containing immense blocks 6 or 8 feet in diameter, many of which must have come from Scandinavia, and above the unstratified till are a series of contorted beds of sand and gravel. This series may be considered to represent a period of intense glaciation. Above this again comes the middle drift of Mr. Searles Wood, junior, yielding shells which indicate, as is now generally admitted, a comparatively mild condition of climate. Upon this middle drift lies the upper boulder clay, which is well developed in South Norfolk and Suffolk, and which is of unmistakable glacial origin. Newer than all these are the Mundesley fresh-water beds, which lie in a hollow denuded out of the foregoing series. In this formation a black peaty deposit containing seeds of plants, insects, shells, and scales and bones of fishes, has been found, all indicating a mild and temperate condition of climate. Among the shells there is, as in the forest bed, the *Paludina marginata.* And that an arctic condition of things in England followed is believed by Mr. Fisher and others, on the evidence of the "Trail" described by the former observer.

Cave and River Deposits.—Evidence of the existence of warm periods during the glacial epoch is derived from a class of facts which have long been regarded by geologists as very puzzling, namely, the occurrence of mollusca and mammalia of a southern type associated in England and on the continent with those of an extremely arctic character. For example, *Cyrena fluminalis* is a shell which does not live at present in any European river, but inhabits the Nile and parts of Asia, especially Cashmere. *Unio littoralis,* extinct in Britain, is still abundant in the Loire; *Paludina marginata* does not exist in this country. These shells of a southern type have been found in post-tertiary deposits at Gray's Thurrock, in Essex; in the valley of the Ouse, near Bedford; and at Hoxne, in Suffolk, associated with a *Hippopotamus* closely allied to that now inhabiting the Nile, and *Elephas antiquus,* an animal remarkable for its southern range. Amongst other forms of a southern type which have been met with in the cave and river deposits, are the spotted hyæna from Africa, an animal, says Mr. Dawkins, identical, except in size, with the cave hyæna, the African elephant (*E. Africanus*), and the *Elephas meridionalis,* the great beaver (*Trogontherium*), the cave hyæna (*Hyæna spelæa*), the cave lion (*Felis leo,* var. *spelæa*), the lynx (*Felis lynx*), the sabre-toothed tiger (*Machairodus latidens*), the rhinoceros (*Rhinoceros megarhinus* and *R. leptorhinus*). But the most extraordinary thing is that along with these, associated in the same beds, have been found the remains of such animals of an arctic type as the glutton (*Gulo luscus*), the ermine (*Mustela erminea*), the reindeer (*Cervus tarandus*), the musk-ox or musk-sheep (*Ovibos moschatus*), the aurochs (*Bison priscus*), the woolly rhinoceros (*Rhinoceros tichorhinus*), the mammoth (*Elephas primigenius*), and others of a like character. According to Mr. Boyd Dawkins, these southern animals extended as far north as Yorkshire in England, and the northern animals as far south as the latitude of the Alps and Pyrenees.*

The Explanation of the Difficulty.—As an explanation of these puzzling phenomena, I suggested, in the Philosophical Magazine for November, 1868, that these southern animals lived in our island during the warm periods of the glacial epoch, while the northern animals lived during the cold periods. This view I am happy to find has lately been supported by Sir John Lubbock; further, Mr. James Geikie, in his "Great Ice Age," and also in the Geological Magazine, has entered so fully into the subject and brought forward such a body of evidence in support of it, that, in all probability, it will, ere long, be generally accepted. The only objection which has been advanced, so far as I am aware, deserving of serious consideration, is that by Mr. Boyd Dawkins, who holds that if these migrations had been *secular* instead of seasonal, as is supposed by Sir Charles Lyell and himself, the arctic and southern animals would now be found in separate deposits. It is perfectly true that if there had been only one cold and one warm period, each of geologically immense duration, the remains might, of course, be expected to have been found in separate beds; but when we consider that the glacial epoch consisted of a long *succession of alternate cold and warm periods*, of not more than ten or twelve thousand years each, we can hardly expect that in the river deposits belonging to this long cycle we should be able to distinguish the deposits of the cold periods from those of the warm.

Shell Beds.—Evidence of warm inter-glacial periods may be justly inferred from the presence of shells of a southern type which have been found in glacial beds, of which some illustrations follow.

In the southern parts of Norway, from the present sea-level

* Quart. Journ. Geol. Soc., xxviii., p. 435.

up to 500 feet, are found glacial shell beds, similar to those of Scotland. In these beds *Trochus magus, Tapes decussata,* and *Pholas candida* have been found, shells which are distributed between the Mediterranean and the shores of England, but no longer live round the coasts of Norway.

At Capellbacken, near Udevalla, in Sweden, there is an extensive bed of shells 20 to 30 feet in thickness. This formation has been described by Mr. Gwyn Jeffreys.* It consists of several distinct layers, apparently representing many epochs and conditions. Its shells are of a highly arctic character, and several of the species have not been found living south of the arctic circle. But the remarkable circumstance is that it contains *Cypræa lurida,* a Mediterranean shell, which Mr. Jeffreys, after some hesitation, believed to belong to the bed. Again, at Lilleherstehagen, a short distance from Capellbacken, another extensive deposit is exposed. "Here the upper layer," says Mr. Jeffreys, "gives a singular result. Mixed with the universal *Trophon clathratus* (which is a high northern species, and found living only within the arctic circle) are many shells of a southern type, such are *Ostrea edulis, Tapes pullastra, Corbula gibba,* and *Aporrhais pes-pelicani.*

At Kempsey, near Worcester, a shell bed is described by Sir R. Murchison in his "Silurian System" (p. 533), in which *Bulla ampulla* and a species of *Oliva,* shells of a southern type, have been found.

A case somewhat similar to the above is recorded by the Rev. Mr. Crosskey as having been met with in Scotland at the Kyles of Bute. "Among the Clyde beds, I have found," he says, "a layer containing shells, in which those of a more southern type appear to exist in greater profusion and perfection than even in our present seas. It is an open question," he continues, "whether our climate was not slightly warmer than it is now between the glacial epoch and the present day."†

* Brit. Assoc. Report, 1863.

† Trans. Glasgow Nat. Hist. Soc., vol. i., p. 115.

In a glacial bed near Greenock, Mr. A. Bell found the fry of living Mediterranean forms, viz., *Conus Mediterraneus* and *Cardita trapezia.*

Although deposits containing shells of a temperate or of a southern type in glacial beds have not been often recorded, it by no means follows that such deposits are actually of rare occurrence. That glacial beds should contain deposits indicating a temperate or a warm condition of climate is a thing so contrary to all preconceived opinions regarding the sequence of events during the glacial epoch, that most geologists, were they to meet with a shell of a southern type in one of those beds, would instantly come to the conclusion that its occurrence there was purely accidental, and would pay no special attention to the matter.

Evidence derived from "Borings."—With the view of ascertaining if additional light would be cast on the sequence of events, during the formation of the boulder clay, by an examination of the journals of bores made through a great depth of surface deposits, I collected, during the summer of 1867, about two hundred and fifty such records, put down in all parts of the mining districts of Scotland. An examination of these bores shows most conclusively that the opinion that the boulder clay, or lower till, is one great undivided formation, is wholly erroneous.

These two hundred and fifty bores represent a total thickness of 21,348 feet, giving 86 feet as the mean thickness of the deposits passed through. Twenty of these have one boulder clay, with beds of stratified sand or gravel beneath the clay; twenty-five have *two* boulder clays, with stratified beds of sand and gravel between; ten have *three* boulder clays; one has *four* boulder clays; two have *five* boulder clays; and no one has fewer than *six* separate masses of boulder clay, with stratified beds of sand and gravel between; sixteen have two or three separate boulder clays, differing altogether in colour and hardness, without any stratified beds between. We have, therefore, out of two hundred and fifty bores, seventy-five of them representing a condition of things wholly different from that exhibited to the geologist in ordinary sections.

The full details of the character of the deposits passed through by these bores, and their bearing on the history of the glacial epoch, have been given by Mr. James Bennie, in an interesting paper read before the Glasgow Geological Society,* to which I would refer all those interested in the subject of surface geology.

The evidence afforded by these bores of the existence of warm inter-glacial periods will, however, fall to be considered in a subsequent chapter.†

Another important and unexpected result obtained from these bores to which we shall have occasion to refer, was the evidence which they afforded of a Continental Period.

Striated Pavements.—It has been sometimes observed that in horizontal sections of the boulder clay, the stones and boulders are all striated in one uniform direction, and this has been effected over the original markings on the boulders. It has been inferred from this that a pause of long duration must have taken place in the formation of the boulder clay, during which the ice disappeared and the clay became hardened into a solid mass. After which the old condition of things returned, glaciers again appeared, passed over the surface of the hardened clay with its imbedded boulders, and ground it down in the same way as they had formerly done the solid rocks underneath the clay.

An instance of striated pavements in the boulder clay was observed by Mr. Robert Chambers in a cliff between Portobello and Fisherrow. At several places a narrow train of blocks was observed crossing the line of the beach, somewhat like a quay or mole, but not more than a foot above the general level. All the blocks *had flat sides uppermost, and all the flat sides were striated in the same direction* as that of the rocky surface through-

* Trans. of the Geol. Soc. of Glasgow, vol. iii., p. 133. See also "Great Ice Age," chaps. xii. and xiii.

† Chap. XXIX.

out the country. A similar instance was also observed between Leith and Portobello. "There is, in short," says Mr. Chambers, "a surface of the boulder clay, deep down in the entire bed, which, to appearance, has been in precisely the same circumstances as the fast rock surface below had previously been. It has had in its turn to sustain the weight and abrading force of the glacial agent, in whatever form it was applied; and the additional deposits of the boulder clay left over this surface may be presumed to have been formed by the agent on that occasion."*

Several cases of a similar character were observed by Mr. James Smith, of Jordanhill, on the beach at Row, and on the shore of the Gareloch.† Between Dunbar and Cockburnspath, Professor Geikie found along the beach, for a space of 30 or 40 square yards, numbers of large blocks of limestone with flattened upper sides, imbedded in a stiff red clay, and all striated in one direction. On the shores of the Solway he found another example.‡

The cases of striated pavements recorded are, however, not very numerous. But this by no means shows that they are of rare occurrence in the boulder clay. These pavements, of course, are to be found only in the interior of the mass, and even there they can only be seen along a horizontal section. But sections of this kind are rarely to be met with, for river channels, quarries, railway cuttings, and other excavations of a similar character which usually lay open the boulder clay, exhibit vertical sections only. It is therefore only along the sea-shore, as Professor Geikie remarks, where the surface of the clay has been worn away by the action of the waves, that opportunities have hitherto been presented to the geologist for observing them.

There can be little doubt that during the warm periods of the glacial epoch our island would be clothed with a luxuriant

* Edin. New Phil. Journ., vol. liv., p. 272.
† "Newer Pliocene Geology," p. 129. John Gray & Co., Glasgow.
‡ "Glacial Drift of Scotland," p. 67.

flora. At the end of a cold period, when the ice had disappeared, the whole face of the country would be covered over to a considerable depth with a confused mass of stones and boulder clay. A surface thus wholly destitute of every seed and germ would probably remain for years without vegetation. But through course of time life would begin to appear, and during the thousands of years of perpetual summer which would follow, the soil, uncongenial as it no doubt must have been, would be forced to sustain a luxuriant vegetation. But although this was the case, we need not wonder that now scarcely a single vestige of it remains; for when the ice sheet again crept over the island everything animate and inanimate would be ground down to powder. We are certain that prior to the glacial epoch our island must have been covered with life and vegetation. But not a single vestige of these are now to be found; no, not even of the very soil on which the vegetation grew. The solid rock itself upon which the soil lay has been ground down to mud by the ice sheet, and, to a large extent, as Professor Geikie remarks, swept away into the adjoining seas.* It is now even more difficult to find a trace of the ancient soil *under* the boulder clay than it is to find remains of the soil of the warm periods *in* that clay. As regards Scotland, cases of old land surfaces under the boulder clay are as seldom recorded as cases of old land surfaces in it. In so far as geology is concerned, there is as much evidence to show that our island was clothed with vegetation during the glacial epoch as there is that it was so clothed prior to that epoch.

* "Glacial Drift of Scotland," p. 12.

CHAPTER XVI.

WARM INTER-GLACIAL PERIODS IN ARCTIC REGIONS.

Cold Periods best marked in Temperate, and Warm Periods in Arctic, Regions. —State of Arctic Regions during Glacial Period.—Effects of Removal of Ice from Arctic Regions.—Ocean-Currents; Influence on Arctic Climate.—Reason why Remains of Inter-glacial Period are rare in Arctic Regions. —Remains of Ancient Forests in Banks's Land, Prince Patrick's Island, &c. —Opinions of Sir R. Murchison, Captain Osborn, and Professor Haughton. —Tree dug up by Sir E. Belcher in lat. 75° N.

IN the temperate regions the cold periods of the glacial epoch would be far more marked, than the warm inter-glacial periods. The condition of things which prevailed during the cold periods would differ far more widely from that which now prevails than would the condition of things during the warm periods. But as regards the polar regions the reverse would be the case; there the warm inter-glacial periods would be far more marked than the cold periods. The condition of things prevailing in those regions during the warm periods would be in strongest contrast to what now obtains, but this would not hold true in reference to the cold periods; for during the latter, matters there would be pretty much the same as at present, only a good deal more severe. The reason of this may be seen from what has already been stated in Chapter IV.; but as it is a point of considerable importance in order to a proper understanding of the physical state of things prevailing in polar regions during the glacial epoch, I shall consider this part of the subject more fully.

During the cold periods, our island, and nearly all places in the northern temperate regions down to about the same latitude, would be covered with snow and ice, and all animal and vegetable life within the glaciated area would to a great extent be destroyed. The presence of the ice would of itself, for reasons already explained, lower the mean annual temperature to near the freezing-point. The summers, notwithstanding the proximity of the sun, would not be warm, on the contrary their temperature would rise little above the freezing-point. An excess of evaporation would no doubt take place, owing to the increase in the intensity of the sun's rays, but this result would only tend to increase the snowfall.*

During the warm periods our country and the regions under consideration would experience conditions not differing much from those of the present, but the climate would probably be somewhat warmer and more equable. The proximity of the sun during winter would prevent snow from falling. The summers, owing to the greater distance of the sun, would probably be somewhat colder than they are now. But the loss of heat during summer would be to a large extent compensated for by two causes to which we must here refer. (1.) The much greater amount of heat conveyed by ocean-currents than at present. (2.) Our summers are now cooled to a considerable extent by cold aërial currents from the ice-covered regions of the north. But during the period in question there would be little or no ice in arctic regions, consequently the winds would be comparatively warm, whatever direction they came from.

Let us next direct our attention to the state of things in the arctic regions during the glacial epoch. At present Greenland and other parts of the arctic regions occupied by land are almost wholly covered with ice, and as a consequence nearly destitute of vegetable life. During the cold periods of the glacial epoch the quantity of snow falling would doubtless be greater and the ice thicker, but as regards organic life, matters would not probably be much worse than they are at present. In fact, so far as Greenland and the antarctic continent are concerned, they are about as destitute of plant life as they can be. Although an increase in the thickness of the arctic ice would not greatly alter the present state of matters in those regions,

* See Chapter IV.

yet what a transformation would ensue upon the disappearance of the ice! This would not only raise the summer temperature some twenty degrees or so, but would afford the necessary conditions for the existence of abundant animal and plant life. The severity of the climate of Greenland is due to a very considerable extent, as we have already seen, to the presence of ice. Get rid of the permanent ice, and the temperature of the country, *cæteris paribus,* would instantly rise. That Greenland should ever have enjoyed a temperate climate, capable of supporting abundant vegetation, has often been matter of astonishment, but this wonder diminishes when we reflect that during the warm periods it would be in the arctic regions that the greatest heating effect would take place, this being due mainly to the transference of nearly all the warm inter-tropical waters to one hemisphere.

It has been shown in Chapter II. that the heating effects at present resulting from the transference of heat by ocean-currents increase as we approach the poles. As a consequence of this it follows that during the warm periods, when the quantity of warm water transferred would be nearly doubled, the *increase of heat resulting from this cause would itself increase* as the warm pole was approached. This effect, combined with the shortness of the winter in perihelion and the nearness of the sun during that season, would prevent the accumulation of snow. During summer, the sun, it is true, would be at a much greater distance from the earth than at present, but it must be borne in mind that for a period of three months the quantity of heat received from the sun at the north pole would be greater than that received at the equator. Consequently, after the winter's snow was melted, this great amount of heat would go to raise the temperature, and the arctic summer could not be otherwise than hot. It is not hot at present, but this, be it observed, is because of the presence of the ice. When we take all these facts into consideration we need not be surprised that Greenland once enjoyed a condition of climate totally different from that which now obtains in that region.

It is, therefore, in the arctic and antarctic regions where we ought to find the most marked and decided evidence of warm inter-glacial periods. And doubtless such evidence would be abundantly forthcoming had these regions not been subjected to such intense denudation since the glacial epoch, and were so large a portion of the land not still buried beneath an icy covering, and therefore beyond the geologist's reach. Only on islands and such outlying places as are not shrouded in snow and ice can we hope to meet with any trace of the warm periods of the glacial epoch: and we may now proceed to consider what relics of these warm periods have actually been discovered in arctic regions.

Evidence of Warm Periods in Arctic Regions.—The fact that stumps, &c., of full-grown trees have been found in places where at present nothing is to be met with but fields of snow and ice, and where the mean annual temperature scarcely rises above the zero of the Fahrenheit thermometer, is good evidence to show that the climate of the arctic regions was once much warmer than now. The remains of an ancient forest were discovered by Captain McClure, in Banks's Land, in latitude 74° 48′. He found a great accumulation of trees, from the sea-level to an elevation of upwards of 300 feet. "I entered a ravine," says Captain McClure, "some miles inland, and found the north side of it, for a depth of 40 feet from the surface, composed of one mass of wood similar to what I had before seen."* In the ravine he observed a tree protruding about 8 feet, and 3 feet in circumference. And he further states that, "*From the perfect state of the bark*, and the position of the trees so far from the sea, there can be but little doubt that they grew originally in the country." A cone of one of these fir-trees was brought home, and was found to belong apparently to the genus *Abies*, resembling *A.* (*Pinus*) *alba*.

In Prince Patrick's Island, in latitude 76° 12′ N., longitude 122° W., near the head of Walker Inlet, and a considerable distance in the interior in one of the ravines, a tree protruding

* "Discovery of the North-West Passage," p. 213.

about 10 feet from a bank was discovered by Lieutenant Mecham. It proved to be 4 feet in circumference. In its neighbourhood several others were seen, all of them similar to some he had found at Cape Manning; each of them measured 4 feet round and 30 feet in length. The carpenter stated that the trees resembled larch. Lieutenant Mecham, from their appearance and position, concluded that they must have grown in the country.*

Trees under similar conditions were also found by Lieutenant Pim on Prince Patrick's Island, and by Captain Parry on Melville Island, all considerably above the present sea-level and at a distance from the shore. On the coast of New Siberia, Lieutenant Anjou found a cliff of clay containing stems of trees still capable of being used for fuel.

"This remarkable phenomenon," says Captain Osborn, "opens a vast field for conjecture, and the imagination becomes bewildered in trying to realise that period of the world's history when the absence of ice and a milder climate allowed forest trees to grow in a region where now the ground-willow and dwarf-birch have to struggle for existence."

Sir Roderick Murchison came to the conclusion that all those trees were drifted to their present position when the islands of the arctic archipelago were submerged. But it was the difficulty of accounting for the growth of trees in such a region which led him to adopt this hypothesis. His argument is this: "If we imagine," he says, "that the timber found in those latitudes grew on the spot we should be driven to adopt the anomalous hypothesis that, notwithstanding physical relations of land and water similar to those which now prevail, trees of large size grew on such *terra firma* within a few degrees of the north pole!—a supposition which I consider to be wholly incompatible with the data in our possession, and at variance with the laws of the isothermal lines."† This reasoning of Sir Roderick's may be quite correct, on the supposition that

* "Voyage of the *Resolute*," p. 294.
† Quart. Journ. Geol. Soc., vol. xi., p. 540.

changes of climate are due to changes in the distribution of sea and land, as advocated by Sir Charles Lyell. But these difficulties disappear if we adopt the views advocated in the foregoing chapters. As Captain Osborn has pointed out, however, Sir Roderick's hypothesis leaves the real difficulty untouched. "A very different climate," he says, "must then have existed in those regions to allow driftwood so perfect as to retain its bark to reach such great distances; and perhaps it may be argued that if that sea was sufficiently clear of ice to allow such timber to drift unscathed to Prince Patrick's Land, that that *very absence of a frozen sea would allow fir-trees to grow in a soil naturally fertile.*" *

As has been already stated, all who have seen those trees in arctic regions agree in thinking that they grew *in situ.* And Professor Haughton, in his excellent account of the arctic archipelago appended to McClintock's "Narrative of Arctic Discoveries," after a careful examination of the entire evidence on the subject, is distinctly of the same opinion; while the recent researches of Professor Heer put it beyond doubt that the drift theory must be abandoned.

Undoubtedly the arctic archipelago was submerged to an extent that could have admitted of those trees being floated to their present positions. This, as we shall see, follows from theory; but submergence, without a warmer condition of climate, would not enable trees to reach those regions with their bark entire.

But in reality we are not left to theorise on the subject, for we have a well-authenticated case of one of those trees being got by Captain Belcher standing erect in the position in which it grew. It was found immediately to the northward of the narrow strait opening into Wellington Sound, in lat. 75° 32′ N. long. 92° W., and about a mile and a half inland. The tree was dug up out of the frozen ground, and along with it a portion of the soil which was immediately in contact with the roots. The whole was packed in canvas and brought to

* "McClure's North-West Passage," p. 214. Second Edition.

England. Near to the spot several knolls of peat mosses about nine inches in depth were found, containing the bones of the lemming in great numbers. The tree in question was examined by Sir William Hooker, who gave the following report concerning it, which bears out strongly the fact of its having grown *in situ.*

"The piece of wood brought by Sir Edward Belcher from the shores of Wellington Channel belongs to a species of pine, probably to the *Pinus* (*Abies*) *alba,* the most northern conifer. The structure of the wood of the specimen brought home differs remarkably in its anatomical character from that of any other conifer with which I am acquainted. Each concentric ring (or annual growth) consists of two zones of tissue; one, the outer, that towards the circumference, is broader, of a pale colour, and consists of ordinary tubes of fibres of wood, marked with discs common to all coniferæ. These discs are usually opposite one another when more than one row of them occur in the direction of the length of the fibre; and, what is very unusual, present radiating lines from the central depression to the circumference. Secondly, the inner zone of each annual ring of wood is narrower, of a dark colour, and formed of more slender woody fibres, with thicker walls in proportion to their diameter. These tubes have few or no discs upon them, but are covered with spiral striæ, giving the appearance of each tube being formed of a twisted band. The above characters prevail in all parts of the wood, but are slightly modified in different rings. Thus the outer zone is broader in some than in others, the disc-bearing fibres of the outer zone are sometimes faintly marked with spiral striæ, and the spirally marked fibres of the inner zone sometimes bear discs. These appearances suggest the annual recurrence of some special cause that shall thus modify the first and last formed fibres of each year's deposit, so that that first formed may differ in amount as well as in kind from that last formed; and the peculiar conditions of an arctic climate appear to afford an adequate solution. The inner, or first-formed zone, must be regarded as imperfectly developed, being deposited at a season when the functions of the plant are very intermittently exercised, and when a few short hours of sunshine are daily succeeded by many of extreme cold. As the season advances the sun's heat and light are continuous during the greater part of the twenty-four hours, and the newly formed wood fibres are hence more perfectly developed, they are much longer, present no signs of striæ, but are studded with discs of a more highly organized structure than are usual in the natural order to which this tree belongs." *

Another circumstance which shows that the tree had grown where it was found is the fact that in digging up the roots portions of the leaves were obtained. It may also be mentioned that near this place was found an old river channel cut deeply into the rock, which, at some remote period, when the climate must have been less rigorous than at present, had been occupied by a river of considerable size.

Now, it is evident that if a tree could have grown at Wellington Sound, there is no reason why one might not have grown at Banks's Land, or at Prince Patrick's Island. And, if the climatic condition of the country would allow one tree to grow, it would equally as well allow a hundred, a thousand, or a whole forest. If this, then, be the case, Sir Roderick's objection to the theory of growth *in situ* falls to the ground.

Another circumstance which favours the idea that those trees grew during the glacial epoch is the fact that although they are recent, geologically speaking, and belong to the drift series, yet they are, historically speaking, very old. The wood, though not fossilized, is so hardened and changed by age that it will scarcely burn.

* "British Association Report for 1855," p. 381. "The Last of the Arctic Voyages," vol. i., p. 381.

CHAPTER XVII.

FORMER GLACIAL EPOCHS.—REASON OF THE IMPERFECTION OF GEOLOGICAL RECORDS IN REFERENCE TO THEM.

Two Reasons why so little is known of Glacial Epochs.—Evidence of Glaciation to be found on Land-surfaces.—Where are all our ancient Land-surfaces?—The stratified Rocks consist of a Series of old Sea-bottoms.—Transformation of a Land-surface into a Sea-bottom obliterates all Traces of Glaciation.—Why so little remains of the Boulder Clays of former Glacial Epochs.—Records of the Glacial Epoch are fast disappearing.—Icebergs do not striate the Sea-bottom.—Mr. Campbell's Observations on the Coast of Labrador.—Amount of Material transported by Icebergs much exaggerated.—Mr. Packard on the Glacial Phenomena of Labrador.—Boulder Clay the Product of Land-ice.—Palæontological Evidence.—Paucity of Life characteristic of a Glacial Period.—Warm Periods better represented by Organic Remains than cold.—Why the Climate of the Tertiary Period was supposed to be warmer than the present.—Mr. James Geikie on the Defects of Palæontological Evidence.—Conclusion.

Two Reasons why so little is known of former Glacial Epochs.—If the glacial epoch resulted from the causes discussed in the foregoing chapters, then such epochs must have frequently supervened. We may, therefore, now proceed to consider what evidence there is for the former occurrence of excessive conditions of climate during previous geological ages. When we begin our inquiry, however, we soon find that the facts which have been recorded as evidence in favour of the action of ice in former geological epochs are very scanty indeed. Two obvious reasons for this may be given, namely, (1) The imperfection of the geological records themselves, and (2) the little attention hitherto paid toward researches of this kind. The notion, once so prevalent, that the climate of our earth was much warmer in the earlier geological ages than it is now, and that it has ever since been gradually becoming cooler, was wholly at variance with the idea of former ice-periods. And this conviction of the *a priori* improbability of cold periods having obtained during Palæozoic and Mesozoic ages tended to prevent due attention being paid to such facts as seemed to bear upon the subject. But our limited knowledge of former glacial epochs must no doubt be attributed chiefly to the actual imperfection of the geological records. So great is this imperfection that the mere absence of direct geological evidence cannot reasonably be regarded as sufficient proof that the conclusions derived from astronomical and physical considerations regarding former ice-periods are improbable. Nor is this all. The geological records of ancient glacial conditions are not only imperfect, but, as I shall endeavour to show, this imperfection *follows as a natural consequence from the principles of geology itself.* There are not merely so many blanks or gaps in the records, but a reason exists in the very nature of geological evidence why such breaks in the record might reasonably be expected to occur.

Evidence of Glaciation to be found chiefly on Land-surfaces.—It is on a land-surface that the principal traces of the action of ice during a glacial epoch are left, for it is there that the stones are chiefly striated, the rocks ground down, and the boulder clay formed. But where are all our ancient land-surfaces? They are not to be found. The total thickness of the stratified rocks of Great Britain is, according to Professor Ramsay, nearly fourteen miles. But from the top to the bottom of this enormous pile of deposits there is hardly a single land-surface to be detected. True patches of old land-surfaces of a local character exist, such, for example, as the dirt-beds of Portland; but, with the exception of coal-seams, every general formation from top to bottom has been accumulated under water, and none but the under-clays *ever existed as a land*-surface. And it is here, in such a formation, that the geologist has to collect all his information regarding the existence of former glacial epochs. The entire stratified rocks of the globe, with the exception of the coal-beds and under-clays (in neither of which would one expect to find traces of ice-action), consist almost entirely of a *series of old sea-bottoms*, with here and there an occasional fresh-water deposit. Bearing this in mind, what is the sort of evidence which we can now hope to find in these old sea-bottoms of the existence of former ice-periods?

Every geologist of course admits that the stratified rocks are not old land-surfaces, but a series of old sea-bottoms formed out of the accumulated material derived from the degradation of primeval land-surfaces. And it is true that all land-surfaces once existed as sea-bottoms; but the stratified rocks consist of a series of old sea-bottoms which never were land-surfaces. Many of them no doubt have been repeatedly above the sea-level, and may once have possessed land-surfaces; but these, with the exception of the under-clays of the various coal measures, the dirt-beds of Portland, and one or two more patches, have all been denuded away. The important bearing which this consideration has on the nature of the evidence which we can now expect to find of the existence of former glacial epochs has certainly been very much overlooked.

If we examine the matter fully we shall be led to conclude that the *transformation of a land-surface into a sea-bottom* will probably completely obliterate every trace of glaciation which that land-surface may once have presented. We cannot, for example, expect to meet with polished and striated stones belonging to a former land glaciation; for such stones are not carried down bodily and unchanged by our rivers and deposited in the sea. They become broken up by subaërial agencies into gravel, sand, and clay, and in this condition are transported seawards. Nor even if we supposed it possible that the stones and boulders derived from a mass of till could be carried down to sea by river-action, could we at the same time fail to admit that such stones would be deprived of all their ice-markings, and become water-worn and rounded on the way.*

* Mr. James Geikie informs me that the great accumulations of gravel which occur so abundantly in the low grounds of Switzerland, and which are, undoubtedly, merely the re-arranged materials originally brought down from the Alps as till and as moraines by the glaciers during the glacial epoch, rarely or never yield a single scratched or glaciated stone. The action of the rivers escaping

Nor can we expect to find boulder clay among the stratified rocks, for boulder clay is not carried down as such and deposited in the sea, but under the influence of the denuding agents becomes broken up into soft mud, clay, sand, and gravel, as it is gradually peeled off the land and swept seawards. Patches of boulder clay may have been now and again forced into the sea by ice and eventually become covered up; but such cases are wholly exceptional, and their absence in any formation cannot fairly be adduced as a proof that that formation does not belong to a glacial period.

The only evidence of the existence of land-ice during former periods which we can reasonably expect to meet with in the stratified rocks, consists of erratic blocks which may have been transported by icebergs and dropped into the sea. But unless the glaciers of such epochs reached the sea, we could not possibly possess even this evidence. Traces in the stratified rocks of the effects of land-ice during former epochs must, in the very nature of things, be rare indeed. The only sort of evidence which, as a general rule, we may expect to detect, is the presence of large erratic blocks imbedded in strata which from their constitution have evidently been formed in still water. But this is quite enough; for it proves the existence of ice at the time the strata were being deposited as conclusively as though we saw the ice floating with the blocks upon it. This sort of evidence, when found in low latitudes, ought to be received as conclusive of the existence of former glacial epochs; and, no doubt, would have been so received had it not been for the idea that, if these blocks had been transported by ice, there ought in addition to have been found striated stones, boulder clay, and other indications of the agency of land-ice.

Of course all erratics are not necessarily transported by

from the melting ice has succeeded in obliterating all trace of striæ. It is the same, he says, with the heaps of gravel and sand in the lower grounds of Sweden and Norway, Scotland and Ireland. These deposits are evidently in the first place merely the materials carried down by the swollen rivers that issued from the gradually melting ice-fields and glaciers. The stones of the gravel derived from the demolition of moraines and till, have lost all their striæ and become in most cases well water-worn and rounded.

masses of ice broken from the terminal front of glaciers. The "ice foot," formed by the freezing of the sea along the coasts of the higher latitudes of Greenland, carries seawards immense quantities of blocks and *débris*. And again stones and boulders are frequently frozen into river-ice, and when the ice breaks up in spring are swept out to sea, and may be carried some little distance before they are dropped. But both these cases can occur only in regions where the winters are excessive; nor is it at all likely that such ice-rafts will succeed in making a long voyage. If, therefore, we could assure ourselves that the erratics occasionally met with in certain old geological formations in low latitudes were really transported from the land by an ice-foot or a raft of river-ice, we should be forced to conclude that very severe climatic conditions must have obtained in such latitudes at the time the erratics were dispersed.

The reason why we now have, comparatively speaking, so little direct evidence of the existence of former glacial periods will be more forcibly impressed upon the mind, if we reflect on how difficult it would be in a million or so of years hence to find any trace of what we now call the glacial epoch. The striated stones would by that time be all, or nearly all, disintegrated, and the till washed away and deposited in the bottom of the sea as stratified sands and clays. And when these became consolidated into rock and were raised into dry land, the only evidence that we should probably then have that there ever had been a glacial epoch would be the presence of large blocks of the older rocks, which would be found imbedded in the upraised formation. We could only infer that there had been ice at work from the fact that by no other known agency could we conceive such blocks to have been transported and dropped in a still sea.

Probably few geologists believe that during the Middle Eocene and the Upper Miocene periods our country passed through a condition of glaciation as severe as it has done during the Post-pliocene period; yet when we examine the subject carefully, we find that there is actually no just ground to conclude that it has not. For, in all probability, throughout the strata to be eventually formed out of the destruction of the now existing land-surfaces, evidence of ice-action will be as scarce as in Eocene or Miocene strata.

If the stratified rocks forming the earth's crust consisted of a series of old land-surfaces instead (as they actually do) of a series of old sea-bottoms, then probably traces of many glacial periods might be detected.

Nearly all the evidence which we have regarding the glacial epoch has been derived from what we find on the now existing land-surfaces of the globe. But probably not a vestige of this will exist in the stratified beds of future ages, formed out of the destruction of the present land-surfaces. Even the very arctic shell-beds themselves, which have afforded to the geologist such clear proofs of a frozen sea during the glacial epoch, will not be found in those stratified rocks; for they must suffer destruction along with everything else which now exists above the sea-level. There is probably not a single relic of the glacial epoch which has ever been seen by the eye of man that will be treasured up in the stratified rocks of future ages. Nothing that does not lie buried in the deeper recesses of the ocean will escape complete disintegration and appear imbedded in those formations. It is only those objects which lie in our existing sea-bottoms that will remain as monuments of the glacial epoch of the Post-tertiary period. And, moreover, it will only be those portions of the sea-bottoms that may happen to be upraised into dry land that will be available to the geologist of future ages. The point to be determined now is this:—*Is it probable that the geologist of the future will find in the rocks formed out of the now existing sea-bottoms more evidence of a glacial epoch during Post-tertiary times than we now do of one during, say, the Miocene, the Eocene, or the Permian period?* Unless this can be proved to be the case, we have no ground whatever to conclude that the cold periods of the Miocene, Eocene, and Permian periods were not as severe as that of the glacial epoch. This is evident, for the only relics which now remain of the glacial epochs of those periods are simply what happened to be protected in the then existing sea-bottoms. Every vestige that lay on the land would in all probability be destroyed by subaërial agency and carried into the sea in a sedimentary form. But before we can determine whether or not there is more evidence of the glacial epoch in our now existing sea-bottoms than there is of former glacial epochs in the stratified rocks (which are in reality the sea-bottoms belonging to ancient epochs), we must first ascertain what is the nature of those marks of glaciation which are to be found in a sea-bottom.

Icebergs do not striate the Sea-bottom.—We know that the rocky face of the country was ground down and striated during the glacial epoch; and this is now generally believed to have been done by land-ice. But we have no direct evidence that the floor of the ocean, beyond where it may have been covered with land-ice, was striated. Beyond the limits of the land-ice it could be striated only by means of icebergs. But do icebergs striate the rocky bed of the ocean? Are they adapted for such work? It seems to be often assumed that they are. But I have been totally unable to find any rational grounds for such a belief. Clean ice can have but little or no erosive power, and never could scratch a rock. To do this it must have grinding materials in the form of sand, mud, or stones. But the bottoms of icebergs are devoid of all such materials. Icebergs carry the grinding materials on their backs, not on their bottoms. No doubt, when the iceberg is launched into the deep, great masses of sand, mud, and stones will be adhering to its bottom. But no sooner is the berg immersed, than a melting process commences at its sides and lower surface in contact with the water; and the consequence is, the materials adhering to the lower surface soon drop off and sink to the bottom of the sea. The iceberg, divested of these materials, can now do very little harm to the rocky sea-bottom over which it floats. It is true that an iceberg moving with a velocity of a few miles an hour, if it came in contact with the sea-bottom, would, by the mere force of concussion, tear up loose and disjointed rocks, and hurl some of the loose materials to a distance; but it would do but little in the way of grinding down the rock against which it struck. But even supposing the bottom of the iceberg were properly shod with the necessary grinding materials, still it would be but a very inefficient grinding agent; for a *floating* iceberg would not be in contact with the sea-bottom. And if it were in contact with the sea-bottom, it would soon become stranded and, of course, motionless, and under such conditions could produce no effect.

It is perfectly true that although the bottom of the berg may be devoid of grinding materials, yet these may be found lying on the surface of the submarine rock over which the ice moves. But it must be borne in mind that the same current which will move the icebergs over the surface of the rock will move the sand, mud, and other materials over it also; so that the markings effected by the ice would in all probability be erased by the current. In the deep recesses of the ocean the water has been found to have but little or no motion. But icebergs always follow the path of currents; and it is very evident that at the comparatively small depth of a thousand feet or so reached by icebergs the motion of the water will be considerable; and the continual shifting of the small particles of the mud and sand will in all probability efface the markings which may be made now and again by a passing berg.

Much has been said regarding the superiority of icebergs as grinding and striating agents in consequence of the great velocity of their motion in comparison with that of land-ice. But it must be remembered that it is while the iceberg is floating, and before it touches the rock, that it possesses high velocity. When the iceberg runs aground, its motion is suddenly arrested or greatly reduced. But if the iceberg advancing upon a sloping sea-bottom is raised up so as to exert great pressure, it will on this account be the more suddenly arrested, the motion will be slow, and the distance passed over short, before the berg becomes stranded. If it exerts but little pressure on the sea-bottom, it may retain a considerable amount of motion and advance to a considerable distance before it is brought to a stand; but, exerting little pressure, it can perform but little work. Land-ice moves slowly, but then it exerts enormous pressure. A glacier 1,000 feet in thickness has a pressure on its rocky bed equal to about 25 tons on the square foot; but an iceberg a mile in thickness, forced up on a sloping sea-bottom to an elevation of 20 feet (and this is perhaps more than any ocean-current could effect), would only exert a pressure of about half a ton on the square foot, or about 1-50th part of the pressure of the glacier 1,000 feet in thickness. A great deal has been said about the erosive and crushing power of icebergs of enormous thickness, as if their thickness gave them any additional pressure. An iceberg 100 feet in thickness will exert just as much pressure as one a mile in thickness. The pressure of an iceberg is not like that of a glacier, in proportion to its thickness, but to the height to which it is raised out of the water. An iceberg 100 feet in thickness raised 10 feet will exert exactly the same pressure as one a mile in thickness raised to an equal height.

To be an efficient grinding agent, steadiness of motion, as well as pressure, is essential. A rolling or rocking motion is ill-adapted for grinding down and striating a rock. A steady rubbing motion under pressure is the thing required. But an iceberg is not only deficient in pressure, but also deficient in steadiness of motion. When an iceberg moving with considerable velocity comes on an elevated portion of the sea-bottom, it does not move steadily onwards over the rock, unless the pressure of the berg on the rock be trifling. The resistance being entirely at the bottom of the iceberg, its momentum, combined with the pressure of the current, applied wholly above the point of resistance, tends to make the berg bend forward, and in some cases upset (when it is of a cubical form). The momentum of the moving berg, instead of being applied in forcing it over the rock against which it comes in contact, is probably all consumed in work against gravitation in raising the berg upon its front edge. After the momentum is consumed, unless the berg be completely upset, it will fall back under the force of gravitation to its original position. But the momentum which it acquires from gravitation in falling backwards carries it beyond its position of repose in an opposite direction. It will thus continue to rock backwards and forwards until the friction of the water brings it to rest. The momentum of the berg, instead of being applied to the work of grinding and striating the sea-bottom, will chiefly be consumed in heat in the agitation of the water. But if the berg does advance, it will do so with a rocking unsteady motion, which, as Mr. Couthouy * and Professor Dana † observe, will tend rather to obliterate striations than produce them.

A floating berg moves with great steadiness; but a berg that has run aground cannot advance with a steady motion. If the rock over which the berg moves offers little resistance, it may do so; but in such a case the berg could produce but little effect on the rock.

Dr. Sutherland, who has had good opportunities to witness the effects of icebergs, makes some most judicious remarks on the subject. "It will be well" he says, "to bear in mind that when an iceberg *touches the ground, if that ground be hard and resisting, it must come to a stand*, and the propelling power continuing, a slight leaning over in the water, or yielding motion of the whole mass, may compensate readily for being so suddenly arrested. If, however, the ground be soft, so as not to arrest the motion of the iceberg at once, a moraine will be the result; but the moraine thus raised will tend to bring it to a stand." ‡

There is another cause referred to by Professor Dana, which, to a great extent, must prevent the iceberg from having an opportunity of striating the sea-bottom, even though it were otherwise well adapted for so doing. It is this: the bed of the

* Report on Icebergs, read before the Association of American Geologists, *Silliman's Journal*, vol. xliii., p. 163 (1842).

† "Manual of Geology," p. 677.

‡ Quart. Journ. Geol. Soc., vol: ix., p. 306.

ocean in the track of icebergs must be pretty much covered with stones and rubbish dropped from the melting bergs. And this mass of rubbish will tend to protect the rock.*

If icebergs cannot be shown *à priori*, from mechanical considerations, to be well adapted for striating the sea-bottom, one would naturally expect, from the confident way in which it is asserted that they are so adapted, that the fact has been at least established by actual observation. But, strange as it may appear, we seem to have little or no proof that icebergs actually striate the bed of the ocean. This can be proved from the direct testimony of the advocates of the iceberg theory themselves.

We shall take the testimony of Mr. Campbell, the author of two well-known works in defence of the iceberg theory, viz., "Frost and Fire," and "A Short American Tramp." Mr. Campbell went in the fall of the year 1864 to the coast of Labrador, the Straits of Belle Isle, and the Gulf of St. Lawrence, for the express purpose of witnessing the effects of icebergs, and testing the theory which he had formed, that the ice-markings of the glacial epoch were caused by floating ice and not by land-ice, as is now generally believed.

The following is the result of his observations on the coast of Labrador.

Hanly Harbour, Strait of Belle Isle:—"The water is 37° F. in July. . . . As fast as one island of ice grounds and bursts, another takes its place; and in winter the whole strait is blocked up by a mass which swings bodily up and down, grating along the bottom at all depths. . . . Examined the beaches and rocks at the water-line, especially in sounds. Found the rocks ground smooth, *but not striated,* in the sounds" (*Short American Tramp*, pp. 68, 107).

Cape Charles and Battle Harbour:—"But though these harbours are all frozen every winter, the *rocks at the water-line are not striated*" (p. 68).

At St. Francis Harbour:—"The water-line is much rubbed, smooth, *but not striated*" (p. 72).

* Dana's "Manual of Geology," p. 677.

Cape Bluff:—"Watched the rocks with a telescope, and *failed to make out striæ anywhere;* but the water-line is everywhere rubbed smooth" (p. 75).

Seal Islands:—"*No striæ are to be seen at the land-wash in these sounds or on open sea-coasts near the present water-line*" (p. 76).

He only mentions having here found striations in the three following places along the entire coast of Labrador visited by him; and in regard to two of these, it seems very doubtful that the markings were made by modern icebergs.

Murray's Harbour:—"This harbour was blocked up with ice on the 20th of July. The water-line is rubbed, and in *some places* striated" (p. 69).

Pack Island:—"The water-line in a narrow sound was polished and striated in the direction of the sound, about N.N.W. This seems to be fresh work done by heavy ice drifting from Sandwich Bay; *but, on the other hand, stages with their legs in the sea, and resting on these very rocks, are not swept away by the ice*" (p. 96). If these markings were modern, why did not the "heavy ice" remove the small fir poles supporting the fishing-stages?

Red Bay:—"Landed half-dressed, and found some striæ perfectly fresh at the water-level, but weathered out a short distance *inland*" (p. 107). The striations "inland" could not have been made by modern icebergs; and it does not follow that because the markings at the water-level were not weathered they were produced by modern ice.

These are the evidences which he found that icebergs striate rocks, on a coast of which he says that, during the year he visited it, "the winter-drift was one vast solid raft of floes and bergs more than 150 miles wide, and perhaps 3,000 feet thick at spots, driven by a whole current bodily over one definite course, year after year, since this land was found" (p. 85).

But Mr. Campbell himself freely admits that the floating ice which comes aground along the shores does not produce striæ. "It is sufficiently evident," he says, "*that glacial striæ are not produced by thin bay ice*" (p. 76). And in "Frost and Fire," vol. ii., p. 237, he states that, "from a careful examination of the water-line at many spots, it appears that bay-ice grinds rocks, *but does not produce striation.*"

"It is impossible," he continues, "to get at rocks over which heavy icebergs now move; but a mass 150 miles wide, perhaps 3,000 feet thick in some parts, and moving at the rate of a mile an hour, or more, *appears to be an engine amply sufficient* to account for striæ on rising rocks." And in "American Tramp," p. 76, he says, "*striæ must be made* in deep water by the large masses which seem to pursue the even tenor of their way in the steady current which flows down the coast."

Mr. Campbell, from a careful examination of the sea-bottom along the coast, finds that the small icebergs do not produce striæ, but the large ones, which move over rocks impossible to be got at, "must" produce them. They "appear" to be amply sufficient to do so. If the smaller bergs cannot striate the sea-bottom, why must the larger ones do so? There is no reason why the smaller bergs should not move as swiftly and exert as much pressure on the sea-bottom as the larger ones. And even supposing that they did not, one would expect that the light bergs would effect on a smaller scale what the heavy ones would do on a larger.

I have no doubt that when Mr. Campbell visited Labrador he expected to find the sea-coast under the water-line striated by means of icebergs, and was probably not a little surprised to find that it actually was not. And I have no doubt that were the sea-bottom in the tracks of the large icebergs elevated into view, he would find to his surprise that it was free from striations also.

So far as observation is concerned, we have no grounds from what Mr. Campbell witnessed to conclude that icebergs striate the sea-bottom.

The testimony of Dr. Sutherland, who has had opportunities of seeing the effects of icebergs in arctic regions, leads us to the same conclusion. "Except," he says, "from the evidence afforded by plants and animals at the bottom, we have *no means whatever* to ascertain the effect produced by icebergs upon the rocks.* In the Malegat and Waigat I have seen whole clusters of these floating islands, drawing from 100 to 250 fathoms, moving to and fro with every return and recession of the tides. I looked very earnestly for grooves and scratches left by icebergs and glaciers in the rocks, but always failed to discover any." †

We shall now see whether river-ice actually produces striations or not. If floating ice under any form can striate rocks, one would expect that it ought to be done by river-ice, seeing that such ice is obliged to follow one narrow definite track.

St. John's River, New Brunswick:—"This river," says Mr. Campbell, "is obstructed by ice during five months of the year. When the ice goes, there is wild work on the bank. Arrived at St. John, drove to the suspension-bridge. . . . At this spot, if *anywhere in the world*, river-ice ought to produce striation. The whole drainage of a wide basin and one of the strongest tides in the world, here work continually in one rock-groove; and in winter this water-power is armed with heavy ice. *There are no striæ* about the water-line." ‡

River St. Lawrence:—"In winter the power of ice-floats driven by water-power is tremendous. The river freezes and packs ice till the flow of water is obstructed. The rock-pass at Quebec is like the Narrows at St. John's, Newfoundland. The whole pass, about a mile wide, was paved with great broken slabs and round boulders of worn ice as big as small stacks, piled and tossed, and heaped and scattered upon the level water below and frozen solid. . . . . This kind of ice does NOT *produce striation* at the water-margin at Quebec. At Montreal, when the river 'goes,' the ice goes with it with a vengeance. . . . . The *piers are not yet striated* by river-ice at Montreal. . . . . The rocks at the high-water level have *no trace* of glacial striæ. . . . . The rock at Ottawa is rubbed by

* Quart. Journ. Geol. Soc., vol. ix., p. 306. † "Journal," vol. i., p. 38.

‡ "Short American Tramp," pp. 168, 174.

river-ice every spring, and *always in one direction, but it is not striated.* . . . . The surfaces are all rubbed smooth, and the edges of broken beds are rounded where exposed to the ice; *but there are no striæ.*" *

When Sir Charles Lyell visited the St. Lawrence in 1842, at Quebec he went along with Colonel Codrington "and searched carefully below the city in the channel of the St. Lawrence, at low water, near the shore, for the signs of glacial action at the precise point where the chief pressure and friction of packed ice are exerted every year," but found none.

"At the bridge above the Falls of Montmorenci, over which a large quantity of ice passes every year, the gneiss is polished, and kept perfectly free from lichens, but not more so than rocks similarly situated at waterfalls in Scotland. In none of these places were any long straight grooves observable." †

The only thing in the shape of modern ice-markings which he seems to have met with in North America was a few straight furrows half an inch broad in soft sandstone, at the base of a cliff at Cape Blomidon in the Bay of Fundy, at a place where during the preceding winter "packed" ice 15 feet thick had been pushed along when the tide rose over the sandstone ledges.‡

The very fact that a geologist so eminent as Sir Charles Lyell, after having twice visited North America, and searched specially for modern ice-markings, was able to find only two or three scratches, upon a soft sandstone rock, which he could reasonably attribute to floating ice, ought to have aroused the suspicion of the advocates of the iceberg theory that they had really formed too extravagant notions regarding the potency of floating ice as a striating agent.

There is no reason to believe that the grooves and markings noticed by M. Weibye and others on the Scandinavian coast and other parts of northern Europe were made by icebergs.

* "Short American Tramp," pp. 239—241.
† "Travels in North America," vol. ii., p. 137.
‡ Ibid., vol. ii., p. 174.

Professor Geikie has clearly shown, from the character and direction of the markings, that they are the production of land-ice.* If the floating ice of the St. Lawrence and the icebergs of Labrador are unable to striate and groove the rocks, it is not likely that those of northern Europe will be able to do so.

It will not do for the advocates of the iceberg theory to assume, as they have hitherto done, that, as a matter of course, the sea-bottom is being striated and grooved by means of icebergs. They must prove that. They must either show that, as a matter of fact, icebergs are actually efficient agents in striating the sea-bottom, or prove from mechanical principles that they must be so. The question must be settled either by observation or by reason; mere opinion will not do.

The Amount of Material transported by Icebergs much exaggerated.—The transporting of boulders and rubbish, and not the grinding and striating of rocks, is evidently the proper function of the iceberg. But even in this respect I fear too much has been attributed to it.

In reading the details of voyages in the arctic regions one cannot help feeling surprised how seldom reference is made to stones and rubbish being seen on icebergs. Arctic voyagers, like other people, when they are alluding to the geological effects of icebergs, speak of enormous quantities of stones being transported by them; but in reading the details of their voyages, the impression conveyed is that icebergs with stones and blocks of rock upon them are the exceptions. The greater portion of the narratives of voyages in arctic regions consists of interesting and detailed accounts of the voyager's adventures among the ice. The general appearance of the icebergs, their shape, their size, their height, their colour, are all noticed; but rarely is mention made of stones being seen. That the greater number of icebergs have no stones or rubbish on them is borne out by the positive evidence of geologists who have had opportunities of seeing icebergs.

Mr. Campbell says:—"It is remarkable that up to this

* Proceedings of the Royal Society of Edinburgh, Session 1865—66, p. 537.

time we have only seen a few doubtful stones on bergs which we have passed. . . . . Though no bergs with stones *on them or in them* have been approached during this voyage, many on board the *Ariel* have been close to bergs heavily laden. . . . . A man who has had some experience of ice has *never seen a stone on a berg* in these latitudes. Captain Anderson, of the *Europa*, who is a geologist, has *never seen a stone on a berg* in crossing the Atlantic. *No stones were clearly seen on this trip.*"* Captain Sir James Anderson (who has long been familiar with geology, has spent a considerable part of his life on the Atlantic, and has been accustomed to view the iceberg as a geologist as well as a seaman) has never seen a stone on an iceberg in the Atlantic. This is rather a significant fact.

Sir Charles Lyell states that, when passing icebergs on the Atlantic, he "was most anxious to ascertain whether there was any mud, stones, or fragments of rocks on any one of these floating masses; but after examining about forty of them without perceiving any signs of frozen matter, I left the deck when it was growing dusk."† After he had gone below, one was said to be seen with something like stones upon it. The captain and officers of the ship assured him that they had *never seen a stone upon a berg.*

The following extract from Mr. Packard's "Memoir on the Glacial Phenomena of Labrador and Maine," will show how little is effected by the great masses of floating ice on the Labrador coast either in the way of grinding and striating the rocks, or of transporting stones, clay, and other materials.

"Upon this coast, which during the summer of 1864 was lined with a belt of floe-ice and bergs probably two hundred miles broad, and which extended from the Gulf of St. Lawrence at Belles Amours to the arctic seas, this immense body of floating ice seemed *directly* to produce but little alteration in its physical features. If we were to ascribe the grooving and polishing of rocks to the action of floating ice-floes and bergs,

* "Short American Tramp," pp. 77, 81, 111.
† "Second Visit," vol. ii., p. 367.

how is it that the present shores far above (500), and at least 250 feet below, the water-line are often jagged and angular, though constantly stopping the course of masses of ice impelled four to six miles an hour by the joint action of tides, currents, and winds? No boulders, or gravel, or mud were seen upon any of the bergs or masses of shore-ice. They had dropped all burdens of this nature nearer their points of detachment in the high arctic regions." . . . .

"This huge area of floating ice, embracing so many thousands of square miles, was of greater extent, and remained longer upon the coast, in 1864, than for forty years previous. It was not only pressed upon the coast by the normal action of the Labrador and Greenland currents, which, in consequence of the rotatory motion of the earth, tended to force the ice in a south-westerly direction, but the presence of the ice caused the constant passage of cooler currents of air from the sea over the ice upon the heated land, giving rise during the present season to a constant succession of north-easterly winds from March until early in August, which further served to crowd the ice into every harbour and recess upon the coast. It was the universal complaint of the inhabitants that the easterly winds were more prevalent, and the ice 'held' later in the harbours this year than for many seasons previous. Thus the fisheries were nearly a failure, and vegetation greatly retarded in its development. But so far as polishing and striating the rocks, depositing drift material, and thus modifying the contour of the surface of the present coast, this modern mass of bergs and floating ice effected comparatively little. Single icebergs, when small enough, entered the harbours, and there stranding, soon pounded to pieces upon the rocks, melted, and disappeared. From Cape Harrison, in lat. 55°, to Caribo Island, was an interrupted line of bergs stranded in 80 to 100 or more fathoms, often miles apart, while others passed to the seaward down by the eastern coast of Newfoundland, or through the Straits of Belle Isle."*

* "Memoirs of Boston Society of Natural History," vol. i. (1867), p. 228.

Boulder Clay the Product of Land-ice.—There is still another point connected with icebergs to which we must allude, viz., the opinion that great masses of the boulder clay of the glacial epoch were formed from the droppings of icebergs. If boulder clay is at present being accumulated in this manner, then traces of the boulder clay deposits of former epochs might be expected to occur. It is perfectly obvious that *unstratified* boulder clay could not have been formed in this way. Stones, gravel, sand, clay, and mud, the ingredients of boulder clay, tumbled all together from the back of an iceberg, could not sink to the bottom of the sea without separating. The stones would reach the bottom first, then the gravel, then the sand, then the clay, and last of all the mud, and the whole would settle down in a stratified form. But, besides, how could the *clay* be derived from icebergs? Icebergs derive their materials from the land before they are launched into the deep, and while they are in the form of land-ice. The materials which are found on the backs of icebergs are what fell upon the ice from mountain tops and crags projecting above the ice. Icebergs are chiefly derived from continental ice, such as that of Greenland, where the whole country is buried under one continuous mass, with only a lofty mountain peak here and there rising above the surface. And this is no doubt the chief reason why so few icebergs have stones upon their backs. The continental ice of Greenland is not, like the glaciers of the Alps, covered with loose stones. Dr. Robert Brown informs me that no moraine matter has ever been seen on the inland ice of Greenland. It is perfectly plain that clay does not fall upon the ice. What falls upon the ice is stones, blocks of rocks, and the loose *débris.* Clay and mud we know, from the accounts given by arctic voyagers, are sometimes washed down upon the coast-ice; but certainly very little of either can possibly get upon an iceberg. Arctic voyagers sometimes speak of seeing clay and mud upon bergs; but it is probable that if they had been near enough they would have found that what they took for clay and mud were merely dust and rubbish.

Undoubtedly the boulder clay of many places bears unmistakable evidence of having been formed under water; but it does not on that account follow that it was formed from the droppings of icebergs. The fact that the boulder clay in every case *is chiefly composed of materials derived from the country on which the clay lies,* proves that it was not formed from matter transported by icebergs. The clay, no doubt, contains stones and boulders belonging to other countries, which in some cases may have been transported by icebergs; but the clay itself has not come from another country. But if the clay itself has been derived from the country on which it lies, then it is absurd to suppose that it was deposited from icebergs. The clay and materials which are found on icebergs are derived from the land on which the iceberg is formed; but to suppose that icebergs, after floating about upon the ocean, should always return to the country which gave them birth, and there deposit their loads, is rather an extravagant supposition.

From the facts and considerations adduced we are, I would venture to presume, warranted to conclude that, with the exception of what may have been produced by land-ice, very little in the shape of boulder clay or striated rocks belonging to the glacial epoch lies buried under the ocean—and that when the now existing land-surfaces are all denuded, probably scarcely a trace of the glacial epoch will then be found, except the huge blocks that were transported by icebergs and dropped into the sea. It is therefore probable that we have as much evidence of the existence of a glacial epoch during former periods as the geologists of future ages will have of the existence of a glacial epoch during the Post-tertiary period, and that consequently we are not warranted in concluding that the glacial epoch was something unique in the geological history of our globe.

Palæontological Evidence.—It might be thought that if glacial epochs have been numerous, we ought to have abundance of palæontological evidence of their existence. I do not know if this necessarily follows. Let us take the glacial epoch itself for example, which is quite a modern affair. Here we do not require to go and search in the bottom of the sea for the evidence of its existence; for we have the surface of the land in almost identically the same state in which it was when the ice left it, with the boulder clay and all the wreck of the ice lying upon it. But what geologist, with all these materials before him, would be able to find out from palæontological evidence alone that there had been such an epoch? He might search the whole, but would not be able to find fossil evidence from which he could warrantably infer that the country had ever been covered with ice. We have evidence in the fossils of the Crag and other deposits of the existence of a colder condition of climate prior to the true glacial period, and in the shell-beds of the Clyde and other places of a similar state of matters after the great ice-sheets had vanished away. But in regard to the period of the true boulder clay or till, when the country was enveloped in ice, palæontology has almost nothing whatever to tell us. "Whatever may be the cause," says Sir Charles Lyell, "the fact is certain that over large areas in Scotland, Ireland, and Wales, I might add throughout the northern hemisphere on both sides of the Atlantic, the stratified drift of the glacial period is very commonly devoid of fossils." *

In the "flysch" of the Eocene of the Alps, to which we shall have occasion to refer in the next chapter, in which the huge blocks are found which prove the existence of ice-action during that period, few or no fossils have been found. So devoid of organic remains is that formation, that it is only from its position, says Sir Charles, that it is known to belong to the middle or "nummulitic" portion of the great Eocene series. Again, in the conglomerates at Turin, belonging to the Upper Miocene period, in which the angular blocks of limestone are found which prove that during that period Alpine glaciers reached the sea-level in the latitude of Italy, not a single organic remain has been found. It would seem that an extreme paucity of organic life is a characteristic of a glacial period,

* "Antiquity of Man," p. 268. Third Edition.

which warrants us in concluding that the absence of organic remains in any formation otherwise indicative of a cold climate cannot be regarded as sufficient evidence that that formation does not belong to a cold period.

In the last chapter it was shown why so little evidence of the warm periods of the glacial epoch is now forthcoming. The remains of the *faunas* and *floras* of those periods were nearly wholly destroyed and swept into the adjoining seas by the ice-sheet that covered the land. It is upon the present land-surface that we find the chief evidence of the last glacial epoch, but the traces of the warm periods of that epoch are hardly now to be met with in that position since they have nearly all been obliterated or carried into the sea.

In regard to former glacial epochs, however, ice-marked rocks, scratched stones, moraines, till, &c., no longer exist; the land-surfaces of those old times have been utterly swept away. The only evidence, therefore, of such ancient glacial epochs, that we can hope to detect, must be sought for in the deposits that were laid down upon the sea-bottom; where also we may expect to find traces of the warm periods that alternated during such epochs with glacial conditions. It is plain, moreover, that the palæontological evidence in favour of warm periods will always be the most abundant and satisfactory.

Judging from geological evidence alone, we naturally conclude that, as a general rule, the climate of former periods was somewhat warmer than it is at the present day. It is from fossil remains that the geologist principally forms his estimate of the character of the climate during any period. Now, in regard to fossil remains, the warm periods will always be far better represented than the cold; for we find that, as a *general rule, those formations which geologists are inclined to believe indicate a cold condition of climate are remarkably devoid of fossil remains.* If a geologist does not keep this principle in view, he will be very apt to form a wrong estimate of the general character of the climate of a period of such enormous length as say the Tertiary.

Suppose that the presently existing sea-bottoms, which have been forming since the commencement of the glacial epoch, were to become consolidated into rock and thereafter to be elevated into dry land, we should then have a formation which might be properly designated the Post-pliocene. It would represent the time which has elapsed from the beginning of the glacial epoch to the present day. Suppose one to be called upon as a geologist to determine from that formation what was the general character of the climate during the period in question, what would probably be the conclusion at which he would arrive? He would probably find here and there patches of boulder clay containing striated and ice-worn stones. Now and again he would meet with bones of the mammoth and the reindeer, and shells of an arctic type. He would likewise stumble upon huge blocks of the older rocks imbedded in the formation, from which he would infer the existence of icebergs and glaciers reaching the sea-level. But, on the whole, he would perceive that the greater portion of the fossil remains met with in this formation implied a warm and temperate condition of climate. At the lower part of the formation, corresponding to the time of the true boulder-clay, there would be such a scarcity of organic remains that he would probably feel at a loss to say whether the climate at that time was cold or hot. But if the intense cold of the glacial epoch was not continuous, but broken up by intervening warm periods during which the ice, to a considerable extent at least, disappeared for a long period of time (and there are few geologists who have properly studied the subject who will positively deny that such was the case), then the country would no doubt during those warm periods possess an abundance of plant and animal life. It is quite true that we may almost search in vain on the present land-surface for the organic remains which belonged to those inter-glacial periods; for they were nearly all swept away by the ice which followed. But no doubt in the deep recesses of the ocean, buried under hundreds of feet of sand, mud, clay, and gravel, lie multitudes of the plants and animals which then flourished on the land, and were carried down by rivers into the sea. And along with these lie the skeletons, shells, and other exuviæ of the creatures which flourished in the warm seas of those periods. Now looking at the great abundance of fossils indicative of warm and genial conditions which the lower portions of this formation would contain, the geologist might be in danger of inferring that the earlier part of the Post-pliocene period was a warmer period, whereas we, at the present day, looking at the matter from a different standpoint, declare that part to have been characterized by cold or glacial conditions. No doubt, if the beds formed during the cold periods of the glacial epoch could be distinguished from those formed during the warm periods, the fossil remains of the one would indicate a cold condition of climate, and those of the other a warm condition; but still, taking the entire epoch as a whole, the percentage of fossil remains indicative of a warm condition would probably so much exceed that indicative of a cold condition, that we should come to the conclusion that the character of the climate, as a whole, during the epoch in question was warm and equable.

As geologists we have, as a rule, no means of arriving at a knowledge of the character of the climate of any given period but through an examination of the sea-bottoms belonging to that period; for these contain all the evidence upon the subject. But unless we exercise caution, we shall be very apt, in judging of the climate of such a period, to fall into the same error that we have just now seen one might naturally fall into were he called upon to determine the character of the climate during the glacial epoch from the nature of the organic remains which lie buried in our adjoining seas. On this point Mr. J. Geikie's observations are so appropriate, that I cannot do better than introduce them here. "When we are dealing," says this writer, "with formations so far removed from us in time, and in which the animal and plant remains depart so widely from existing forms of life, we can hardly expect to derive much aid from the fossils in our attempts to detect traces of cold climatic conditions. The arctic shells in our Post-tertiary clays are convincing proofs of the former existence in our latitude of a severe climate; but when we go so far back as Palæozoic ages, we have no such clear evidence to guide us. All that palæontologists can say regarding the fossils belonging to these old times is simply this, that they seem to indicate, generally speaking, mild, temperate, or genial, and even sometimes tropical, conditions of climate. Many of the fossils, indeed, if we are to reason from analogy at all, could not possibly have lived in cold seas. But, for aught that we know, there may have been alternations of climate during the deposition of each particular formation; and these changes may be marked by the presence or absence, or by the greater or less abundant development, of certain organisms at various horizons in the strata. Notwithstanding all that has been done, our knowledge of the natural history of these ancient seas is still very imperfect; and therefore, in the present state of our information, we are not entitled to argue, from the general aspect of the fossils in our older formations, that the temperature of the ancient seas was never other than mild and genial."*

Conclusion.—From what has already been stated it will, I trust, be apparent that, assuming glacial epochs during past geological ages to have been as numerous and as severe as the Secular theory demands, still it would be unreasonable to expect to meet with abundant traces of them. The imperfection of the geological record is such that we ought not to be astonished that so few relics of former ice ages have come down to us. It will also be apparent that the palæontological evidence of a warm condition of climate having obtained during any particular age, is no proof that a glacial epoch did not also supervene during the same cycle of time. Indeed it is quite the reverse; for the warm conditions of which we have proof may indicate merely the existence of an inter-glacial period. Furthermore, if the Secular theory of changes of climate be admitted, then evidence of a warm condition of climate having

* "Great Ice Age," p. 512.

prevailed in arctic regions during any past geological age may be regarded as presumptive proof of the existence of a glacial epoch; that is to say, of an epoch during which cold and warm conditions of climate alternated. Keeping these considerations in view, we shall now proceed to examine briefly what evidence we at present have of the former existence of glacial epochs.

CHAPTER XVIII.

FORMER GLACIAL EPOCHS; GEOLOGICAL EVIDENCE OF.

Cambrian Conglomerate of Islay and North-west of Scotland.—Ice-action in Ayrshire and Wigtownshire during Silurian Period.—Silurian Limestones in Arctic Regions.—Professor Ramsay on Ice-action during Old Red Sandstone Period.—Warm Climate in Arctic Regions during Old Red Sandstone Period.—Professor Geikie and Mr. James Geikie on a Glacial Conglomerate of Lower Carboniferous Age.—Professor Haughton and Professor Dawson on Evidence of Ice-action during Coal Period.—Mr. W. T. Blanford on Glaciation in India during Carboniferous Period.—Carboniferous Formations of Arctic Regions.—Professor Ramsay on Permian Glaciers.—Permian Conglomerate in Arran.—Professor Hull on Boulder Clay of Permian Age.—Permian Boulder Clay of Natal.—Oolitic Boulder Conglomerate in Sutherlandshire.—Warm Climate in North Greenland during Oolitic Period.—Mr. Godwin-Austen on Ice-action during Cretaceous Period.—Glacial Conglomerates of Eocene Age in the Alps.—M. Gastaldi on the Ice-transported Limestone Blocks of the Superga.—Professor Heer on the Climate of North Greenland during Miocene Period.

CAMBRIAN PERIOD.

Island of Islay.—Good evidence of ice-action has been observed by Mr. James Thomson, F.G.S.,* in strata which he believes to be of Cambrian age. At Port Askaig, Island of Islay, below a precipitous cliff of quartzite 70 feet in height, there is a mass of arenaceous talcose schist containing fragments of granite, some angular, but most of them rounded, and of all sizes, from mere particles to large boulders. As there is no granite in the island from which these boulders could have been derived, he justly infers that they must have been transported by the agency of ice. The probability of his conclusion is strengthened by the almost total absence of stratification in the deposit in question.

* Brit. Assoc., 1870, p. 88.

North-west of Scotland.—Mr. J. Geikie tells me that much of the Cambrian conglomerate in the north-west of Scotland strongly reminds him of the coarse shingle beds (Alpine diluvium) which so often crowd the old glacial valleys of Switzerland and Northern Italy. In many places the stones of the Cambrian conglomerate have a subangular, blunted shape, like those of the re-arranged moraine débris of Alpine countries.

SILURIAN PERIOD.

Wigtownshire.—The possibility of glacial action so far back as the Silurian age has been suggested. In beds of slate and shales in Wigtownshire of Lower Silurian age Mr. J. Carrick Moore found beds of conglomerate of a remarkable character. The fragments generally vary from the size of one inch to a foot in diameter, but in some of the beds, boulders of 3, 4, and even 5 feet in diameter occur. There are no rocks in the neighbourhood from which any of these fragments could have been derived. The matrix of this conglomerate is sometimes a green trappean-looking sandstone of exceeding toughness, and sometimes an indurated sandstone indistinguishable from many common varieties of greywacke.*

Ayrshire.—Mr. James Geikie states that in Glenapp, and near Dalmellington, he found embedded in Lower Silurian strata blocks and boulders from one foot to 5 feet in diameter of gneiss, syenite, granite, &c., none of which belong to rocks of those neighbourhoods.† Similar cases have been found in Galway, Ireland, and at Lisbellaw, south of Enniskillen.‡ In America, Professor Dawson describes Silurian conglomerates with boulders 2 feet in diameter.

Arctic Regions.—The existence of warm inter-glacial periods during that age may be inferred from the fact that in the arctic regions we find widespread masses of Silurian limestones containing encrinites, corals, and mollusca, and other fossil

* Quart. Journ. Geol. Soc., vol. v., p. 10. Phil. Mag. for April, 1865, p. 289.
† "Great Ice Age," p. 512.
‡ Jukes' "Manual of Geology," p. 421.

remains, for an account of which see Professor Haughton's geological account of the Arctic Archipelago appended to McClintock's "Narrative of Arctic Discoveries."*

OLD RED SANDSTONE.

North of England.—According to Professor Ramsay and some other geologists the brecciated, subangular conglomerates and boulder beds of the Old Red Sandstone of Scotland and the North of England are of glacial origin. When these conglomerates and the recent boulder clay come together it is difficult to draw the line of demarcation between them.

Professor Ramsay observed some very remarkable facts in connection with the Old Red Sandstone conglomerates of Kirkby Lonsdale, and Sedburgh, in Westmoreland and Yorkshire. I shall give the results of his observations in his own words.

"The result is, that we have found many stones and blocks distinctly scratched, and on others the ghosts of scratches nearly obliterated by age and chemical action, probably aided by pressure at a time when these rocks were buried under thousands of feet of carboniferous strata. In some cases, however, the markings were probably produced within the body of the rock itself by pressure, accompanied by disturbance of the strata; but in others the longitudinal and cross striations convey the idea of glacial action. The shapes of the stones of these conglomerates, many of which are from 2 to 3 feet long, their flattened sides and subangular edges, together with the confused manner in which they are often arranged (like stones in the drift), have long been enough to convince me of their ice-borne character; and the scratched specimens, when properly investigated, may possibly convince others."†

Isle of Man.—The conglomerate of the Old Red Sandstone in the Isle of Man has been compared by Mr. Cumming to "a consolidated ancient boulder clay." And he remarks, "Was it so that those strange trilobitic-looking fishes of that era had to

* See also Quarterly Journal Geological Society, vol. xi., p. 510.

† The *Reader* for August 12, 1865.

endure the buffeting of ice-waves, and to struggle amidst the wreck of ice-floes and the crush of bergs?"*

Australia.—A conglomerate similar to that of Scotland has been found in Victoria, Australia, by Mr. Selwyn, at several localities. Along the Wild Duck Creek, near Heathcote, and also near the Mia-Mia, Spring Plains, Redesdale, localities in the Colony of Victoria, where it was examined by Messrs. Taylor and Etheridge, Junior, this conglomerate consists of a mixture of granite pebbles and boulders of various colours and textures, porphyries, indurated sandstone, quartz, and a peculiar flint-coloured rock in a matrix of bluish-grey very hard mud-cement.† Rocks similar to the pebbles and blocks composing the conglomerate do not occur in the immediate neighbourhood; and from the curious mixture of large and small angular and waterworn fragments it was conjectured that it might possibly be of glacial origin. Scratched stones were not observed, although a careful examination was made. From similar mud-pebble beds on the Lerderderg River, Victoria, Mr. R. Daintree obtained a few pebbles grooved after the manner of ice-scratched blocks.‡

And the existence of a warm condition of climate during the Old Red Sandstone period is evidenced by the fossiliferous limestones of England, Russia, and America. On the banks of the Athabasca River, Rupert-Land, Sir John Richardson found beds of limestone containing *Producti*, *Spiriferi*, an *Orthis* resembling *O. resupinata*, *Terebratula reticularis*,§ and a *Pleurotomaria*, which, in the opinion of the late Dr. Woodward, who examined the specimens, are characteristic of Devonian rocks of Devonshire.

* "History of the Isle of Man," p. 86. My colleague, Mr. John Horne, in his "Sketch of the Geology of the Isle of Man," Trans. of Edin. Geol. Soc., vol. ii., part iii., considers this conglomerate to be of Lower Carboniferous age.

† See Selwyn, "Phys. Geography and Geology of Victoria." 1866. pp. 15—16; Taylor and Etheridge, *Geol. Survey Vict.*, *Quarter Sheet* 13, *N.E.*

‡ Report on the Geology of the District of Ballan, Victoria. 1866. p. 11.

§ *Atrypa reticularis.*

CARBONIFEROUS PERIOD.

France.—It is now a good many years since Mr. Godwin-Austen directed attention to what he considered evidence of ice-action during the coal period. This geologist found in the carboniferous strata of France large angular blocks which he could not account for without inferring the former action of ice. "Whether from local elevation," he says, "or from climatic conditions, there are certain appearances over the whole which imply that at one time the temperature must have been very low, as glacier-action can alone account for the presence of the large angular blocks which occur in the lowest detrital beds of many of the southern coal-basins."*

Scotland.—In Scotland great beds of conglomerate are met with in various parts, which are now considered by Professor Geikie, Mr. James Geikie, and other officers of the Geological Survey who have had opportunities of examining them, to be of glacial origin. "They are," says Mr. James Geikie, "quite unstratified, and the stones often show that peculiar blunted form which is so characteristic of glacial work." † Many of the stones found by Professor Geikie, several of which I have had an opportunity of seeing, are well striated.

In 1851 Professor Haughton brought forward at the Geological Society of Dublin, a case of angular fragments of granite occurring in the carboniferous limestone of the county of Dublin; and he explained the phenomena by the supposition of the transporting power of ice.

North America.—In one of the North American coal-fields Professor Newberry found a boulder of quartzite 17 inches by 12 inches, imbedded in a seam of coal. Similar facts have also been recorded both in the United States, and in Nova Scotia. Professor Dawson describes what he calls a gigantic esker of Carboniferous age, on the outside of which large

* Quart. Journ. Geol. Soc., vol. xii., p. 58.

† "Great Ice Age," p. 513.

travelled boulders were deposited, probably by drift-ice; while in the swamps within, the coal flora flourished.*

India.—Mr. W. T. Blanford, of the Geological Survey of India, states that in beds considered to be of Carboniferous age are found large boulders, some of them as much as 15 feet in diameter. The bed in which these occur is a fine silt, and he refers the deposition of the boulders to ice-action. Within the last three years his views have received singular confirmation in another part of India, where beds of limestone were found striated below certain overlying strata. The probability that these appearances are due, as Mr. Blanford says, to the action of ice, is strengthened by the consideration that about five degrees farther to the north of the district in question rises the cold and high table-land of Thibet, which during a glacial epoch would undoubtedly be covered with ice that might well descend over the plains of India.†

Arctic Regions.—A glacial epoch during the Carboniferous age may be indirectly inferred from the probable existence of warm inter-glacial periods, as indicated by the limestones with fossil remains found in arctic regions.

That an equable condition of climate extended to near the north pole is proved by the fact that in the arctic regions vast masses of carboniferous limestone, having all the characters of the mountain limestone of England, have been found. "These limestones," says Mr. Isbister, "are most extensively developed in the north-east extremity of the continent, where they occupy the greater part of the coast-line, from the north side of the Kotzebue Sound to within a few miles of Point Barrow, and form the chief constituent of the lofty and conspicuous headlands of Cape Thomson, Cape Lisburn, and Cape Sabine." ‡ Limestone of the same age occurs extensively along the Mackenzie River. The following fossils have been found in these limestones:—*Terebratula resupinata,*§ *Lithostrotion basaltiforme, Cyathophyllum dianthum, C. flexuosum, Turbinolia mitrata, Pro-*

* "Great Ice Age," p. 513.
† Brit. Assoc. Report for 1873.
‡ Quart. Journ. Geol. Soc., vol. xi., p. 519.
§ *Orthis resupinata.*

*ductus Martini,** *Dentalium Sarcinula*, *Spiriferi*, *Orthidæ*, and encrinital fragments in the greatest abundance.

Among the fossils brought home from Depôt Point, Albert Land, by Sir E. Belcher, Mr. Salter found the following, belonging to the Carboniferous period:—*Fusulina hyperborea*, *Stylastrea inconferta*, *Zaphrentis ovibos*, *Clisiophyllum tumulus*, *Syringopora (Aulopora)*, *Fenestella Arctica*, *Spirifera Keilhavii*, *Productus cora*, *P. semireticulatus*.†

Coal-beds of Carboniferous age are extensively developed in arctic regions. The fuel is of a highly bituminous character, resembling, says Professor Haughton, the gas coals of Scotland. The occurrence of coal in such high latitudes indicates beyond doubt that a mild and temperate condition of climate must, during some part of the Carboniferous age, have prevailed up to the very pole.

"In the coal of Jameson's Land, on the east side of Greenland, lying in latitude 71°, and in that of Melville Island, in latitude 75° N., Professor Jameson found plants resembling fossils of the coal-fields of Britain."‡

PERMIAN PERIOD.

England.—From the researches of Professor Ramsay in the Permian breccias, we have every reason to believe that during a part of the Permian age our country was probably covered with glaciers reaching to the sea. These brecciated stones, he states, are mostly angular or subangular, with flattened sides and but very slightly rounded at the edges, and are imbedded in a deep red marly paste. At Abberley Hill some of the masses are from 2 to 3 feet in diameter, and in one of the quarries, near the base of Woodbury Hill, Professor Ramsay saw one 2 feet in diameter. Another was observed at Woodbury Rock, 4 feet long, 3 feet broad, and 1½ feet thick. The boulders were found in South Staffordshire, Enville, in

* *Prod. semireticulatus* var. *Martini*. Sow.

† "Belcher's Voyage," vol. ii., p. 377.

‡ "Journal of a Boat Voyage through Rupert-Land," vol. ii., p. 208.

Abberley and Malvern Hills, and other places. "They seem," he says, "to have been derived from the conglomerate and green, grey, and purple Cambrian grits of the Longmynd, and from the Silurian quartz-rocks, slates, felstones, felspathic ashes, greenstones, and Upper Caradoc rocks of the country between the Longmynd and Chirbury. But then," he continues, "the south end of the Malvern Hills is from forty to fifty miles, the Abberleys from twenty-five to thirty-five miles, Enville from twenty to thirty miles, and South Staffordshire from thirty-five to forty miles distant from that country."*

It is physically impossible, Professor Ramsay remarks, that these blocks could have been transported to such distances by any other agency than that of ice. Had they been transported by water, supposing such a thing possible, they would have been rounded and water-worn, whereas many of these stones are flat slabs, and most of them have their edges but little rounded. And besides many of them are highly polished, and others grooved and finely striated, exactly like those of the ancient glaciers of Scotland and Wales. Some of these specimens are to be seen in the Museum of Practical Geology, Jermyn Street.

Scotland.—In the Island of Arran, Mr. E. A. Wunsch and Mr. James Thomson found a bed of conglomerate which they considered of Permian age, and probably of glacial origin. This conglomerate enclosed angular fragments of various schistose, volcanic, and limestone rocks, and contained carboniferous fossils.

Ireland.—At Armagh, Ireland, Professor Hull found boulder beds of Permian age, containing pebbles and boulders, sometimes 2 feet in diameter. Some of the boulders must have been transported from a region lying about 30 miles to the north-west of the locality in which they now occur. It is difficult to conceive, says Professor Hull, how rock fragments of such a size could have been carried to their present position by any other agency than that of floating ice. This boulder-bed is

* Quart. Journ. Geol. Soc., vol. xi., p. 197.

overlaid by a recent bed of boulder clay. Professor Ramsay, who also examined the section, agrees with Professor Hull that the bed is of Permian age, and unquestionably of ice-formation.*

Professor Ramsay feels convinced that the same conclusions which he has drawn in regard to the Permian breccia of England will probably yet be found to hold good in regard to much of that of North Germany.† And there appears to be some ground for concluding that the cold of that period even reached to India.‡

South Africa.—An ancient boulder clay, supposed to be either of Permian or Jurassic age, has been extensively found in Natal, South Africa. This deposit, discovered by Dr. Sutherland, the Surveyor-General of the colony, is thus described by Dr. Mann:—

"The deposit itself consists of a greyish-blue argillaceous matrix, containing fragments of granite, gneiss, graphite, quartzite, greenstone, and clay-slate. These imbedded fragments are of various size, from the minute dimensions of sand-grains up to vast blocks measuring 6 feet across, and weighing from 5 to 10 tons. They are smoothed, as if they had been subject to a certain amount of attrition in a muddy sediment; but they are not rounded like boulders that have been subjected to sea-breakers. The fracture of the rock is not conchoidal, and there is manifest, in its substance, a rude disposition towards wavy stratification."

"Dr. Sutherland inclines to think that the transport of vast massive blocks of several tons' weight, the scoring of the subjacent surfaces of sandstone, and the simultaneous deposition of minute sand-grains and large boulders in the same matrix, all point to one agency as the only one which can be rationally admitted to account satisfactorily for the presence of this remarkable formation in the situations in which it is found. He believes that the boulder-bearing clay of Natal is of

* Explanation Memoir to Sheet 47, "Geological Survey of Ireland."

† Phil. Mag., vol. xxix., p. 290.

‡ "Memoirs of the Geological Survey of India," vol. i., part i.

analogous nature to the great Scandinavian drift, to which it is certainly intimately allied in intrinsic mineralogical character; that it is virtually a vast moraine of olden time; and that ice, in some form or other, has had to do with its formation, at least so far as the deposition of the imbedded fragments in the amorphous matrix are concerned." *

In the discussion which followed the reading of Dr. Sutherland's paper, Professor Ramsay pointed out that in the Natal beds enormous blocks of rock occurred, which were 60 or 80 miles from their original home, and still remained angular; and there was a difficulty in accounting for the phenomena on any other hypothesis than that suggested.

Mr. Stow, in his paper on the Karoo beds, has expressed a similar opinion regarding the glacial character of the formation.†

But we have in the Karoo beds evidence not only of glaciation, but of a much warmer condition of things than presently exists in that latitude. This is shown from the fact that the shells of the *Trigona*-beds indicate a tropical or subtropical condition of climate.

Arctic Regions.—The evidence which we have of the existence of a warm climate during the Permian period is equally conclusive. The close resemblance of the *flora* of the Permian period to that of Carboniferous times evidently points to the former prevalence of a warm and equable climate. And the existence of the magnesian limestone in high latitudes seems to indicate that during at least a part of the Permian period, just as during the accumulution of the carboniferous limestone, a warm sea must have obtained in those latitudes.

OOLITIC PERIOD.

North of Scotland.—There is not wanting evidence of something like the action of ice during the Oolitic period.‡

In the North of Scotland Mr. James Geikie says there is a

* Quart. Journ. Geol. Soc., vol. xxvi., p. 514.
† Ibid., vol. xxvii., p. 544.
‡ Phil. Mag., vol. xxix., p. 290.

coarse boulder conglomerate associated with the Jurassic strata in the east of Sutherland, the possibly glacial origin of which long ago suggested itself to Professor Ramsay and other observers. Mr. Judd believes the boulders to have been floated down by ice from the Highland mountains at the time the Jurassic strata were being accumulated.

North Greenland.—During the Oolitic period a warm condition of climate extended to North Greenland. For example, in Prince Patrick's Island, at Wilkie Point, in lat. 76° 20′ N., and long. 117° 20′ W., Oolitic rocks containing an ammonite (*Ammonites McClintocki*, Haughton), like *A. concavus* and other shells of Oolitic species, were found by Captain McClintock.* In Katmai Bay, near Behring's Straits, the following Oolitic fossils were discovered—*Ammonites Wasnessenskii*, *A. biplex*, *Belemnites paxillosus*, and *Unio liassinus*.† Captain McClintock found at Point Wilkie, in Prince Patrick's Island, lat. 76° 20′, a bone of *Ichthyosaurus*, and Sir E. Belcher found in Exmouth Island, lat. 76° 16′ N., and long. 96° W., at an elevation of 570 feet above the level of the sea, bones which were examined by Professor Owen, and pronounced to be those of the same animal.‡ Mr. Salter remarks that at the time that these fossils were deposited, "a condition of climate something like that of our own shores was prevailing in latitudes not far short of 80° N."§ And Mr. Jukes says that during the Oolitic period, "in latitudes where now sea and land are bound in ice and snow throughout the year, there formerly flourished animals and plants similar to those living in our own province at that time. The questions thus raised," continues Mr. Jukes, "as to the climate of the globe when cephalopods and reptiles such as we should expect to find only in warm or temperate seas, could live in such high latitudes, are not easy to answer."‖ And

* Journal of the Royal Dublin Society for February, 1857.

† Quart. Journ. Geol. Soc., vol. xi., p. 519.

‡ "The Last of the Arctic Voyages," by Captain Sir E. Belcher, vol. ii., p. 389. Appendix Brit. Assoc. Report for 1855, p. 79.

§ Ibid., vol. ii., p. 379. Appendix.

‖ "Manual of Geology," pp. 395, 493.

Professor Haughton remarks, that he thinks it highly improbable that any change in the position of land and water could ever have produced a temperature in the sea at 76° north latitude which would allow of the existence of ammonites, especially species so like those that lived at the same time in the tropical warm seas of the South of England and France at the close of the Liassic, and commencement of the Lower Oolitic period.*

The great abundance of the limestone and coal of the Oolitic system shows also the warm and equable condition of the climate which must have then prevailed.

CRETACEOUS PERIOD.

Croydon.—A large block of crystalline rock resembling granite was found imbedded in a pit, on the side of the old London and Brighton road near Purley, about two miles south of Croydon. Mr. Godwin-Austen has shown conclusively that it must have been transported there by means of floating ice. This boulder was associated with loose sea-sand, coarse shingle, and a smaller boulder weighing twenty or twenty-five pounds, and all water-worn. These had all sunk together without separating. Hence they must have been firmly held together, both during the time that they were being floated away, and also whilst sinking to the bottom of the cretaceous sea. Mr. Godwin-Austen supposes the whole to have been carried away frozen to the bottom of a mass of ground-ice. When the ice from melting became unable to float the mass attached to it, the whole would then sink to the bottom together.†

Dover.—While the workmen were employed in cutting the tunnel on the London, Chatham, and Dover Railway, between Lydden Hill and Shepherdswell, a few miles from Dover, they came upon a mass of coal imbedded in chalk, at a depth of 180 feet. It was about 4 feet square, and from 4 to 10 inches thick.

* Appendix to McClintock's "Arctic Discoveries."

† Quart. Journ. Geol. Soc., vol. xiv., p. 262. Brit. Assoc. Report for 1857, p. 62.

The coal was friable and highly bituminous. It resembled some of the Wealden or Jurassic coal, and was unlike the true coal of the coal-measures. The specific gravity of the coal precluded the supposition that it could have floated away of itself into the cretaceous sea. "Considering its friability," says Mr. Godwin-Austen, "I do not think that the agency of a floating tree could have been engaged in its transport; but, looking at its flat, angular form, it seems to me that its history may agree with what I have already suggested with reference to the boulder in the chalk at Croydon. We may suppose that during the Cretaceous period some bituminous beds of the preceding Oolitic period lay so as to be covered with water near the sea-margin, or along some river-bank, and from which portions could be carried off by ice, and so drifted away, until the ice was no longer able to support its load."*

Mr. Godwin-Austen then mentions a number of other cases of blocks being found in the chalk. In regard to those cases he appropriately remarks that, as the cases where the occurrence of such blocks has been observed are likely to be far less numerous than those which have escaped observation, or failed to have been recorded, and as the chalk exposed in pits and quarries bears only a most trifling proportion to the whole horizontal extent of the formation, we have no grounds to conclude that the above are exceptional cases.

Boulders have also been found in the cretaceous strata of the Alps by Escher von der Linth.†

The existence of warm periods during the Cretaceous age is plainly shown by the character of the flora and fauna of that age. The fact that chalk is of organic origin implies that the climate must have been warm and genial, and otherwise favourable to animal life. This is further manifested by such plants as *Cycas* and *Zamia*, which betoken a warm climate, and by the corals and huge sauroid reptiles which then inhabited our waters.

* Quart. Journ. Geol. Soc., vol. xvi., p. 327. *Geologist*, 1860, p. 38.
† Phil. Mag., vol. xxix., p. 290.

It is, in fact, the tropical character of the fauna of that period which induced Sir Charles Lyell to reject Mr. Godwin-Austen's idea that the boulders found in the chalk had been transported by floating ice. Such a supposition, implying a cold climate, "is," Sir Charles says, "inconsistent with the luxuriant growth of large chambered univalves, numerous corals, and many fish, and other fossils of tropical forms."

The recent discovery of the Cretaceous formation in Greenland shows that during that period a mild and temperate condition of climate must have prevailed in that continent up to high latitudes. "This formation in Greenland," says Dr. Robert Brown, "has only been recently separated from the Miocene formation, with which it is associated and was supposed to be a part of. It is, as far as we yet know, only found in the vicinity of Kome or Koke, near the shores of Omenak Fjord, in about 70° north latitude, though traces have been found elsewhere on Disco, &c. The fossils hitherto brought to Europe have been very few, and consist of plants which are now preserved in the Stockholm and Copenhagen Museums. From these there seems little doubt that the age assigned to this limited deposit (so far as we yet know) by the celebrated palæontologist, Professor Oswald Heer, of Zurich, is the correct one."* Dr. Brown gives a list of the Cretaceous flora found in Greenland.

EOCENE PERIOD.

Switzerland.—In a coarse conglomerate belonging to the "*flysch*" of Switzerland, an Eocene formation, there are found certain immense blocks, some of which consist of a variety of granite which is not known to occur *in situ* in any part of the Alps. Some of the blocks are 10 feet and upwards in length, and one at Halekeren, at the Lake of Thun, is 105 feet in length, 90 feet in breadth, and 45 feet in height. Similar blocks are found in the Apennines. These unmistakably

* Trans. Geol. Soc. of Glasgow, vol. v., p. 64.

indicate the presence of glaciers or floating ice. This conclusion is further borne out by the fact that the "*flysch*" is destitute of organic remains. But the hypothesis that these huge masses were transported to their present sites by glaciers or floating ice has been always objected to, says Sir Charles Lyell, "on the ground that the Eocene strata of Nummulitic age in Switzerland, as well as in other parts of Europe, contain genera of fossil plants and animals characteristic of a warm climate. And it has been particularly remarked," he continues, "by M. Desor that the strata most nearly associated with the '*flysch*' in the Alps are rich in echinoderms of the *Spatangus* family which have a decided tropical aspect."*

But according to the theory of Secular Changes of Climate, the very fact that the "*flysch*" is immediately associated with beds indicating a warm or even tropical condition of climate, is one of the strongest proofs which could be adduced in favour of its glacial character, for the more severe a cold period of a glacial epoch is, the warmer will be the periods which immediately precede and succeed. These crocodiles, tortoises, and tropical flora probably belong to a warm Eocene inter-glacial period.

MIOCENE PERIOD.

Italy.—We have strong evidence in favour of the opinion that a glacial epoch existed during the Miocene period. It has been shown by M. Gastaldi, that during that age Alpine glaciers extended to the sea-level.

Near Turin there is a series of hills, rising about 500 or 600 feet above the valleys, composed of beds of Miocene sandstone, marl, and gravel, and loose conglomerate. These beds have been carefully examined and described by M. Gastaldi.† The hill of the Luperga has been particularly noticed by him. Many of the stones in these beds are striated in a manner similar to those found in the true till or boulder clay of this

* "Principles," vol. i., p. 209. Eleventh Edition.

† "Memoirs of the Royal Academy of Science of Turin," Second Series, vol. xx. I am indebted for the above particulars to Professor Ramsay, who visited the spot along with M. Gastaldi.

country. But what is most remarkable is the fact that large erratic blocks of limestone, many of them from 10 to 15 feet in diameter, are found in abundance in these beds. It has been shown by Gastaldi that these blocks have all been derived from the outer ridge of the Alps on the Italian side, namely, from the range extending from Ivrea to the Lago Maggiore, and consequently they must have travelled from twenty to eighty miles. So abundant are these large blocks, that extensive quarries have been opened in the hills for the sake of procuring them. These facts prove not only the existence of glaciers on the Alps during the Miocene period, but of glaciers extending to the sea and breaking up into icebergs; the stratification of the beds amongst which the blocks occur sufficiently indicating aqueous action and the former presence of the sea.

That the glaciers of the Southern Alps actually reached to the sea, and sent their icebergs adrift over what are now the sunny plains of Northern Italy, is sufficient proof that during the cold period of Miocene times the climate must have been very severe. Indeed, it may well have been as severe as, if not even more excessive than, the intensest severity of climate experienced during the last great glacial epoch.

Greenland.—Of the existence of warm conditions during Miocene times, geology affords us abundant evidence. I shall quote the opinion of Sir Charles Lyell on this point:—

"We know," says Sir Charles, "that Greenland was not always covered with snow and ice; for when we examine the tertiary strata of Disco Island (of the Upper Miocene period), we discover there a multitude of fossil plants which demonstrate that, like many other parts of the arctic regions, it formerly enjoyed a mild and genial climate. Among the fossils brought from that island, lat. 70° N., Professor Heer has recognised *Sequoia Landsdorfii*, a coniferous species which flourished throughout a great part of Europe in the Miocene period. The same plant has been found fossil by Sir John Richardson within the Arctic Circle, far to the west on the Mackenzie River, near the entrance of Bear River; also by some Danish naturalists in

Iceland, to the east. The Icelandic surturband or lignite, of this age, has also yielded a rich harvest of plants, more than thirty-one of them, according to Steenstrup and Heer, in a good state of preservation, and no less than fifteen specifically identical with Miocene plants of Europe. Thirteen of the number are arborescent; and amongst others is a tulip-tree (*Liriodendron*), with its fruit and characteristic leaves, a plane (*Platanus*), a walnut, and a vine, affording unmistakable evidence of a climate in the parallel of the Arctic Circle which precludes the supposition of glaciers then existing in the neighbourhood, still less any general crust of continental ice like that of Greenland."*

At a meeting of the British Association, held at Nottingham in August 1866, Professor Heer read a valuable paper on the "Miocene Flora of North Greenland." In this paper some remarkable conclusions as to the probable temperature of Greenland during the Miocene period were given.

Upwards of sixty different species brought from Atanekerdluk, a place on the Waigat opposite Disco, in lat. 70° N., have been examined by him.

A steep hill rises on the coast to a height of 1,080 feet, and at this level the fossil plants are found. Large quantities of wood in a fossilized or carbonized condition lie about. Captain Inglefield observed one trunk thicker than a man's body standing upright. The leaves, however, are the most important portion of the deposit. The rock in which they are found is a sparry iron ore, which turns reddish brown on exposure to the weather. In this rock the leaves are found, in places packed closely together, and many of them are in a very perfect condition. They give us a most valuable insight into the nature of the vegetation which formed this primeval forest.

He arrives at the following conclusions:—

1. *The fossilized plants of Atanekerdluk cannot have been drifted from any great distance. They must have grown on the spot where they were found.*

* "Antiquity of Man," Second Edition, p. 237.

This is shown—

(*a*) By the fact that Captain Inglefield and Dr. Ruik observed trunks of trees standing upright.

(*b*) By the great abundance of the leaves, and the perfect state of preservation in which they are found.

(*c*) By the fact that we find in the stone both fruits and seeds of the trees whose leaves are also found there.

(*d*) By the occurrence of insect remains along with the leaves.

2. *The flora of Atanekerdluk is Miocene.*

3. *The flora is rich in species.*

4. *The flora proves without a doubt that North Greenland, in the Miocene epoch, had a climate much warmer than its present one. The difference must be at least* 29° F.

Professor Heer discusses at considerable length this proposition. He says that the evidence from Greenland gives a final answer to those who objected to the conclusions as to the Miocene climate of Europe drawn by him on a former occasion. It is quite impossible that the trees found at Atanekerdluk could ever have flourished there if the temperature were not far higher than it is at present. This is clear from many of the species, of which we find the nearest living representative 10° or even 20° of latitude to the south of the locality in question.

The trees of Atanekerdluk were not, he says, all at the extreme northern limit of their range, for in the Miocene flora of Spitzbergen, lat. 78° N., we find the beech, plane, hazelnut, and some other species identical with those from Greenland, and we may conclude, he thinks, that the firs and poplars which we meet at Atanekerdluk and Bell Sound, Spitzbergen, must have reached up to the North Pole if land existed there in the tertiary period.

"The hills of fossilized wood," he adds, "found by McClure and his companions in Banks's Land (lat. 74° 27′ N.), are therefore discoveries which should not astonish us, they only confirm the evidence as to the original vegetation of the polar regions which we have derived from other sources."

The *Sequoia Landsdorfii* is the most abundant of .the trees of Atanekerdluk. The *Sequoia sempairveus* is its present representative. This tree has its extreme northern limit about lat. 53° N. For its existence it requires a summer temperature of 59° or 61° F. Its fruit requires a temperature of 64° for ripening. The winter temperature must not fall below 34°, and that of the whole year must be at least 49°. The temperature of Atanekerdluk during the time that the Miocene flora grew could not have been under the above.*

Professor Heer concludes his paper as follows :—

"I think these facts are convincing, and the more so that they are not insulated, but confirmed by the evidence derivable from the Miocene flora of Iceland, Spitzbergen, and Northern Canada. These conclusions, too, are only links in the grand chain of evidence obtained from the examination of the Miocene flora of the whole of Europe. They prove to us that we could not by any re-arrangement of the relative positions of land and water produce for the northern hemisphere a climate which would explain the phenomena in a satisfactory manner. We must only admit that we are face to face with a problem, whose solution in all probability must be attempted, and, we doubt not, completed by the astronomer."

* Dr. Robert Brown, in a recent Memoir on the Miocene Beds of the Disco District (Trans. Geol. Soc. Glasg., vol. v., p. 55), has added considerably to our knowledge of these deposits. He describes the strata in detail, and gives lists of the plant and animal remains discovered by himself and others, and described by Professor Heer. Professor Nordenskjöld has likewise increased the data at our command (Transactions of the Swedish Academy, 1873); and still further evidence in favour of a warm climate having prevailed in Greenland during Miocene times has been obtained by the recent second German polar expedition.

CHAPTER XIX.

GEOLOGICAL TIME.—PROBABLE DATE OF THE GLACIAL EPOCH.

Geological Time measurable from Astronomical Data.—M. Leverrier's Formulæ. —Tables of Eccentricity for 3,000,000 Years in the Past and 1,000,000 Years in the Future.—How the Tables have been computed.—Why the Glacial Epoch is more recent than had been supposed.—Figures convey a very inadequate Conception of immense Duration.—Mode of representing a Million of Years.—Probable Date of the Glacial Epoch.

If those great Secular variations of climate which we have been considering be indirectly the result of changes in the eccentricity of the earth's orbit, then we have a means of determining, at least so far as regards recent epochs, when these variations took place. If the glacial epoch be due to the causes assigned, we have a means of ascertaining, with tolerable accuracy, not merely the date of its commencement, but the length of its duration. M. Leverrier has not only determined the superior limit of the eccentricity of the earth's orbit, but has also given formulæ by means of which the extent of the eccentricity for any period, past or future, may be computed.

A well-known astronomer and mathematician, who has specially investigated the subject, is of opinion that these formulæ give results which may be depended upon as approximately correct for *four millions of years* past and future. An eminent physicist has, however, expressed to me his doubts as to whether the results can be depended on for a period so enormous. M. Leverrier in his Memoir has given a table of the eccentricity for 100,000 years before and after 1800 A.D., computed for intervals of 10,000 years. This table, no doubt, embraces a period sufficiently great for ordinary astronomical purposes, but it is by far too limited to afford information in regard to geological epochs.

With the view of ascertaining the probable date of the glacial epoch, as well as the character of the climate for a long course of ages, Table I. was computed from M. Leverrier's formulæ.* It shows the eccentricity of the earth's orbit and longitude of the perihelion for 3,000,000 of years back, and 1,000,000 of years to come, at periods 50,000 years apart.

On looking over the table it will be seen that there are three principal periods when the eccentricity rose to a very high value, with a few subordinate maxima between. It will be perceived also that during each of those periods the eccentricity does not remain at the same uniform value, but rises and falls, in one case twice, and in the other two cases three times. About 2,650,000 years back we have the eccentricity almost at its inferior limit. It then begins to increase, and fifty thousand years afterwards, namely at 2,600,000 years ago, it reaches 0660; fifty thousand years after this period it has diminished to ·0167, which is about its present value. It then begins to increase, and in another fifty thousand years, namely at 2,500,000 years ago, it approaches to almost the superior limit, its value being then ·0721. It then begins to diminish, and at 2,450,000 years ago it has diminished to ·0252. These two maxima, separated by a minimum and extending over a period

* The following are M. Leverrier's formulæ for computing the eccentricity of the earth's orbit, given in his "Memoir" in the *Connaissance des Temps* for 1843:—

Eccentricity in (t) years after January 1, 1800 $= \sqrt{h^2 + l^2}$ where

$$h = 0{\cdot}000526 \text{ Sin } (gt + \beta) + 0{\cdot}016611 \text{ Sin } (g_1t + \beta_1) + 0{\cdot}002366 \text{ Sin } (g_2t + \beta_2)$$
$$+ 0{\cdot}010622 \text{ Sin } (g_3t + \beta_3) - 0{\cdot}018925 \text{ Sin } (g_4t + \beta_4)$$
$$+ 0{\cdot}011782 \text{ Sin } (g_5t + \beta_5) - 0{\cdot}016913 \text{ Sin } (g_6t + \beta_6)$$

and

$$l = 0{\cdot}000526 \text{ Cos } (gt + \beta) + 0{\cdot}016611 \text{ Cos } (g_1t + \beta_1) + 0{\cdot}002366 \text{ Cos } (g_2t + \beta_2)$$
$$+ 0{\cdot}010622 \text{ Cos } (g_3t + \beta_3) - 0.018925 \text{ Cos } (g_4t + \beta_4)$$
$$+ 0{\cdot}011782 \text{ Cos } (g_5t + \beta_5) - 0{\cdot}016913 \text{ Cos } (g_6t + \beta_6)$$

$g = 2''{\cdot}25842$ $\qquad \beta = 126°\ 43'\ 15''$
$g_1 = 3''{\cdot}71364$ $\qquad \beta_1 = 27\ 21\ 26$
$g_2 = 22''{\cdot}4273$ $\qquad \beta_2 = 126\ 44\ 8$
$g_3 = 5''{\cdot}2989$ $\qquad \beta_3 = 85\ 47\ 45$
$g_4 = 7''{\cdot}5747$ $\qquad \beta_4 = 35\ 38\ 43$
$g_5 = 17''{\cdot}1527$ $\qquad \beta_5 = -25\ 11\ 33$
$g_6 = 17''{\cdot}8633$ $\qquad \beta_6 = -45\ 28\ 59$

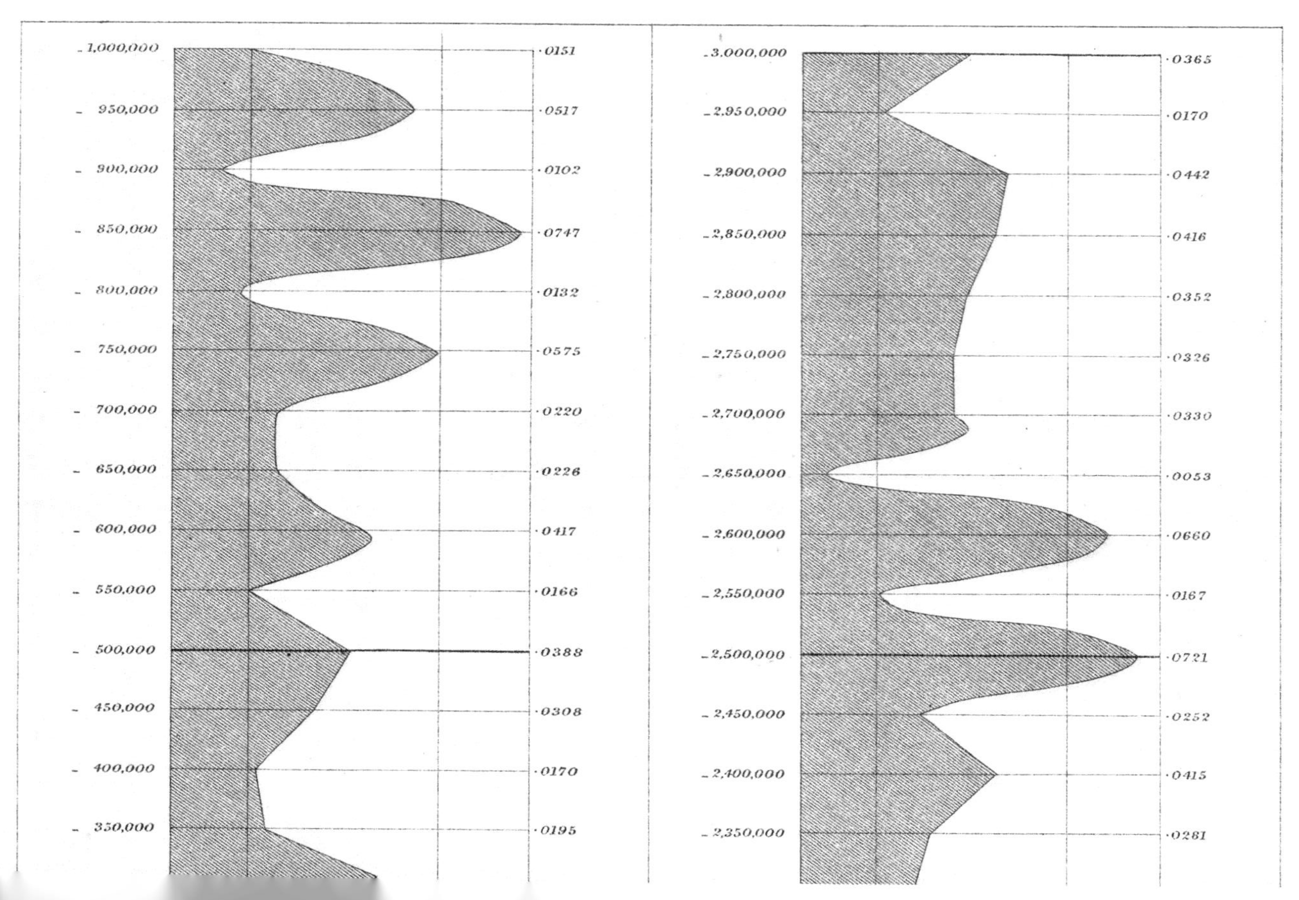

DIAGRAM REPRESENTING THE VARIATIONS IN TH

The Or

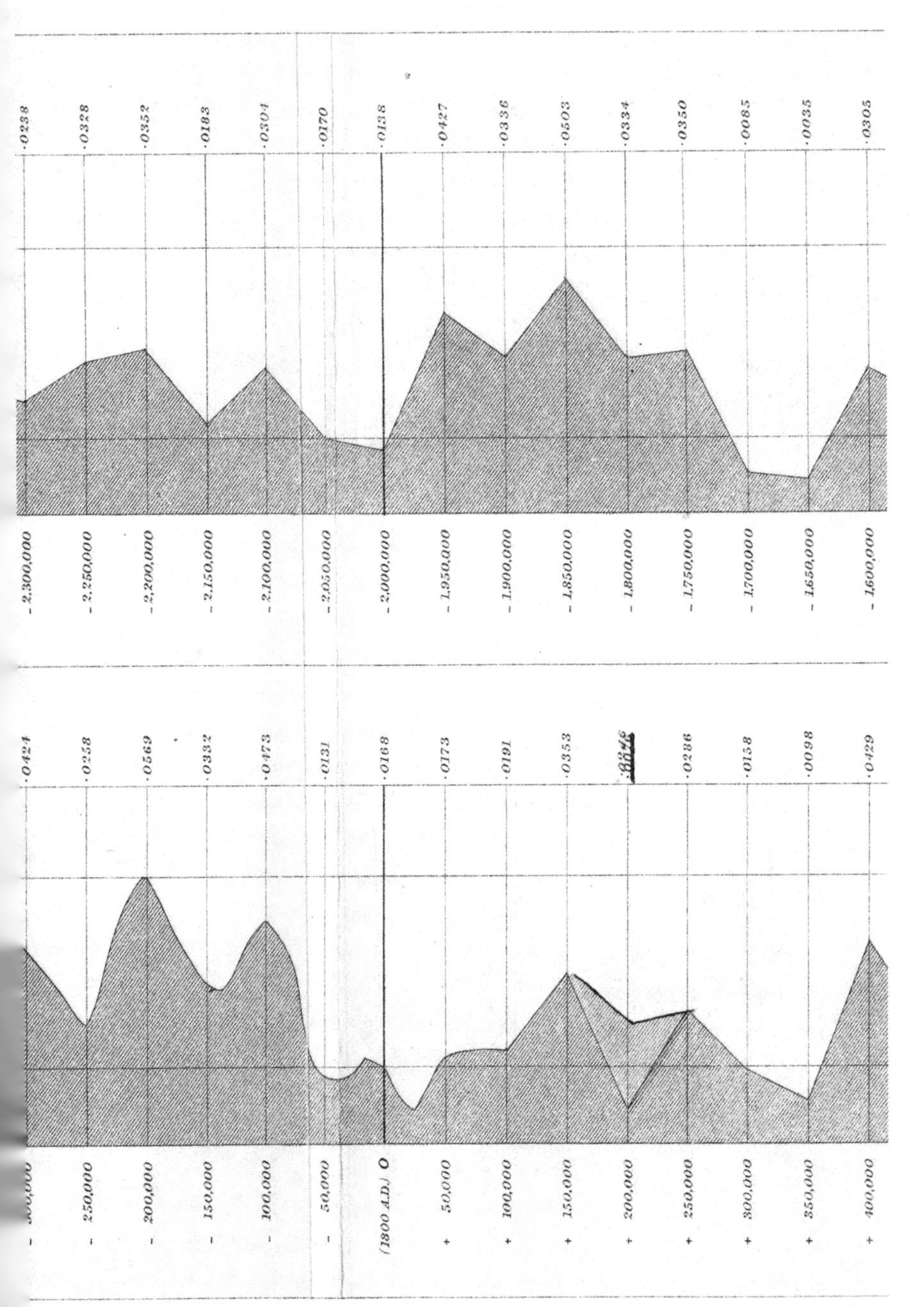

ECCENTRICITY OF THE EARTH'S ORBIT FOR THREE MILLION OF YEARS BEFORE

nates are joined by straight lines where the values, at intervals of 10,000 years, between them have not been dete

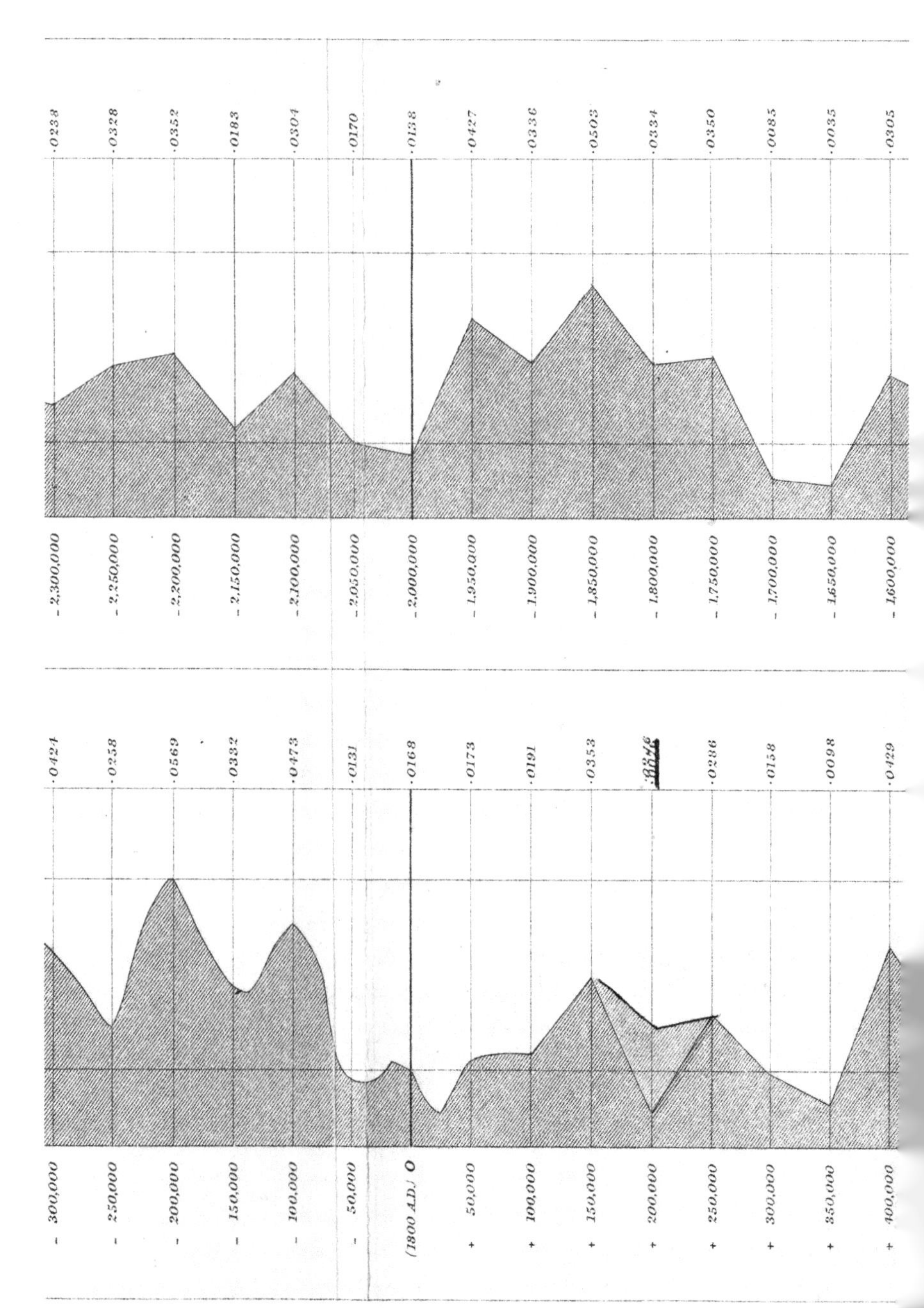

E ECCENTRICITY OF THE EARTH'S ORBIT FOR THREE MILLION OF YEARS BEFOR

inates are joined by straight lines where the values, at intervals of 10,000 years, between them have not been d

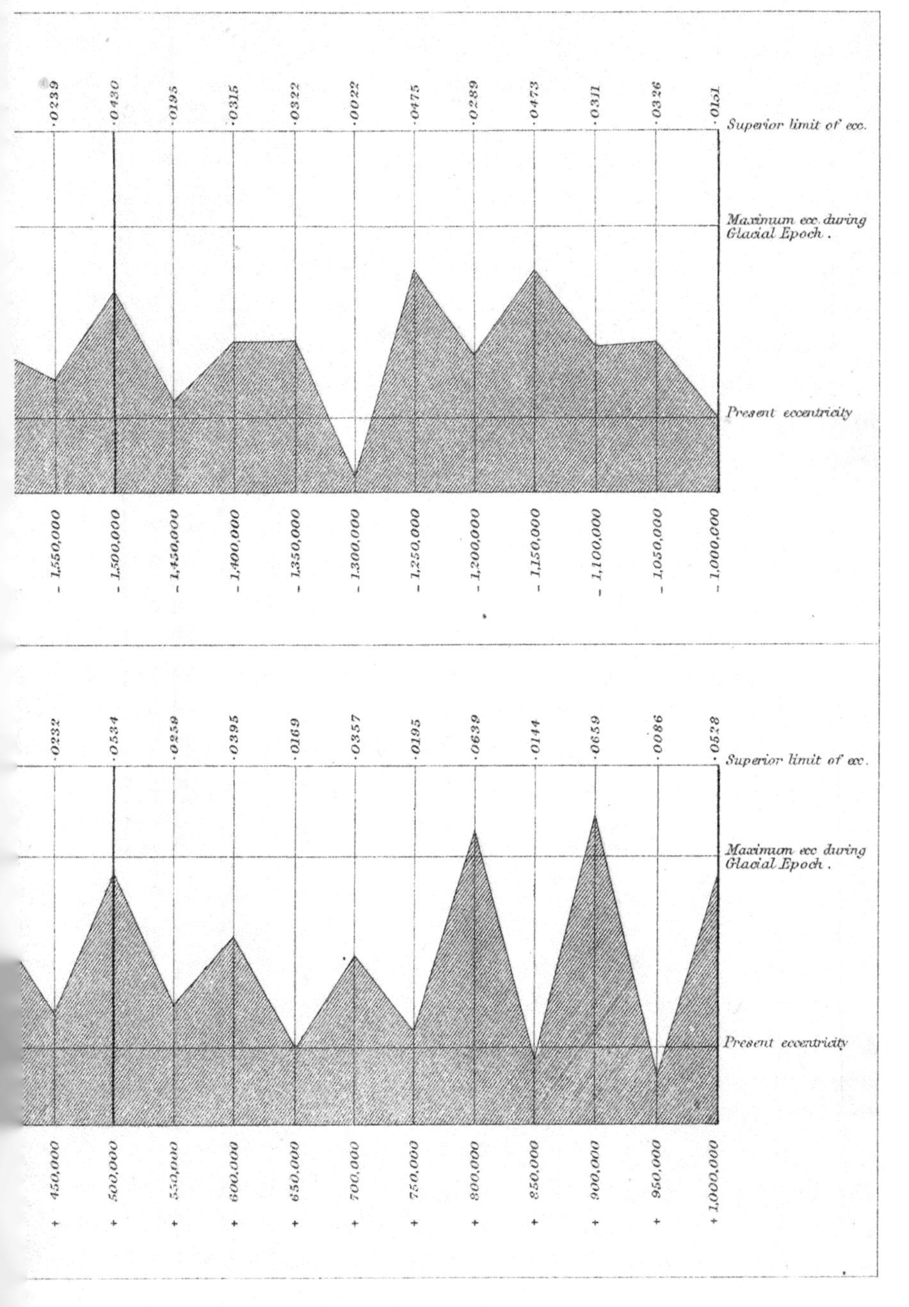

W & A.K. Johnston, Edin

800 A.D. AND ONE MILLION OF YEARS AFTER IT.

mined.

of 200,000 years, constitute the first great period of high eccentricity. We then pass onwards for upwards of a million and a half years, and we come to the second great period. It consists of three maxima separated by two minima. The first maximum occurred at 950,000 years ago, the second or middle one at 850,000 years ago, and the third and last at 750,000 years ago—the whole extending over a period of nearly 300,000 years. Passing onwards for another million and half years, or to about 800,000 years in the future, we come to the third great period. It also consists of three maxima one hundred thousand years apart. These occur at the periods 800,000, 900,000, and 1,000,000 years to come, respectively, separated also by two minima. Those three great periods, two of them in the past and one of them in the future, included in the Table, are therefore separated from each other by an interval of upwards of 1,700,000 years.

In this Table there are seven periods when the earth's orbit becomes nearly circular, four in the past and three in the future.

The Table shows also four or five subordinate periods of high eccentricity, the principal one occurring 200,000 years ago.

The variations of eccentricity during the four millions of years, are represented to the eye diagrammitically in Plate IV.

In order to determine with more accuracy the condition of the earth's orbit during the three periods of great eccentricity included in Table I., I computed the values for periods of ten thousand years apart, and the results are embodied in Tables II., III., and IV.

There are still eminent astronomers and physicists who are of opinion that the climate of the globe never could have been seriously affected by changes in the eccentricity of its orbit. This opinion results, no doubt, from viewing the question as a purely astronomical one. Viewed from an astronomical standpoint, as has been already remarked, there is actually nothing from which any one could reasonably conclude with certainty whether a change of eccentricity would seriously affect climate or not. By means of astronomy we ascertain the extent of the

TABLE I.

THE ECCENTRICITY AND LONGITUDE OF THE PERIHELION OF THE EARTH'S ORBIT FOR 3,000,000 YEARS IN THE PAST AND 1,000,000 YEARS IN THE FUTURE, COMPUTED FOR INTERVALS OF 50,000 YEARS.

PAST TIME.			PAST TIME.		
Number of years before epoch 1800.	Eccentricity.	Longitude of perihelion.	Number of years before epoch 1800.	Eccentricity.	Longitude of perihelion.
— 3,000,000	0·0365	39° 30′	— 1,950,000	0·0427	120° 32′
— 2,950,000	0·0170	210 39	— 1,900,000	0·0336	188 31
— 2,900,000	0·0442	200 52	— 1,850,000	0·0503	272 14
— 2,850,000	0·0416	0 18	— 1,800,000	0·0334	354 52
— 2,800,000	0·0352	339 14	— 1,750,000	0·0350	65 25
— 2,750,000	0·0326	161 22	— 1,700,000	0·0085	95 13
— 2,700,000	0·0330	65 37	— 1,650,000	0·0035	168 23
— 2,650,000	0·0053	318 40	— 1,600,000	0·0305	158 42
— 2,600,000	0·0660	190 4	— 1,550,000	0·0239	225 57
— 2,550,000	0·0167	298 34	— 1,500,000	0·0430	303 29
— 2,500,000	0·0721	338 36	— 1,450,000	0·0195	57 11
— 2,450,000	0·0252	109 33	— 1,400,000	0·0315	97 35
— 2,400,000	0·0415	116 40	— 1,350,000	0·0322	293 38
— 2,350,000	0·0281	308 23	— 1,300,000	0·0022	0 48
— 2,300,000	0·0238	195 25	— 1,250,000	0·0475	105 50
— 2,250,000	0·0328	141 18	— 1,200,000	0·0289	239 34
— 2,200,000	0·0352	307 6	— 1,150,000	0·0473	250 27
— 2,150,000	0·0183	307 5	— 1,100,000	0·0311	55 24
— 2,100,000	0·0304	98 40	— 1,050,000	0·0326	4 8
— 2,050,000	0·0170	334 46			
— 2,000,000	0·0138	324 4			

TABLE I.—*Continued.*

THE ECCENTRICITY AND LONGITUDE OF THE PERIHELION OF THE EARTH'S ORBIT FOR 3,000,000 YEARS IN THE PAST AND 1,000,000 YEARS IN THE FUTURE, COMPUTED FOR INTERVALS OF 50,000 YEARS.

PAST TIME.			FUTURE TIME.		
Number of years before epoch 1800.	Eccentricity.	Longitude of perihelion.	Number of years after epoch 1800.	Eccentricity.	Longitude of perihelion.
—1,000,000	0·0151	248° 22′	A.D 1800	0·0168	99° 30′
— 950,000	0·0517	97 51	+ 50,000	0·0173	38 12
— 900,000	0·0102	135 2	+ 100,000	0·0191	114 50
— 850,000	0·0747	239 28	+ 150,000	0·0353	201 57
— 800,000	0·0132	343 49	+ 200,000	0·0246	279 41
— 750,000	0·0575	27 18	+ 250,000	0·0286	350 54
— 700,000	0·0220	208 13	+ 300,000	0·0158	172 29
— 650,000	0·0226	141 29	+ 350,000	0·0098	201 40
— 600,000	0·0417	32 34	+ 400,000	0·0429	6 9
— 550,000	0·0166	251 50	+ 450,000	0·0231	98 37
— 500,000	0·0388	193 56	+ 500,000	0·0534	157 26
— 450,000	0·0308	356 52	+ 550,000	0·0259	287 31
— 400,000	0·0170	290 7	+ 600,000	0·0395	285 43
— 350,000	0·0195	182 50	+ 650,000	0·0169	144 3
— 300,000	0·0424	23 29	+ 700,000	0·0357	17 12
— 250,000	0·0258	59 39	+ 750,000	0·0195	0 53
— 200,000	0·0569	168 18	+ 800,000	0·0639	140 38
— 150,000	0·0332	242 56	+ 850,000	0·0144	176 41
— 100,000	0·0473	316 18	+ 900,000	0·0659	291 16
— 50,000	0·0131	50 14	+ 950,000	0·0086	115 13
			+1,000,000	0·0528	57 31

TABLE II.

ECCENTRICITY, LONGITUDE OF THE PERIHELION, &c., &c., FOR INTERVALS OF 10,000 YEARS, FROM 2,650,000 TO 2,450,000 YEARS AGO.

THE GLACIAL EPOCH OF THE *Eocene period* IS PROBABLY COMPREHENDED WITHIN THIS TABLE.

I.	II.	III.	IV.	Winter occurring in aphelion.			
Number of years before A.D. 1800.	Eccentricity of orbit.	Longitude of perihelion.	Number of degrees passed over by the perihelion. Motion retrograde at periods marked R.	V. Excess of winter over summer, in days.	VI. Midwinter intensity of the sun's heat. Present intensity =1000.	VII. Number of degrees by which the midwinter temperature is lowered.	VIII. Midwinter temperature of Great Britain.
2,650,000	0·0053	318° 40					
2,640,000	0·0173	54 25	95° 45′			F.	F.
2,630,000	0·0331	93 37	39 12	15·4	906	26°·2	12°·8
2,620,000	0·0479	127 12	33 35	22·2	884	33·3	5·7
2,610,000	0·0591	158 36	31 24	27·4	862	38·3	0·7
2,600,000	0·0660	190 4	31 28	30·6	851	41·5	—2·5
2,590,000	0·0666	220 28	30 24	30·9	850	41·8	—2·8
2,580,000	0·0609	249 56	29 28	28·3	859	39·2	—0·2

2,570,000	0·0492	277 24	27 28	22·9	878	33·9	5·1
2,560,000	0·0350	305 2	27 38	16·2	902	27·1	11·9
2,550,000	0·0167	298 34	R 6 28				
2,540,000	0·0192	253 58	R 44 36				
2,530,000	0·0369	259 19	5 21	17·1	899	28·0	11·0
2,520,000	0·0537	283 7	23 48	25·0	871	35·9	3·1
2,510,000	0·0660	310 4	26 57	30·6	851	41·5	—2·5
2,500,000	0·0721	338 36	28 32	33·5	841	44·2	—5·2
2,490,000	0·0722	7 36	29 0	33·6	841	44·3	—5·3
2,480,000	0·0662	35 46	28 10	30·8	850	41·7	—2·7
2,470,000	0·0553	63 26	27 40	25·7	868	36·6	2·4
2,460,000	0·0410	89 13	25 47	19·1	892	30·0	9·0
2,450,000	0·0252	109 33	20 20	11·7			

TABLE III.

ECCENTRICITY, LONGITUDE OF THE PERIHELION, &c., &c., FOR INTERVALS OF 10,000 YEARS, FROM 1,000,000 TO 700,000 YEARS AGO.

THE GLACIAL EPOCH OF THE *Miocene period* IS PROBABLY COMPREHENDED WITHIN THIS TABLE.

I.	II.	III.	VI.	Winter occurring in aphelion.			
Number of years before A.D. 1800.	Eccentricity of orbit.	Longitude of perihelion.	Number of degrees passed over by the perihelion. Motion retrograde at periods marked R.	V. Excess of winter over summer, in days.	VI. Midwinter intensity of the sun's heat. Present intensity =1000.	VII. Number of degrees by which the midwinter temperature is lowered.	VIII. Midwinter temperature of Great Britain.
1,000,000	0·0151	248° 22					
990,000	0·0224	313 50	65° 28′			F.	F.
980,000	0·0329	358 2	44 12	15·3	906	26°·1	12°·9
970,000	0·0441	32 40	34 38	20·5	887	31·5	7·5
960,000	0·0491	66 49	34 9	22·8	878	33·8	5·2
950,000	0·0517	97 51	31 2	24·0	874	35·0	4·0
940,000	0·0495	127 42	29 51	23·0	878	34·0	5·0
930,000	0·0423	156 11	28 29	19·7	890	30·6	8·4
920,000	0·0305	181 40	25 29	14·2	910	25·0	14·0
910,000	0·0156	194 15	12 35				
900,000	0·0102	135 2	R 59 13				

890,000	0·0285	127 1	R 8 1				
880,000	0·0456	152 33	25 32	21·2	884	32·2	6·8
870,000	0·0607	180 23	27 50	28·2	859	39·0	0·0
860,000	0·0708	209 41	29 18	32·9	843	43·6	—4·6
850,000	0·0747	239 28	29 47	34·7	837	45·3	—6·3
840,000	0·0698	269 14	29 46	32·4	845	43·2	—4·2
830,000	0·0623	298 28	29 14	29·0	857	40·0	—1·0
820,000	0·0476	326 4	27 36	22·1	881	33·1	5·9
810,000	0·0296	348 30	22 26				
800,000	0·0132	343 49	R 4 41				
790,000	0·0171	293 19	R 50 30				
780,000	0·0325	303 37	10 18	15·2	907	26·0	13·0
770,000	0·0455	328 38	25 1	21·2	884	32·2	6·8
760,000	0·0540	357 12	28 34	25·1	870	36·0	3·0
750,000	0·0575	27 18	30 6	26·7	864	37·7	1·3
740,000	0·0561	58 30	31 12	26·1	867	37·0	2·0
730,000	0·0507	90 55	32 25	23·6	876	34·6	4·4
720,000	0·0422	125 14	34 19	19·6	890	30·6	8·4
710,000	0·0307	177 26	52 12	14·3	910	25·0	14·0
700,000	0·0220	208 13	30 47				

TABLE IV.

ECCENTRICITY, LONGITUDE OF THE PERIHELION, &c., &c., FOR INTERVALS OF 10,000 YEARS, FROM 250,000 YEARS AGO TO THE PRESENT DTAE.

THE *Glacial epoch* IS PROBABLY COMPREHENDED WITHIN THIS TABLE.

I.	II.	III.	IV.	Winter occurring in aphelion.			
Number of years before A.D. 1800.	Eccentricity of orbit.	Longitude of perihelion.	Number of degrees passed over by the perihelion. Motion retrograde at periods marked R.	V. Excess of winter over summer, in days.	VI. Midwinter intensity of the sun's heat. Present intensity = 1000.	VII. Number of degrees by which the mid-winter temperature is lowered.	VIII. Midwinter temperature of Great Britain.
		° ′	° ′			F.	F.
250,000	0·0258	59 39					
240,000	0·0374	74 58	15 19	17·4	898	28·3	10·7
S 230,000	0·0477	102 49	27 51	22·2	885	33·2	5·8
S 220,000	0·0497	124 33	21 44	23·2	877	34·1	4·9
S 210,000	0·0575	144 55	20 22	26·7	864	37·7	1·3
200,000	0·0569	168 18	23 23	26·5	865	37·4	1·6
S 190,000	0·0532	190 4	21 46	24·7	871	35·7	3·3
S 180,000	0·0476	209 22	19 18	22·1	881	33·1	5·9
S 170,000	0·0437	228 7	18 45	20·3	887	31·3	7·7

160,000	0·0364	236 38	8 31	16·9	900	27·8	11·2
150,000	0·0332	242 56	6 18	15·4	905	26·2	12·8
140,000	0·0346	246 29	3 33	16·1	903	26·9	12·1
130,000	0·0384	259 34	13 5	17·8	896	28·8	10·2
120,000	0·0431	274 47	15 13	20·1	888	31·0	8·0
110,000	0·0460	293 48	19 1	21·4	883	32·4	6·6
100,000	0·0473	316 18	22 30	22·0	881	33·0	6·0
L 90,000	0·0452	340 2	23 44	21·0	885	32·0	7·0
L 80,000	0·0398	4 13	24 11	18·5	894	29·4	9·6
L 70,000	0·0316	27 22	23 9	14·7	908	25·5	13·5
L 60,000	0·0218	46 8	18 46				
50,000	0·0131	50 14	4 6				
L 40,000	0·0109	28 36	R 21 38				
L 30,000	0·0151	5 50	R 22 46				
L 20,000	0·0188	44 0	38 10				
L 10,000	0·0187	78 28	34 28				
A.D. 1800	0·0168	99 30	21 2				

eccentricity at any given period, how much the winter may exceed the summer in length (or the reverse), how much the sun's heat is increased or decreased by a decrease or an increase of distance, and so forth; but we obtain no information whatever regarding how these will actually affect climate. This, as we have already seen, must be determined wholly from physical considerations, and it is an exceedingly complicated problem. An astronomer, unless he has given special attention to the physics of the question, is just as apt to come to a wrong conclusion as any one else. The question involves certain astronomical elements; but when these are determined everything then connected with the matter is purely physical. Nearly all the astronomical elements of the question are comprehended in the accompanying Tables.

In Tables II., III., and IV., column I. represents the dates of the periods, column II. the eccentricity, column III. the longitude of the perihelion. In Table IV the eccentricity and the longitude of the perihelion of the six periods marked with an S are copied from a letter of Mr. Stone to Sir Charles Lyell, published in the Supplement of the Phil. Mag. for June, 1865; the eight periods marked L are copied from M. Leverrier's Table, to which reference has been made. For the correctness of everything else, both in this Table and in the other three, I alone am responsible.

Column IV. gives the number of degrees passed over by the perihelion during each 10,000 years. From this column it will be seen how irregular is the motion of the perihelion. At four different periods it had a retrograde motion for 20,000 years. Column V. shows the number of days by which the winter exceeds the summer when the winter occurs in aphelion. Column VI. shows the intensity of the sun's heat during mid-winter, when the winter occurs in aphelion, the present mid-winter intensity being taken at 1,000. These six columns comprehend all the astronomical part of the Tables. Regarding the correctness of the principles upon which these columns are constructed, there is no diversity of opinion. But these columns afford no direct information as to the character of the climate, or how much the temperature is increased or diminished. To find this we pass on to columns VII. and VIII., calculated on physical principles. Now, unless the physical principles upon which these three columns are calculated be wholly erroneous, change of eccentricity must undoubtedly very seriously affect climate. Column VII. shows how many degrees Fahrenheit the temperature is lowered by a decrease in the intensity of the sun's heat corresponding to column VI. For example, 850,000 years ago, if the winters occurred then in aphelion, the direct heat of the sun during midwinter would be only $\frac{837}{1000}$ of what it is at present at the same season of the year, and column VII. shows that this decrease in the intensity of the sun's heat would lower the temperature 45°·3 F.

The principle upon which this result is arrived at is this:—The temperature of space, as determined by Sir John Herschel, is —239° F. M. Pouillet, by a different method, arrived at almost the same result. If we take the midwinter temperature of Great Britain at 39°, then 239°+39°=278° will represent the number of degrees of rise due to the sun's heat at midwinter; in other words, it takes a quantity of sun-heat which we have represented by 1000 to maintain the temperature of the earth's surface in Great Britain 278° above the temperature of space. Were the sun extinguished, the temperature of our island would sink 278° below its present midwinter temperature, or to the temperature of space. But 850,000 years ago, as will be seen from Table III., if the winters occurred in aphelion, the heat of the sun at midwinter would only equal 837 instead of 1000 as at present. Consequently, if it takes 1,000 parts of heat to maintain the temperature 278° above the temperature of space, 837 parts of heat will only be able to maintain the temperature 232°·7 above the temperature of space; for 232°·7 is to 278 as 837 is to 1,000. Therefore, if the temperature was then only 232°·7 above that of space, it would be 45°·3 below what it is at present. This is what the temperature would be on the supposition, of course, that it depended wholly on the sun's intensity and was not modified by other causes. This method has already been discussed at some length in Chapter II. But whether these values be too high or too low, one thing is certain, that a very slight increase or a very slight decrease in the quantity of heat received from the sun must affect temperature to a considerable extent. The direct heat of the moon, for example, cannot be detected by the finest instruments which we possess; yet from 238,000 observations made at Prague during 1840-66, it would seem that the temperature is sensibly affected by the mere change in the lunar perigee and inclination of the moon's orbit.*

Column VIII. gives the midwinter temperature. It is found by subtracting the numbers in column VII. from 39°, the present midwinter temperature.

I have not given a Table showing the temperature of the summers at the corresponding periods. This could not well be done; for there is no relation at the periods in question between the intensity of the sun's heat and the temperature of the summers. One is apt to suppose, without due consideration, that the summers ought to be then as much warmer than they are at present, as the winters were then colder than now. Sir Charles Lyell, in his "Principles," has given a column of summer temperatures calculated from my table upon this principle. Astronomically the principle is correct, but physically, as was shown in Chapter IV., it is totally erroneous, and calculated to convey a wrong impression regarding the whole subject of geological climate. The summers at those periods, instead of being much warmer than they are at present, would in reality be much colder, notwithstanding the great increase in the intensity of the sun's heat resulting from the diminished distance of the sun.

What, then, is the date of the glacial epoch? It is perfectly obvious that if the glacial epoch resulted from a high state of eccentricity, it must be referred either to the period included

* See Professor C. V. Zenger's paper "On the Periodic Change of Climate caused by the Moon," Phil. Mag. for June, 1868.

in Table III. or to the one in Table IV. In Table III. we have a period extending from about 980,000 to about 720,000 years ago, and in Table IV. we have a period beginning about 240,000 years ago, and extending down to about 80,000 years ago. As the former period was of greater duration than the latter, and the eccentricity also attained to a higher value, I at first felt disposed to refer the glacial epoch proper (the time of the till and boulder clay) to the former period; and the latter period, I was inclined to believe, must have corresponded to the time of local glaciers towards the close of the glacial epoch, the evidence for which (moraines) is to be found in almost every one of our Highland glens. On this point I consulted several eminent geologists, and they all agreed in referring the glacial epoch to the former period; the reason assigned being that they considered the latter period to be much too recent and of too short duration to represent that epoch.

Pondering over the subject during the early part of 1866, reasons soon suggested themselves which convinced me that the glacial epoch must be referred to the latter and not to the former period. Those reasons I shall now proceed to state at some length, since they have a direct bearing, as will be seen, on the whole question of geological time.

It is the modern and philosophic doctrine of uniformity that has chiefly led geologists to over-estimate the length of geological periods. This philosophic school teaches, and that truly, that the great changes undergone by the earth's crust must have been produced, not by convulsions and cataclysms of nature, but by those ordinary agencies that we see at work every day around us, such as rain, snow, frost, ice, and chemical action, &c. It teaches that the valleys were not produced by violent dislocations, nor the hills by sudden upheavals, but that they were actually carved out of the solid rock by the silent and gentle agency of chemical action, frost, rain, ice, and running water. It teaches, in short, that the rocky face of our globe has been carved into hill and dale, and ultimately worn down to the sea-level, by means of these apparently trifling agents, not only once or twice, but probably dozens of times over during past ages. Now, when we reflect that with such extreme slowness do these agents perform their work, that we might watch their operations from year to year, and from century to century, if we could, without being able to perceive that they make any very sensible advance, we are necessitated to conclude that geological periods must be enormous. And the conclusion at which we thus arrive is undoubtedly correct. It is, in fact, impossible to form an adequate conception of the length of geological time. It is something too vast to be fully grasped by our minds. But here we come to the point where the fundamental mistake arises; Geologists do not err in forming too great a conception of the extent of geological periods, *but in the mode in which they represent the length of these periods in numbers.* When we speak of units, tens, hundreds, thousands, we can form some notion of what these quantities represent; but when we come to millions, tens of millions, hundreds of millions, thousands of millions, the mind is then totally unable to follow, and we can only use these numbers as representations of quantities that turn up in calculation. We know, from the way in which they do turn up in our process of calculation, whether they are correct representations of things in actual nature or not; but we could not, from a mere comparison of these quantities with the thing represented by them, say whether they were actually too small or too great.

At present, geological estimates of time are little else than mere conjectures. Geological science has hitherto afforded no trustworthy means of estimating the positive length of geological epochs. Geological phenomena tell us most emphatically that these periods must be long; but how long they have hitherto failed to inform us. Geological phenomena represent time to the mind under a most striking and imposing form. They present to the eye, as it were, a sensuous representation of time; the mind thus becomes deeply impressed with a sense of immense duration; and when one under these feelings is called upon to put down in figures what he believes will represent that duration, he is very apt to be deceived. If, for example, a million of years as represented by geological phenomena and a million of years as represented by figures were placed before our eyes, we should certainly feel startled. We should probably feel that a unit with six ciphers after it was really something far more formidable than we had hitherto supposed it to be. Could we stand upon the edge of a gorge a mile and a half in depth that had been cut out of the solid rock by a tiny stream, scarcely visible at the bottom of this fearful abyss, and were we informed that this little streamlet was able to wear off annually only $\frac{1}{10}$ of an inch from its rocky bed, what would our conceptions be of the prodigious length of time that this stream must have taken to excavate the gorge? We should certainly feel startled when, on making the necessary calculations, we found that the stream had performed this enormous amount of work in something less than a million of years.

If, for example, we could possibly form some adequate conception of a period so prodigious as one hundred millions of years, we should not then feel so dissatisfied with Sir W. Thomson's estimate that the age of the earth's crust is not greater than that.

Here is one way of conveying to the mind some idea of what a million of years really is. Take a narrow strip of paper an inch broad, or more, and 83 feet 4 inches in length, and stretch it along the wall of a large hall, or round the walls of an apartment somewhat over 20 feet square. Recall to memory the days of your boyhood, so as to get some adequate conception of what a period of a hundred years is. Then mark off from one of the ends of the strip $\frac{1}{10}$ of an inch The $\frac{1}{10}$ of the inch will then represent one hundred years, and the entire length of the strip a million of years. It is well worth making the experiment, just in order to feel the striking impression that it produces on the mind.

The latter period, which we have concluded to be that of the glacial epoch, extended, as we have seen, over a period of 160,000 years. But as the glaciation was only on one hemisphere at a time, 80,000 years or so would represent the united length of the cold periods. In order to satisfy ourselves that this period is sufficiently long to account for all the amount of denudation effected during the glacial epoch, let us make some rough estimate of the probable rate at which the surface of the country would be ground down by the ice. Suppose the ice to grind off only one-tenth of an inch annually this would give upwards of 650 feet as the quantity of rock removed during the time. But it is probable that it did not amount to one-fourth part of that quantity. Whether one-tenth of an inch per annum be an over-estimate or an under-estimate of the rate of denudation by the ice, it is perfectly evident that the period in question is sufficiently long, so far as denudation is concerned, to account for the phenomena of the glacial epoch.

But admitting that the period under consideration is sufficiently *long* to account for all the denudation which took place *during* the glacial epoch, we have yet to satisfy ourselves that it is also sufficiently *remote* to account for all the denudation which has taken place *since* the glacial epoch. Are the facts of geology consistent with the idea that the close of the glacial epoch does not date back beyond 80,000 years?

This question could be answered if we knew the present rate of subaërial denudation, for the present rate evidently does not differ greatly from that which has obtained since the close of the glacial epoch.

CHAPTER XX.

GEOLOGICAL TIME.—METHOD OF MEASURING THE RATE OF SUBAËRIAL DENUDATION.

Rate of Subaërial Denudation a Measure of Time.—Rate determined from Sediment of the Mississippi.—Amount of Sediment carried down by the Mississippi; by the Ganges.—Professor Geikie on Modern Denudation.—Professor Geikie on the Amount of Sediment conveyed by European Rivers.—Rate at which the Surface of the Globe is being denuded.—Alfred Tylor on the Sediment of the Mississippi.—The Law which determines the Rate of Denudation.—The Globe becoming less oblate.—Carrying Power of our River Systems the true Measure of Denudation.—Marine Denudation trifling in comparison to Subaërial.—Previous Methods of measuring Geological Time. —Circumstances which show the recent Date of the Glacial Epoch.—Professor Ramsay on Geological Time.

It is almost self-evident that the rate of subaërial denudation must be equal to the rate at which the materials are carried off the land into the sea, but the rate at which the materials are carried off the land is measured by the rate at which sediment is carried down by our river systems. *Consequently, in order to determine the present rate of subaërial denudation, we have only to ascertain the quantity of sediment annually carried down by the river systems.*

Knowing the quantity of sediment transported by a river, say annually, and the area of its drainage, we have the means of determining the rate at which the surface of this area is being lowered by subaërial denudation. And if we know this in reference to a few of the great continental rivers draining immense areas in various latitudes, we could then ascertain with tolerable correctness the rate at which the surface of the globe is being lowered by subaërial denudation, and also the length of time which our present continents can remain above the sea-level. Explaining this to Professor Ramsay during the winter of 1865, I learned from him that accurate measurements had been made of the amount of sediment annually carried down by the Mississippi River, full particulars of which investigations were to be found in the Proceedings of the American Association for the Advancement of Science for 1848. These proceedings contain a report by Messrs Brown and Dickeson, which unfortunately over-estimated the amount of sediment transported by the Mississippi by nearly four times what was afterwards found by Messrs Humphreys and Abbot to be the actual amount. From this estimate, I was led to the conclusion that if the Mississippi is a fair representative of rivers in general, our existing continents would not remain longer than one million and a half years above the sea-level.* This was a conclusion so startling as to excite suspicion that there must have been some mistake in reference to Messrs. Brown and Dickeson's data. It showed beyond doubt, however, that the rate of subaërial denudation, when accurately determined by this method, would be found to be enormously greater than had been supposed. Shortly afterwards, on estimating the rate from the data furnished by Humphreys and Abbot, I found the rate of denudation to be about one foot in 6,000 years. Taking the mean elevation of all the land as ascertained by Humboldt to be 1,000 feet, the whole would therefore be carried down into the ocean by our river systems in about 6,000,000 of years if no elevation of the land took place.† The following are the data and mode of computation by which this conclusion was arrived at. It was found by Messrs. Humphreys and Abbot that the average amount of sediment held in suspension in the waters of the Mississippi is about $\frac{1}{1500}$ of the weight of the water, or $\frac{1}{2900}$ by bulk. The annual discharge of the river is 19,500,000,000,000 cubic feet of water. The quantity of sediment carried down into the Gulf of Mexico amounts to 6,724,000,000 cubic feet. But besides that which is held in suspension, the river pushes down into the sea about 750,000,000 cubic feet of earthy matter, making in all a total of 7,474,000,000 cubic feet

* Phil. Mag. for February, 1867. † Phil. Mag. for May, 1868.

transferred from the land to the sea annually. Where does this enormous mass of material come from? Unquestionably it comes from the ground drained by the Mississippi. The area drained by the river is 1,244,000 square miles. Now 7,474,000,000 cubic feet removed off 1,224,000 square miles of surface is equal to $\frac{1}{4566}$ of a foot off that surface per annum, or one foot in 4,566 years. The specific gravity of the sediment is taken at 1·9, that of rock is about 2·5; consequently the amount removed is equal to one foot of rock in about 6,000 years. The average height of the North American continent above the sea-level, according to Humboldt, is 748 feet; consequently, at the present rate of denudation, the whole area of drainage will be brought down to the sea-level in less than 4,500,000 years, if no elevation of the land takes place.

Referring to the above, Sir Charles Lyell makes the following appropriate remarks:—"There seems no danger of our overrating the mean rate of waste by selecting the Mississippi as our example, for that river drains a country equal to more than half the continent of Europe, extends through twenty degrees of latitude, and therefore through regions enjoying a great variety of climate, and some of its tributaries descend from mountains of great height. The Mississippi is also more likely to afford us a fair test of ordinary denudation, because, unlike the St. Lawrence and its tributaries, there are no great lakes in which the fluviatile sediment is thrown down and arrested on its way to the sea."*

The rate of denudation of the area drained by the river Ganges is much greater than that of the Mississippi. The annual discharge of that river is 6,523,000,000,000 cubic feet of water. The sediment held in suspension is equal $\frac{1}{510}$ by weight; area of drainage 432,480 square miles. This gives one foot of rock in 2,358 years as the amount removed.

Rough estimates have been made of the amount of sediment carried down by some eight or ten European rivers; and although those estimates cannot be depended upon as being

* Student's "Elements of Geology," p. 91. Second Edition.

anything like perfectly accurate, still they show (what there is very little reason to doubt) that it is extremely probable that the European continent is being denuded about as rapidly as the American.

For a full account of all that is known on this subject I must refer to Professor Geikie's valuable memoir on Modern Denudation (Transactions of Geological Society of Glasgow, vol. iii.; also Jukes and Geikie's "Manual of Geology," chap. xxv.) It is mainly through the instrumentality of this luminous and exhaustive memoir that the method under consideration has gained such wide acceptance amongst geologists.

Professor Geikie finds that at the present rate of erosion the following is the number of years required by the under-mentioned rivers to remove one foot of rock from the general surface of their basins. Professor Geikie thus shows that the rate of denudation, as determined from the amount of sediment carried down the Mississippi, is certainly not too high.

Danube	6,846 years.
Mississippi	6,000 ,,
Nith	4,723 ,,
Ganges	2,358 ,,
Rhone	1,528 ,,
Hoang Ho	1,464 ,,
Po	729 ,,

By means of subaërial agencies continents are being cut up into islands, the islands into smaller islands, and so on till the whole ultimately disappears.

No proper estimate has been made of the quantity of sediment carried down into the sea by our British rivers. But, from the principles just stated, we may infer that it must be as great in proportion to the area of drainage as that carried down by the Mississippi. For example, the river Tay, which drains a great portion of the central Highlands of Scotland, carries to the sea three times as much water in proportion to its area of drainage as is carried by the Mississippi. And any one who has seen this rapidly running river during a flood, red and turbid with sediment, will easily be convinced that the quantity of solid material carried down by it into the German Ocean must be very great. Mr. John Dougall has found that the waters of the Clyde during a flood hold in suspension $\frac{1}{800}$ by bulk of sediment. The observations were made about a mile above the city of Glasgow. But even supposing the amount of sediment held in suspension by the waters of the Tay to be only one-third (which is certainly an under-estimate) of that of the Mississippi, viz. $\frac{1}{4500}$ by weight, still this would give the rate of denudation of the central Highlands at one foot in 6,000 years, or 1,000 feet in 6 millions of years.

It is remarkable that although so many measurements have been made of the amount of fluviatile sediment being transported seawards, yet that the bearing which this has on the broad questions of geological time and the rate of subaërial denudation should have been overlooked. One reason for this, no doubt, is that the measurements were made, not with a view to determine the rate at which the river basins are being lowered, but mainly to ascertain the age of the river deltas and the rate at which these are being formed.*

The Law which determines the Rate at which any Country is being denuded.—By means of subaërial agencies continents are being cut up into islands, the islands into smaller islands, and so on till the whole ultimately disappears.

So long as the present order of things remains, the rate of denudation will continue while land remains above the sea-level; and we have no warrant for supposing that the rate was during past ages less than it is at the present day. It will not do to object that, as a considerable amount of the sediment carried down by rivers is boulder clay and other materials belonging to

* In an interesting memoir, published in the Phil. Mag. for 1850, Mr. Alfred Tylor estimated that the basin of the Mississippi is being lowered at the rate of one foot in 10,000 years by the removal of the sediment; and he proceeds further, and reasons that one foot removed off the general surface of the land during that period would raise the sea-level three inches. Had it not been that Mr. Tylor's attention was directed to the effects produced by the removal of sediment in raising the level of the ocean rather than in lowering the level of the land, he could not have failed to perceive that he was in possession of a key to unfold the mystery of geological time.

the Ice age, the total amount removed by the rivers is on that account greater than it would otherwise be. Were this objection true, it would follow that, prior to the glacial period, when it is assumed that there was no boulder clay, the face of the country must have consisted of bare rock; for in this case no soil could have accumulated from the disintegration and decomposition of the rocks, *since, unless the rocks of a country disintegrate more rapidly than the river systems are able to carry the disintegrated materials to the sea, no surface soil can form on that country.* The rate at which rivers carry down sediment is evidently not determined by the rate at which the rocks are disintegrated and decomposed, but by the quantity of rain falling, and the velocity with which it moves off the face of the country. Every river system possesses a definite amount of carrying-power, depending upon the slope of the ground, the quantity of rain falling per annum, the manner in which the rain falls, whether it falls gradually or in torrents, and a few other circumstances. When it so happens, as it generally does, that the amount of rock disintegrated on the face of the country is greater than the carrying-power of the river systems can remove, then a soil necessarily forms. But when the reverse is the case no soil can form on that country, and it will present nothing but barren rock. This is no doubt the reason why in places like the Island of Skye, for example, where the rocks are exceedingly hard and difficult to decompose and separate, the ground steep, and the quantity of rain falling very great, there is so much bare rock to be seen. If, prior to the glacial epoch, the rocks of the area drained by the Mississippi did not produce annually more material from their destruction under atmospheric agency than was being carried down by that river, then it follows that the country must have presented nothing but bare rock, if the amount of rain falling then was as great as at present.

But, after all, one foot removed off the general level of the country since the creation of man, according to Mosaic chronology, is certainly not a very great quantity. No person but one who had some preconceived opinions to maintain, would ever think of concluding that one foot of soil during 6,000 years was an extravagant quantity to be washed off the face of the country by rain and floods during that long period. Those who reside in the country and are eye-witnesses of the actual effects of heavy rains upon the soil, our soft country roads, ditches, brooks, and rivers, will have considerable difficulty in actually believing that only one foot has been washed away during the past 6,000 years.

Some may probably admit that a foot of soil may be washed off during a period so long as 6,000 years, and may tell us that what they deny is not that a foot of loose and soft soil, but a foot of solid rock can be washed away during that period. But a moment's reflection must convince them that, unless the rocks of the country were disintegrating and decomposing as rapidly into soil as the rain is carrying the soil away, the surface of the country would ultimately become bare rock. It is true that the surface of our country in many places is protected by a thick covering of boulder clay; but when this has once been removed, the rocks will then disintegrate far more rapidly than they are doing at present.

But slow as is the rate at which the country is being denuded, yet when we take into consideration a period so enormous as 6 millions of years, we find that the results of denudation are really startling. One thousand feet of solid rock during that period would be removed from off the face of the country. But if the mean level of the country would be lowered 1,000 feet in 6 millions of years, how much would our valleys and glens be deepened during that period? This is a problem well worthy of the consideration of those who treat with ridicule the idea that the general features of our country have been carved out by subaërial agency.

In consequence of the retardation of the earth's rotation, occasioned by the friction of the tidal-wave, the sea-level must be slowly sinking at the equator and rising at the poles. But it is probable that the land at the equator is being lowered by denudation as rapidly as the sea-level is sinking. *Nearly one mile must have been worn off the equator during the past* 12 *millions of years,* if the rate of denudation all along the equator be equal to that of the basin of the Ganges. It therefore follows that we cannot infer from the present shape of our globe what was its form, or the rate at which it was rotating, at the time when its crust became solidified. Although it had been as oblate as the planet Jupiter, denudation must in time have given it its present form.

There is another effect which would result from the denudation of the equator and the sinking of the ocean at the equator and its rise at the poles. This, namely, that it would tend to increase the rate of rotation; or, more properly, it would tend to *lessen* the rate of tidal retardation.

But if the rate of denudation be at present so great, what must it have been during the glacial epoch? It must have been something enormous. At present, denudation is greatly retarded by the limited power of our river systems to remove the loose materials resulting from the destruction of the rocks. These materials accumulate and form a thick soil over the surface of the rocks, which protects them, to a great extent, from the weathering effects of atmospheric agents. So long as the amount of rock disintegrated exceeds that which is being removed by the river systems, the soil will continue to accumulate till the amount of rock destroyed per annum is brought to equal that which is being removed. It therefore follows from this principle that the CARRYING-POWER OF OUR RIVER SYSTEMS IS THE TRUE MEASURE OF DENUDATION. But during the glacial epoch the thickness of the soil would have but little effect in diminishing the waste of the rocks; for at that period the rocks were not decomposed by atmospheric agency, but were ground down by the mechanical friction of the ice. But the presence of a thick soil at this period, instead of retarding the rate of denudation, would tend to increase it tenfold, for the soil would then be used as grinding-material for the ice-sheet. In places where the ice was, say, 2,000 feet in thickness, the soil would be forced along over the rocky face of the country, exerting a pressure on the rocks equal to 50 tons on the square foot.

It is true that the rate at which many kinds of rocks decompose and disintegrate is far less than what has been concluded to be the mean rate of denudation of the whole country. This is evident from the fact which has been adduced by some writers, that inscriptions on stones which have been exposed to atmospheric agency for a period of 2,000 years or so, have not been obliterated. But in most cases epitaphs on monuments and tombstones, and inscriptions on the walls of buildings, 200 years old, can hardly be read. And this is not all: the stone on which the letters were cut has during that time rotted in probably to the depth of several inches; and during the course of a few centuries more the whole mass will crumble into dust.

The facts which we have been considering show also how trifling is the amount of denudation effected by the sea in comparison with that by subaërial agents. The entire sea-coast of the globe, according to Dr. A. Keith Johnston, is 116,531 miles. Suppose we take the average height of the coast-line at 25 feet, and take also the rate at which the sea is advancing on the land at one foot in 100 years, then this gives 15,382,500,000 cubic feet of rock as the total amount removed in 100 years by the action of the sea. The total amount of land is 57,600,000 square miles, or 1,605,750,000,000,000 square feet; and if one foot is removed off the surface in 6,000 years, then 26,763,000,000,000 cubic feet is removed by subaërial agency in 100 years, or about 1,740 times as much as that removed by the sea. Before the sea could denude the globe as rapidly as the subaërial agents, it would have to advance on the land at the rate of upwards of 17 feet annually.

It will not do, however, to measure marine denudation by the rate at which the sea is advancing on the land. There is no relation whatever between the rate at which the sea is *advancing* on the land and the rate at which the sea is *denuding* the land. For it is evident that as the subaërial agents bring the coast down to the sea-level, all that the sea has got to do is simply to advance, or at most to remove the loose materials which may lie in its path. The amount of denudation which has been effected by the sea during past geological ages, compared with what has been effected by subaërial agency, is evidently but trifling. Denudation is not the proper function of the sea. The great denuding agents are land-ice, frost, rain, running-water, chemical agency, &c. The proper work which belongs to the sea is the transporting of the loose materials carried down by the rivers, and the spreading of these out so as to form the stratified beds of future ages.

Previous Methods of measuring Geological Time unreliable.—The method which has just been detailed of estimating the rate of subaërial denudation seems to afford the only reliable means of a geological character of determining geological time in absolute measure. The methods which have hitherto been adopted not only fail to give the positive length of geological periods, but some of them are actually calculated to mislead.

The common method of calculating the length of a period from the thickness of the stratified rocks belonging to that period is one of that class. Nothing whatever can be inferred from the thickness of a deposit as to the length of time which was required to form it. The thickness of a deposit will depend upon a great many circumstances, such as whether the deposition took place near to land or far away in the deep recesses of the ocean, whether it occurred at the mouth of a great river or along the sea-shore, or at a time when the sea-bottom was rising, subsiding, or remaining stationary. Stratified formations 10,000 feet in thickness, for example, may, under some conditions, have been formed in as many years, while under other conditions it may have required as many centuries. Nothing whatever can be safely inferred as to the absolute length of a period from the thickness of the stratified formations belonging to that period. Neither will this method give us a trustworthy estimate of the *relative* lengths of geological periods. Suppose we find the average thickness of the Cambrian rocks to be, say, 26,000 feet, the Silurian to be 28,000 feet, the Devonian to be 6,000 feet, and the Tertiary to be 10,000 feet, it would not be safe to assume, as is sometimes done, that the relative duration of those periods must have corresponded to these numbers. Were we sure that we had got the correct average thickness of all the rocks belonging to each of those formations, we might probably be able to arrive at the relative lengths of those periods; but we can never be sure of this. Those formations all, at one time, formed sea-bottoms; and we can only measure such deposits as are now raised above the sea-level. But is not it probable that the relative positions of sea and land during the Cambrian, Silurian, Old Red Sandstone, Carboniferous, and other early periods of the earth's history, differed more from the present than the distribution of sea and land during the Tertiary period differed from that which obtains now? May not the greater portion of the Tertiary deposits be still under the sea-bottom? And if this be the case, it may yet be found at some day in the distant future, when these deposits are elevated into dry land, that they are much thicker than we now conclude them to be. Of course, it is by no means asserted that this is so, but only that they *may* be thicker for anything we know to the contrary; and the possibility that they may, destroys our confidence in the accuracy of this method of determining the relative lengths of geological periods.

Neither does palæontology afford any better mode of measuring geological time. In fact, the palæontological method of estimating geological time, either absolute or relative, from the rate at which species change, appears to be even still more unsatisfactory. If we could ascertain by some means or other the time that has elapsed from some given epoch (say, for example, the glacial) till the present day, and were we sure at the same time that species have changed at a uniform rate during all past ages, then, by ascertaining the percentage of change that has taken place since the glacial epoch, we should have a means of making something like a rough estimate of the length of the various periods. But without some such period to start with, the palæontological method is useless. It will not do to take the historic period as a base-line. It is far too short to be used with safety in determining the distance of periods so remote as those which concern the geologist. But even supposing the palæontologist had a period of sufficient length measured off correctly to begin with, his results would still be unsatisfactory; for it is perfectly obvious, that unless the climatic conditions of the globe during the various periods were nearly the same, the rate at which the species change would certainly not be uniform; but such has not been the case, as an examination of the Tables of eccentricity will show. Take, for example, that long epoch of 260,000 years, beginning about 980,000 years ago and terminating about 720,000 years ago. During that long period the changes from cold to warm conditions of climate every 10,000 or 12,000 years must have been of the most extreme character. Compare that period with the period beginning, say, 80,000 years ago, and extending to nearly 150,000 years into the future, during which there will be no extreme variations of climate, and how great is the contrast! How extensive the changes in species must have been during the first period as compared with those which are likely to take place during the latter!

Besides, it must also be taken into consideration that organization was of a far more simple type in the earlier Palæozoic ages than during the Tertiary period, and would probably on this account change much more slowly in the former than in the latter.

The foregoing considerations render it highly probable, if not certain, that the rate at which the general surface of the globe is being lowered by subaërial denudation cannot be much under one foot in 6,000 years. Now, if we assign the glacial epoch to that period of high eccentricity beginning 980,000 years ago, and terminating 720,000 years ago, then we must conclude that as much as 120 feet must have been denuded off the face of the country since the close of the glacial epoch.

But if as much as this had been carried down by our rivers into the sea, hardly a patch of boulder clay, or any trace of the glacial epoch, should be now remaining on the land. It is therefore evident that the glacial epoch cannot be assigned to that remote period, but ought to be referred to the period terminating about 80,000 years ago. We have, in this latter case, 13 feet, equal to about 18 feet of drift, as the amount removed from the general surface of the country since the glacial epoch. This amount harmonizes very well with the direct evidence of geology on this point. Had the amount of denudation since the close of the glacial epoch been much greater than this, the drift deposits would not only have been far less complete, but the general appearance and outline of the surface of all glaciated countries would have been very different from what they really are.

Circumstances which show the Recent Date of the Glacial Epoch.—One of the circumstances to which I refer is this. When we examine the surface of any glaciated country, such as Scotland, we can easily satisfy ourselves that the upper surface of the ground differs very much from what it would have been had its external features been due to the action of rain and rivers and the ordinary agencies which have been at work since the close of the Ice period. Go where one will in the Lowlands of Scotland, and he shall hardly find a single acre whose upper surface bears the marks of being formed by the denuding agents which are presently in operation. He will observe everywhere mounds and hollows, the existence of which cannot be accounted for by the present agencies at work. In fact these agencies are slowly denuding pre-existing heights and silting up pre-existing hollows. Everywhere one comes upon patches of alluvium which upon examination prove to be simply old glacially formed hollows silted up. True, the main rivers, streams, and even brooks, occupy channels which have been formed by running water, either since or prior to the glacial epoch, but, in regard to the general surface of the country, the present agencies may be said to be just beginning to carve a new line of features out of the old glacially formed surface.

But so little progress has yet been made, that the kames, gravel mounds, knolls of boulder clay, &c., still retain in most cases their original form. Now, when we reflect that more than a foot of drift is being removed from the general surface of the country every 5,000 years or so, it becomes perfectly obvious that the close of the glacial epoch must be of comparatively recent date.

There is another circumstance which shows that the glacial epoch must be referred to the latest period of great eccentricity. If we refer the glacial epoch to the penultimate period of extreme eccentricity, and place its commencement one million of years back, then we must also lengthen out to a corresponding extent the entire geological history of the globe. Sir Charles Lyell, who is inclined to assign the glacial epoch to this penultimate period, considers that when we go back as far as the Lower Miocene formations, we arrive at a period when the marine shells differed as a whole from those now existing. But only 5 per cent. of the shells existing at the commencement of the glacial epoch have since died out. Hence, assuming the rate at which the species change to be uniform, it follows that the Lower Miocene period must be twenty times as remote as the commencement of the glacial epoch. Consequently, if it be one million of years since the commencement of the glacial epoch, 20 millions of years, Sir Charles concludes, must have elapsed since the time of the Lower Miocene period, and 60 millions of years since the beginning of the Eocene period, and about 160 millions of years since the Carboniferous period, and about 240 millions of years must be the time which has elapsed since the beginning of the Cambrian period. But, on the other hand, if we refer the glacial epoch to the latest period of great eccentricity, and take 250,000 years ago as the beginning of that period, then, according to the same mode of calculation, we have 15 millions of years since the beginning of the Eocene period, and 40 millions of years since the Carboniferous period, and 60 millions of years in all since the beginning of the Cambrian period.

If the beginning of the glacial epoch be carried back a million years, then it is probable, as Sir Charles Lyell concludes, that the beginning of the Cambrian period will require to be placed 240 millions of years back. But it is very probable that the length of time embraced by the pre-Cambrian ages of geological history may be as great as that which has elapsed since the close of the Cambrian period, and, if this be so, then we shall be compelled to admit that nearly 500 millions of years have passed away since the beginning of the earth's geological history. But we have evidence of a physical nature which proves that it is absolutely impossible that the existing order of things, as regards our globe, can date so far back as anything like 500 millions of years. The arguments to which I refer are those which have been advanced by Professor Sir William Thomson at various times. These arguments are well known, and to all who have really given due attention to them must be felt to be conclusive. It would be superfluous to state them here; I shall, however, for reasons which will presently appear, refer briefly to one of them, and that one which seems to be the most conclusive of all, viz., the argument derived from the limit to the age of the sun's heat.

Professor Ramsay on Geological Time.—In an interesting suggestive memoir, "On Geological Ages as items of Geological Time,"* Professor Ramsay discusses the comparative values of certain groups of formations as representative of geological time, and arrives at the following general conclusion, viz., "That the local continental era which began with the Old Red Sandstone and closed with the New Red Marl is comparable, in point of geological time, to that occupied in the deposition of the whole of the Mesozoic, or Secondary series, later than the New Red Marl and all the Cainozoic or Tertiary formations, and indeed of all the time that has elapsed since the beginning of the deposition of the Lias down to the present day." This conclusion is derived partly from a comparison of the physical character of the formations constituting each

* Proc. Roy. Soc., No. 152, 1874.

group, but principally from the zoological changes which took place during the time represented by them.

The earlier period represented by the Cambrian and Silurian rocks he also, from the same considerations, considers to have been very long, but he does not attempt to fix its relative length. Of the absolute length of any or all of these great eras of geological time no estimate or guess is given. He believes, however, that the whole time represented by all the fossiliferous rocks, from the earliest Cambrian to the most recent, is, geologically speaking, short compared with that which went before it. After quoting Professor Huxley's enumeration of the many classes and orders of marine life (identical with those still existing), whose remains characterize the lowest Cambrian rocks, he says, "The inference is obvious that in this earliest known varied life we find no evidence of its having lived near the beginning of the zoological series. In a broad sense, compared with what must have gone before, both biologically and physically, all the phenomena connected with this old period seem to my mind to be quite of a recent description, and the climates of seas and lands were of the very same kind as those that the world enjoys at the present day." . . . "In the words of Darwin, when discussing the imperfection of the geological record of this history, 'we possess the last volume alone relating only to two or three countries,' and the reason why we know so little of pre-Cambrian faunas and the physical characters of the more ancient formations as originally deposited, is that below the Cambrian strata we get at once involved in a sort of chaos of metamorphic strata.' "

It seems to me that Professor Ramsay's results lead to the same conclusion regarding the *positive* length of geological periods as those derived from physical considerations. It is true that his views lead us back to an immense lapse of unknown time prior to the Cambrian period, but this practically tends to shorten geological periods. For it is evident that the geological history of our globe must be limited by the age of the sun's heat, no matter how long or short its age may be. This being the case, the greater the length of time which must have elapsed prior to the Cambrian period, the less must be the time which has elapsed since that period. Whatever is added to the one period must be so much taken from the other. Consequently, the longer we suppose the pre-Cambrian periods to have been, the shorter must we suppose the post-Cambrian to be.

CHAPTER XXI.

THE PROBABLE AGE AND ORIGIN OF THE SUN.

Gravitation Theory.—Amount of Heat emitted by the Sun.—Meteoric Theory.—Helmholtz's Condensation Theory.—Confusion of Ideas.—Gravitation not the chief Source of the Sun's Heat.—Original Heat.—Source of Original Heat.—Original Heat derived from Motion in Space.—Conclusion as to Date of Glacial Epoch.—False Analogy.—Probable Date of Eocene and Miocene Periods.

Gravitation Theory of the Origin and Source of the Sun's Heat.—There are two forms in which this theory has been presented: the first, the meteoric theory, propounded by Dr. Meyer, of Heilbronn; and the second, the contraction theory, advocated by Helmholtz.

It is found that 83·4 foot-pounds of heat per second are incident upon a square foot of the earth's surface exposed to the perpendicular rays of the sun. The amount radiated from a square foot of the sun's surface is to that incident on a square foot of the earth's surface as the square of the sun's distance to the square of his radius, or as 46,400 to 1. Consequently 3,869,000 foot-pounds of heat are radiated off every square foot of the sun's surface per second—an amount equal to about 7,000 horse power. The total amount radiated from the whole surface of the sun per annum is $8{,}340 \times 10^{30}$ foot-pounds. To maintain the present rate of radiation, it would require the combustion of about 1,500 lbs. of coal per hour on every square foot of the sun's surface; and were the sun composed of that material, it would be all consumed in less than 5,000 years. The opinion that the sun's heat is maintained by combustion cannot be entertained for a single moment. A pound of coal falling into the sun from an infinite distance would produce by its concussion more than 6,000 times the amount of heat that would be generated by its combustion.

It is well known that the velocity with which a body falling from an infinite distance would reach the sun would be equal to that which would be generated by a constant force equal to the weight of the body at the sun's surface operating through a space equal to the sun's radius. One pound would at the sun's surface weigh about 28 pounds. Taking the sun's radius at 441,000 miles,* the energy of a pound of matter falling into the sun from infinite space would equal that of a 28-pound weight descending upon the earth from an elevation of 441,000 miles, supposing the force of gravity to be as great at that elevation as it is at the earth's surface. It would amount to upwards of 65,000,000,000 foot-pounds. A better idea of this enormous amount of energy exerted by a one-pound weight falling into the sun will be conveyed by stating that it would be sufficient to raise 1,000 tons to a height of 5½ miles. It would project the *Warrior*, fully equipped with guns, stores, and ammunition, over the top of Ben Nevis.

Gravitation is now generally admitted to be the only conceivable source of the sun's heat. But if we attribute the energy of the sun to gravitation as a source, we assign it to a cause the value of which can be accurately determined. Prodigious as is the energy of a single pound of matter falling into the sun, nevertheless a range of mountains, consisting of 176 cubic miles of solid rock, falling into the sun, would maintain his heat for only a single second. A mass equal to that of the earth would maintain the heat for only 93 years, and a mass equal to that of the sun itself falling into the sun would afford but 33,000,000 years' sun-heat.

It is quite possible, however, that a meteor may reach the sun with a velocity far greater than that which it could acquire by gravitation; for it might have been moving in a direct line towards the sun with an original velocity before coming under

* I have taken for the volume and mass of the sun the values given in Professor Sir William Thomson's memoir, Phil. Mag., vol. viii. (1854).

the sensible influence of the sun's attraction. In this case a greater amount of heat would be generated by the meteor than would have resulted from its merely falling into the sun under the influence of gravitation. But then meteors of this sort must be of rare occurrence. The meteoric theory of the sun's heat has now been pretty generally abandoned for the contraction theory advanced by Helmholtz.

Suppose, with Helmholtz, that the sun originally existed as a nebulous mass, filling the entire space presently occupied by the solar system and extending into space indefinitely beyond the outermost planet. The total amount of work in foot-pounds performed by gravitation in the condensation of this mass to an orb of the sun's present size can be found by means of the following formula given by Helmholtz,*

$$\text{Work of condensation} = \frac{3}{5} \cdot \frac{r^2 M^2}{Rm} \cdot g$$

M is the mass of the sun, m the mass of the earth, R the sun's radius, and r the earth's radius. Taking $M = 4230 \times 10^{27}$ lbs., $m = 11{,}920 \times 10^{21}$ lbs., $R = 2{,}328{,}500{,}000$ feet, and $r = 20{,}889{,}272$ feet; we have then for the total amount of work performed by gravitation in foot-pounds,

$$\text{Work} = \frac{3}{5} \cdot \frac{(20{,}889{,}272 \cdot 5)^2 \times (4230 \times 10^{27})^2}{2{,}328{,}500{,}000 \times 11{,}920 \times 10^{21}}$$
$$= 168{,}790 \times 10^{36} \text{ foot-pounds.}$$

The amount of heat thus produced by gravitation would suffice for nearly 20,237,500 years.

These calculations are based upon the assumption that the density of the sun is uniform throughout. But it is highly probable that the sun's density increases towards the centre, in which case the amount of work performed by gravitation would be somewhat more than the above.

Some confusion has arisen in reference to this subject by the introduction of the question of the amount of the sun's specific heat. If we simply consider the sun as an incandescent body

* Phil. Mag., § 4, vol. xi., p. 516 (1856).

in the process of cooling, the question of the amount of the sun's specific heat is of the utmost importance; because the absolute amount of heat which the sun is capable of giving out depends wholly upon his temperature and specific heat. In this case three things only are required: (1), the sun's mass; (2), temperature of the mass; (3), specific heat of the mass. But if we are considering what is the absolute amount of heat which could have been given out by the sun on the hypothesis that gravitation, either according to the meteoric theory suggested by Meyer or according to the contraction theory advocated by Helmholtz, is the only source of his heat, then we have nothing whatever to do with any inquiries regarding the specific heat of the sun. This is evident because the absolute amount of work which gravitation can perform in the pulling of the particles of the sun's mass together, is wholly independent of the specific heat of those particles. Consequently, the amount of energy in the form of heat thus imparted to the particles by gravity must also be wholly independent of specific heat. That is to say, the amount of heat imparted to a particle will be the same whatever may be its specific heat.

Even supposing we limit the geological history of our globe to 100 millions of years, it is nevertheless evident that gravitation will not account for the supply of the sun's heat during so long a period. There must be some other source of much more importance than gravitation. What other source of energy greater than that of gravitation can there be? It is singular that the opinion should have become so common even among physicists, that there is no other conceivable source than gravitation from which a greater amount of heat could have been derived.

The Origin and Chief Source of the Sun's Heat.—According to the foregoing theories regarding the source of the sun's heat, it is assumed that the matter composing the sun, when it existed in space as a nebulous mass, was not originally possessed of temperature, but that the temperature was given to it as the mass became condensed under the force of gravitation. It is supposed that the heat given out was simply the heat of condensation. But it is quite conceivable that the nebulous mass might have been possessed of an original store of heat previous to condensation.

It is quite possible that the very reason why it existed in such a rarefied or gaseous condition was its excessive temperature, and that condensation only began to take place when the mass began to cool down. It seems far more probable that this should have been the case than that the mass existed in so rarefied a condition without temperature. For why should the particles have existed in this separated form when devoid of the repulsive energy of heat, seeing that in virtue of gravitation they had such a tendency to approach to one another? But if the mass was originally in a heated condition, then in condensing it would have to part not only with the heat generated in condensing, but also with the heat which it originally possessed, a quantity which would no doubt much exceed that produced by condensation. To illustrate this principle, let us suppose a pound of air, for example, to be placed in a cylinder and heat applied to it. If the piston be so fixed that it cannot move, 234·5 foot-pounds of heat will raise the temperature of the air 1° C. But if the piston be allowed to rise as the heat is applied, then it will require 330·2 foot-pounds of heat to raise the temperature 1° C. It requires 95·7 foot-pounds more heat in the latter case than in the former. The same amount of energy, viz., 234·5 foot-pounds, in both cases goes to produce temperature; but in the latter case, where the piston is allowed to move, 95·7 foot-pounds of additional heat are consumed in the mechanical work of raising the piston. Suppose, now, that the air is allowed to cool under the same conditions: in the one case 234·5 foot-pounds of heat will be given out while the temperature of the air sinks 1° C.; in the other case, where the piston is allowed to descend, 330·2 foot-pounds will be given out while the temperature sinks 1° C. In the former case, the air in cooling has simply to part with the energy which it possesses in the form of temperature; but in the latter case it has, in addition to this, to part with the energy bestowed upon its molecules by the descending piston. While the temperature of the gas is sinking 1°, 95·7 foot-pounds of energy in the form of heat are being imparted to it by the descending piston; and these have to be got rid of before the temperature is lowered by 1°. Consequently 234·5 foot-pounds of the heat given out previously existed in the air under the form of temperature, and the remaining 95·7 foot-pounds given out were imparted to the air by the descending piston while the gas was losing its temperature. 234·5 foot-pounds represent the energy or heat which the air previously possessed, and 95·7 the energy or heat of condensation.

In the case of the cooling of the sun from a nebulous mass, there would of course be no external force or pressure exerted on the mass analogous to that of the piston on the air; but there would be, what is equivalent to the same, the gravitation of the particles to each other. There would be the pressure of the whole mass towards the centre of convergence. In the case of air, and all perfect gases cooling under pressure, about 234 foot-pounds of the original heat possessed by the gas are given out while 95 foot-pounds are being generated by condensation. We have, however, no reason whatever to believe that in the case of the cooling of the sun the same proportions would hold true. The proportion of original heat possessed by the mass of the sun to that produced by condensation may have been much greater than 234 to 95, or it may have been much less. In the absence of all knowledge on this point, we may in the meantime assume that to be the proportion. The total quantity of heat given out by the sun resulting from the condensation of his mass, on the supposition that the density of the sun is uniform throughout, we have seen to be equal to 20,237,500 years' sun-heat. Then the quantity of heat given out, which previously existed in the mass as original temperature, must have been 49,850,000 years' heat, making in all 70,087,500 years' heat as the total amount.

The above quantity represents, of course, the total amount of heat given out by the mass since it began to condense. But the geological history of our globe must date its beginning at a period posterior to that. For at that time the mass would probably occupy a much greater amount of space than is presently possessed by the entire solar system; and consequently, before it had cooled down to within the limits of the earth's present orbit, our earth could not have had an existence as a separate planet. Previously to that time it must have existed as a portion of the sun's fiery mass. If we assume that it existed as a globe previously to that, and came in from space after the condensation of the sun, then it is difficult to conceive how its orbit should be so nearly circular as it is at present.

Let us assume that by the time that the mass of the sun had condensed to within the space encircled by the orbit of the planet Mercury (that is, to a sphere having, say, a radius of 18,000,000 miles) the earth's crust began to form; and let this be the time when the geological history of our globe dates its commencement. The total amount of heat generated by the condensation of the sun's mass from a sphere of this size to its present volume would equal 19,740,000 years' sun-heat. The amount of original heat given out during that time would equal 48,625,000 years' sun-heat,—thus giving a total of 68,365,000 years' sun-heat enjoyed by our globe since that period. The total quantity may possibly, of course, be considerably more than that, owing to the fact that the sun's density may increase greatly towards his centre. But we should require to make extravagant assumptions regarding the interior density of the sun and the proportion of original heat to that produced by condensation before we could manage to account for anything like the period that geological phenomena are supposed by some to demand.

The question now arises, by what conceivable means could the mass of the sun have become possessed of such a prodigious amount of energy in the form of heat previous to condensation?

What power could have communicated to the mass 50,000,000 years' heat before condensation began to take place?

The Sun's Energy may have originally been derived from Motion in Space.—There is nothing at all absurd or improbable in the supposition that such an amount of energy might have been communicated to the mass. The Dynamical Theory of Heat affords an easy explanation of at least *how* such an amount of energy *may* have been communicated. Two bodies, each one-half the mass of the sun, moving directly towards each other with a velocity of 476 miles per second, would by their concussion generate in a *single moment* the 50,000,000 years' heat. For two bodies of that mass moving with a velocity of 476 miles per second would possess 4149×10^{38} foot-pounds of energy in the form of *vis viva;* and this, converted into heat by the stoppage of their motion, would give an amount of heat which would cover the present rate of the sun's radiation, for a period of 50,000,000 years.

Why may not the sun have been composed of two such bodies? And why may not the original store of heat possessed by him have all been derived from the concussion of these two bodies? Two such bodies coming into collision with that velocity would be dissipated into vapour by such an inconceivable amount of heat as would thus be generated; and when they condensed on cooling, they would form one spherical mass like the sun. It is perfectly true that two such bodies could never attain the required amount of velocity by their mutual gravitation towards each other. But there is no necessity whatever for supposing that their velocities were derived from their mutual attraction alone. They might have been approaching towards each other with the required velocity wholly independent of gravitation.

We know nothing whatever regarding the absolute motion of bodies in space. And beyond the limited sphere of our observation, we know nothing even of their relative motions. There may be bodies moving in relation to our system with inconceivable velocity. For anything that we know to the contrary, were one of these bodies to strike our earth, the shock might be sufficient to generate an amount of heat that would dissipate the earth into vapour, though the striking body might not be heavier than a cannon-ball. There is, however, nothing very extraordinary in the velocity which we have found would be required in the two supposed bodies to generate the 50,000,000 years' heat. A comet, having an orbit extending to the path of the planet Neptune, approaching so near the sun as to almost graze his surface in passing, would have a velocity of about 390 miles per second, which is within 86 miles of the required velocity.

But in the original heating and expansion of the sun into a gaseous mass, an amount of work must have been performed against gravitation equal to that which has been performed by gravitation during his cooling and condensation, a quantity which we have found amounts to about 20,000,000 years' heat. The total amount of energy originally communicated by the concussion must have been equal to 70,000,000 years' sun-heat. A velocity of 563 miles per second would give this amount. It must be borne in mind, however, that the 563 miles per second is the velocity at the moment of collision; about one-half of this velocity would be derived from the mutual attraction of the two bodies in their approach to each other. Suppose each body to be equal in volume to the sun, and of course one-half the density, the amount of velocity which they would acquire by their mutual attraction would be 274 miles per second, consequently we have to assume an original or projected velocity of only 289 miles per second.

If we admit that gravitation is not sufficient to account for the amount of heat given out by the sun during the geological history of our globe, we are compelled to assume that the mass of which the sun is composed existed prior to condensation in a heated condition; and if so, we are further obliged to admit that the mass must have received its heat from some source or other. And as the dissipation of heat into space must have been going on, in all probability, as rapidly before as after condensation took place, we are further obliged to conclude that the heat must have been communicated to the mass immediately before condensation began, for the moment the mass began to lose its heat condensation would ensue. If we confine our speculations to causes and agencies known to exist, the cause which has been assigned appears to be the only conceivable one that will account for the production of such an enormous amount of heat.

The general conclusion to which we are therefore led from physical considerations regarding the age of the sun's heat is, that the entire geological history of our globe must be comprised within less than 100 millions of years, and that consequently the commencement of the glacial epoch cannot date much farther back than 240,000 years.

The facts of geology, more especially those in connection with denudation, seem to geologists to require a period of much longer duration than 100 millions of years, and it is this which has so long prevented them accepting the conclusions of physical science in regard to the age of our globe. But the method of measuring subaërial denudation already detailed seems to me to show convincingly that the geological data, when properly interpreted, are in perfect accord with the deductions of physical science. Perhaps there are now few who have fairly considered the question who will refuse to admit that 100 millions of years are amply sufficient to comprise the whole geological history of our globe.

A false Analogy supposed to exist between Astronomy and Geology.—Perhaps one of the things which has tended to mislead on this point is a false analogy which is supposed to subsist between astronomy and geology, viz., that geology deals with unlimited *time,* as astronomy deals with unlimited *space.* A little consideration, however, will show that there is not much analogy between the two cases.

Astronomy deals with the countless worlds which lie spread out in the boundless infinity of space; but geology deals with only one world. No doubt reason and analogy both favour the idea that the age of the material universe, like its magnitude, is immeasurable; we have no reason, however, to conclude that it is eternal, any more than we have to infer that it is infinite. But when we compare the age of the material universe with its magnitude, we must not take the age of one of its members (say, our globe) and compare it with the size of the universe. Neither must we compare the age of all the presently existing systems of worlds with the magnitude of the universe; but we must compare the past history of the universe as it stretches back into the immensity of bygone *time*, with the presently existing universe as it stretches out on all sides into limitless *space*. For worlds precede worlds in time as worlds lie beyond worlds in space. Each world, each individual, each atom is evidently working out a final purpose, according to a plan prearranged and predetermined by the Divine Mind from all eternity. And each world, like each individual, when it serves the end for which it was called into existence, disappears to make room for others. This is the grand conception of the universe which naturally impresses itself on every thoughtful mind that has not got into confusion about those things called in science the Laws of Nature.*

But the geologist does not pass back from world to world as they stand related to each other in the order of *succession in time*, as the astronomer passes from world to world as they stand related to each other in the order of *coexistence in space*. The researches of the geologist, moreover, are not only confined to one world, but it is only a portion of the history of that one world that can come under his observation. The oldest of existing formations, so far as is yet known, the Laurentian Gneiss, is made up of the waste of previously existing rocks, and it, again, has probably been derived from the degradation of rocks belonging to some still older period. Regarding what succeeds these old Laurentian rocks geology tells us much; but of the formations that preceded, we know nothing whatever. For anything that geology shows to the contrary, the time which may have elapsed from the solidifying of the earth's

* Phil. Mag. for July, 1872, p. 1.

crust to the deposition of the Laurentian strata—an absolute blank—may have been as great as the time that has since intervened.

Probable Date of the Eocene and Miocene Periods.—If we take into consideration the limit which physical science assigns to the age of our globe, and the rapid rate at which, as we have seen, denudation takes place, it becomes evident that the enormous period of 3 millions of years comprehended in the foregoing tables must stretch far back into the Tertiary age. Supposing that the mean rate of denudation during that period was not greater than the present rate of denudation, still we should have no less than 500 feet of rock worn off the face of the country and carried into the sea during these 3 millions of years. This fact shows how totally different the appearance and configuration of the country in all probability was at the commencement of this period from what it is at the present day. If it be correct that the glacial epoch resulted from the causes which we have already discussed, those tables ought to aid us in our endeavour to ascertain *how* much of the Tertiary period may be comprehended within these 3 millions of years.

We have already seen (Chapter XVIII.) that there is evidence of a glacial condition of climate at two different periods during the Tertiary age, namely, about the middle of the Miocene and Eocene periods respectively. As has already been shown, the more severe a glacial epoch is, the more marked ought to be the character of its warm inter-glacial periods; the greater the extension of the ice during the cold periods of a glacial epoch the further should that ice disappear in arctic regions during the corresponding warm periods. Thus the severity of a glacial epoch may in this case be indirectly inferred from the character of the warm periods and the extent to which the ice may have disappeared from arctic regions. Judged by this test, we have every reason to believe that the Miocene glacial epoch was one of extreme severity.

The Eocene conglomerate, devoid of all organic remains, and containing numerous enormous ice-transported blocks, is, as we have seen, immediately associated with nummulitic strata charged with fossils characteristic of a warm climate. Referring to this Sir Charles Lyell says, "To imagine icebergs carrying such huge fragments of stone in so southern a latitude, and at a period immediately preceded and followed by the signs of a warm climate, is one of the most perplexing enigmas which the geologist has yet been called upon to solve."*

It is perfectly true that, according to the generally received theories of the cause of a glacial climate the whole is a perplexing enigma, but if we adopt the Secular theory of change of climate, every difficulty disappears. According to this theory the very fact of the conglomerate being formed at a period immediately preceded and succeeded by warm conditions of climate, is of itself strong presumptive evidence of the conglomerate being a glacial formation. But this is not all, the very highness of the temperature of the preceding and succeeding periods bears testimony to the severity of the intervening glacial period. Despite the deficiency of direct evidence regarding the character of the Miocene and Eocene glacial periods, we are not warranted, for reasons which have been stated in Chapter XVII., to conclude that these periods were less severe than the one which happened in Quaternary times. Judging from indirect evidence, we have some grounds for concluding that the Miocene glacial epoch at least was even more severe and protracted than our recent glacial epoch.

By referring to Table I., or the accompanying diagram, it will be seen that prior to the period which I have assigned as that of the glacial epoch, there are two periods when the eccentricity almost attained its superior limit. The first period occurred 2,500,000 years ago, when it reached 0·0721, and the second period 850,000 years ago, when it attained a still higher value, viz., 0·0747, being within 0·0028 of the superior limit. To the first of these periods I am disposed to assign the glacial epoch of Eocene times, and to the second that of the Miocene age. With the view of determining the character of these

* "Principles," p. 210. Eleventh Edition.

periods Tables II. and III. have been computed. They give the eccentricity and longitude of perihelion at intervals of 10,000 years. It will be seen from Table II. that the Eocene period extends from about 2,620,000 to about 2,460,000 years ago; and from Table III. it will be gathered that the Miocene period lasted from about 980,000 to about 720,000 years ago.

In order to find whether the eccentricity attained a higher value about 850,000 years ago than 0·0747, I computed the values for one or two periods immediately before and after that period, and satisfied myself that the value stated was indeed the highest, as will be seen from the subjoined table:—

851,000	0·07454
850,000	0·074664
849,500	0·07466
849,000	0·07466

How totally different must have been the condition of the earth's climate at that period from what it is at present! Taking the mean distance of the sun to be 91,400,000 miles, his present distance at mid-winter is 89,864,480 miles; but at the period in question, when the winter solstice was in perihelion, his distance at mid-winter would be no less than 98,224,289 miles. But this is not all; our winters are at present shorter than our summers by 7·8 days, but at that period they would be longer than the summers by 34·7 days.

At present the difference between the perihelion and aphelion distance of the sun amounts to only 3,069,580 miles, but at the period under consideration it would amount to no less than 13,648,579 miles!

CHAPTER XXII.

A METHOD OF DETERMINING THE MEAN THICKNESS OF THE SEDIMENTARY ROCKS OF THE GLOBE.

Prevailing Methods defective.—Maximum Thickness of British Rocks.—Three Elements in the Question.—Professor Huxley on the Rate of Deposition.—Thickness of Sedimentary Rocks enormously over-estimated.—Observed Thickness no Measure of mean Thickness.—Deposition of Sediment principally along Sea-margin.—Mistaken Inference regarding the Absence of a Formation.—Immense Antiquity of existing Oceans.

VARIOUS attempts have been made to measure the positive length of geological periods. Some geologists have sought to determine, roughly, the age of the stratified rocks by calculations based upon their probable thickness and the rate at which they may have been deposited. This method, however, is worthless, because the rates which have been adopted are purely arbitrary. One geologist will take the rate of deposit at a foot in a hundred years, while another will assume it to be a foot in a thousand or perhaps ten thousand years; and, for any reasons that have been assigned, the one rate is just as likely to be correct as the other: for if we examine what is taking place in the ocean bed at the present day, we shall find in some places a foot of sediment laid down in a year, while in other places a foot may not be deposited in a thousand years. The stratified rocks were evidently formed at all possible rates. When we speak of the rate of their formation, we must of course refer to the *mean rate*; and it is perfectly true that if we knew the thickness of these rocks and the mean rate at which they were deposited, we should have a ready means of determining their positive age. But there appears to be nearly as great uncertainty regarding the thickness of the sedimentary rocks as regarding the rate at which they were formed. No doubt we can roughly estimate their probable maximum thickness; for instance, Professor Ramsay has found from actual measurement, that the sedimentary formations of Great Britain have a maximum thickness of upwards of 72,000 feet; but all such measurements give us no idea of their mean thickness. What is the mean thickness of the sedimentary rocks of the globe? On this point geology does not afford a definite answer. Whatever the present mean thickness of the sedimentary rocks of our globe may be, it must be small in comparison to the mean thickness of all the sedimentary rocks which have been formed. This is obvious from the fact that the sedimentary rocks of one age are partly formed from the destruction of the sedimentary rocks of former ages. From the Laurentian age down to the present day, the stratified rocks have been undergoing constant denudation.

Unless we take into consideration the quantity of rock removed during past ages by denudation, we cannot—even though we knew the actual mean thickness of the existing sedimentary rocks of the globe, and the rate at which they were formed—arrive at an estimate regarding the length of time represented by these rocks. For if we are to determine the age of the stratified rocks from the rate at which they were formed, we must have, not the present quantity of sedimentary rocks, but the present plus the quantity which has been denuded during past ages. In other words, we must have the absolute quantity formed. In many places the missing beds must have been of enormous thickness. The time represented by beds which have disappeared is, doubtless, as already remarked, much greater than that represented by the beds which now remain. The greater mass of the sedimentary rocks has been formed out of previously existing sedimentary rocks, and these again out of sedimentary rocks still older. As the materials composing our stratified beds may have passed through many cycles of destruction and re-formation, the time required to have deposited at a given rate the present existing mass of sedimentary rocks may be but a fraction of the time required to have deposited at the same rate the total mass that has actually been formed. To measure the age of the sedimentary rocks by the present existing rocks, assumed to be formed at some given rate, even supposing the rate to be correct, is a method wholly fallacious.

"The aggregate of sedimentary strata in the earth's crust," says Sir Charles Lyell, "can never exceed in volume the amount of solid matter which has been ground down and washed away by rivers, waves, and currents. How vast, then, must be the spaces which this abstraction of matter has left vacant! How far exceeding in dimensions all the valleys, however numerous, and the hollows, however vast, which we can prove to have been cleared out by aqueous erosion!" *

I presume there are few geologists who would not admit that if all the rocks which have in past ages been removed by denudation were restored, the mean thickness of the sedimentary rocks of the globe would be at least equal to their present maximum thickness, which we may take at 72,000 feet.

There are three elements in the question; of which if two are known, the third is known in terms of the other two. If we have the mean thickness of all the sedimentary rocks which have been formed and the mean rate of formation, then we have the time which elapsed during the formation; or having the thickness and the time, we have the rate; or, having the rate and the time, we have the thickness.

One of these three, namely, the rate, can, however, be determined with tolerable accuracy if we are simply allowed to assume—what is very probable, as has already been shown—that the present rate at which the sedimentary deposits are being formed may be taken as the mean rate for past ages. If we know the rate at which the land is being denuded, then we know with perfect accuracy the rate at which the sedimentary deposits are being formed in the ocean. This is obvious, because all the materials denuded from the land are deposited in

* "Principles," vol. i., p. 107. Tenth Edition.

the sea; and what is deposited in the sea is just what comes off the land, with the exception of the small proportion of calcareous matter which may not have been derived from the land, and which in our rough estimate may be left out of account.

Now the mean rate of subaërial denudation, we have seen, is about one foot in 6,000 years. Taking the proportion of land to that of water at 576 to 1,390, then one foot taken off the land and spread over the sea-bottom would form a layer 5 inches thick. Consequently, if one foot in 6,000 years represents the mean rate at which the land is being denuded, one foot in 14,400 years represents the mean rate at which the sedimentary rocks are being formed.

Assuming, as before, that 72,000 feet would represent the mean thickness of all the sedimentary rocks which have ever been formed, this, at the rate of one foot in 14,400 years, gives 1,036,800,000 years as the age of the stratified rocks.

Professor Huxley, in his endeavour to show that 100,000,000 years is a period sufficiently long for all the demands of geologists, takes the thickness of the stratified rocks at 100,000 feet, and the rate of deposit at a foot in 1,000 years. One foot of rock per 1,000 years gives, it is true, 100,000 feet in 100,000,000 years. But what about the rocks which have disappeared? If it takes a hundred millions of years to produce a mass of rock equal to that which now exists, how many hundreds of millions of years will it require to produce a mass equal to what has actually been produced?

Professor Huxley adds, "I do not know that any one is prepared to maintain that the stratified rocks may not have been formed on the average at the rate of 1-83rd of an inch per annum." When the rate, however, is accurately determined, it is found to be, not 1-83rd of an inch per annum, but only 1-1200th of an inch, so that the 100,000 feet of rock must have taken 1,440,000,000 years in its formation,—a conclusion which, according to the results of modern physics, is wholly inadmissible.

Either the thickness of the sedimentary rocks has been over-estimated, or the rate of their formation has been under-estimated, or both. If it be maintained that a foot in 14,400 years is too slow a rate of deposit, then it must be maintained that the land must have been denuded at a greater rate than one foot in 6,000 years. But most geologists probably felt surprised when the announcement was first made, that at this rate of denudation the whole existing land of the globe would be brought under the ocean in 6,000,000 of years.

The error, no doubt, consists in over-estimating the thickness of the sedimentary rocks. Assuming, for physical reasons already stated, that 100,000,000 years limits the age of the stratified rocks, and that the proportion of land to water and the rate of denudation have been on the average the same as at present, the mean thickness of sedimentary rocks formed in the 100,000,000 years amounts to only 7,000 feet.

But be it observed that this is the mean thickness on an area equal to that of the ocean. Over the area of the globe it amounts to only 5,000 feet; and this, let it be observed also, is the total mean thickness formed, without taking into account what has been removed by denudation. If we wish to ascertain what is actually the present mean thickness, we must deduct from this 5,000 feet an amount of rock equal to all the sedimentary rocks which have been denuded during the 100,000,000 years; for the 5,000 feet is not the present mean thickness, but the total mean thickness formed during the whole of the 100,000,000 years. If we assume, what no doubt most geologists would be willing to grant, that the quantity of sedimentary rocks now remaining is not over one-half of what has been actually deposited during the history of the globe, then the actual mean thickness of the stratified rocks of the globe is not over 2,500 feet. This startling result would almost necessitate us to suspect that the rate of subaërial denudation is probably greater than one foot in 6,000 years. But, be this as it may, we are apt, in estimating the mean thickness of the stratified rocks of the globe from their ascertained maximum thickness, to arrive at erroneous conclusions. There are considerations which show that the mean thickness of these rocks must be small in proportion to their maximum thickness. The stratified rocks are formed from the sediment carried down by rivers and streamlets and deposited in the sea. It is obvious that the greater quantity of this sediment is deposited near the mouths of rivers, and along a narrow margin extending to no great distance from the land. Did the land consist of numerous small islands equally distributed over the globe, the sediment carried off from these islands would be spread pretty equally over the sea-bottom. But the greater part of the land-surface consists of two immense continents. Consequently, the materials removed by denudation are not spread over the ocean-bottom, but on a narrow fringe surrounding those two continents. Were the materials spread over the entire ocean-bed, a foot removed off the general surface of the land would form a layer of rock only five inches thick. But in the way in which the materials are at present deposited, the foot removed from the land would form a layer of rock many feet in thickness. The greater part of the sediment is deposited within a few miles of the shore.

The entire coast-line of the globe is about 116,500 miles. I should think that the quantity of sediment deposited beyond, say, 100 miles from this coast-line is not very great. No doubt several of the large rivers carry sediment to a much greater distance from their mouths than 100 miles, and ocean currents may in some cases carry mud and other materials also to great distances. But it must be borne in mind that at many places within the 100 miles of this immense coast-line little or no sediment is deposited, so that the actual area over which the sediment carried off the land is deposited is probably not greater than the area of this belt—116,500 miles long and 100 miles broad. This area on which the sediment is deposited, on the above supposition, is therefore equal to about 11,650,000 square miles. The amount of land on the globe is about 57,600,000 square miles. Consequently, one foot of rock, denuded from the surface of the land and deposited on this belt, would make a stratum of rock 5 feet in thickness; but were the sediment spread over the entire bed of the ocean, it would form, as has already been stated, a stratum of rock of only 5 inches in thickness.

Suppose that no subsidence of the land should take place for a period of, say, 3,000,000 of years. During that period 500 feet would be removed by denudation, on an average, off the land. This would make a formation 2,500 feet thick, which some future geologist might call the Post-tertiary formation. But this, be it observed, would be only the mean thickness of the formation on this area; its maximum thickness would evidently be much greater, perhaps twice, thrice, or even four times that thickness. A geologist in the future, measuring the actual thickness of the formation, might find it in some places 10,000 feet in thickness, or perhaps far more. But had the materials been spread over the entire ocean-bed, the formation would have a mean thickness of little more than 200 feet; and spread over the entire surface of the globe, would form a stratum of scarcely 150 feet in thickness. Therefore, in estimating the mean thickness of the stratified rocks of the globe, a formation with a maximum thickness of 10,000 feet may not represent more than 150 feet. A formation with a *mean* thickness of 10,000 feet represents only 600 feet.

It may be objected that in taking the present rate at which the sedimentary deposits are being formed as the mean rate for all ages, we probably under-estimate the total amount of rock formed, because during the many glacial periods which must have occurred in past ages the amount of materials ground off the rocky surface of the land in a given period would be far greater than at present. But, in reply, it must be remembered that although the destruction in ice-covered regions would be greater during these periods than at present, yet the quantity of materials carried down by rivers into the sea would be less. At the present day the greater part of the materials carried down by our rivers is not what is being removed off the rocky face of the country, but the boulder clay, sand, and other materials which were ground off during the glacial epoch. It is therefore possible, on this account, that the rate of deposit may have been less during the glacial epoch than at present.

When any particular formation is wanting in a given area, the inference generally drawn is, that either the formation has been denuded off the area, or the area was a land-surface during the period when that formation was being deposited. From the foregoing it will be seen that this inference is not legitimate; for, supposing that the area had been under water, the chances that materials should have been deposited on that area are far less than are the chances that there should not. There are sixteen chances against one that no formation ever existed in the area.

If the great depressions of the Atlantic, Pacific, and Indian Oceans be, for example, as old as the beginning of the Laurentian period—and they may be so for anything which geology can show to the contrary—then under these oceans little or no stratified rocks may exist. The supposition that the great ocean basins are of immense antiquity, and that consequently only a small proportion of the sedimentary strata can possibly occupy the deeper bed of the sea, acquires still more probability when we consider the great extent and thickness of the Old Red Sandstone, the Permian, and other deposits, which, according to Professor Ramsay and others, have been accumulated in vast inland lakes.

CHAPTER XXIII.

THE PHYSICAL CAUSE OF THE SUBMERGENCE AND EMERGENCE OF THE LAND DURING THE GLACIAL EPOCH.

Displacement of the Earth's Centre of Gravity by Polar Ice-cap.—Simple Method of estimating Amount of Displacement.—Note by Sir W. Thomson on foregoing Method.—Difference between Continental Ice and a Glacier.—Probable Thickness of the Antarctic Ice-cap.—Probable Thickness of Greenland Ice-sheet.—The Icebergs of the Southern Ocean.—Inadequate Conceptions regarding the Magnitude of Continental Ice.

*Displacement of the Earth's Centre of Gravity by Polar Ice-cap.** —In order to represent the question in its most simple elementary form, I shall assume an ice-cap of a given thickness at the pole and gradually diminishing in thickness towards the equator in the simple proportion of the sines of the latitudes, where at the equator its thickness of course is zero. Let us assume, what is actually the case, that the equatorial diameter of the globe is somewhat greater than the polar, but that when the ice-cap is placed on one hemisphere the whole forms a perfect sphere.

I shall begin with a period of glaciation on the southern hemisphere. Let W N E S′ (Fig. 5) be the solid part of the earth, and c its centre of gravity. And let E S W be an ice-cap covering the southern hemisphere. Let us in the first case assume the earth to be of the same density as the cap. The earth with its cap forms now a perfect sphere with its

* The conception of submergence resulting from displacement of the earth's centre of gravity, caused by a heaping-up of ice at one of the poles, was first advanced by M. Adhémar, in his work "*Révolutions de la Mer,*" 1842. When the views stated in this chapter appeared in the *Reader*, I was not aware that M. Adhémar had written on the subject. An account of his mode of viewing the question is given in the Appendix.

centre of gravity at *o*; for W N E S is a circle, and *o* is its centre. Suppose now the whole to be covered with an ocean a few miles deep, the ocean will assume the spherical form, and will be of uniform depth. Let the southern winter solstice begin now to move round from the aphelion. The ice-cap will also commence gradually to diminish in thickness, and another cap will begin to make its appearance on the northern hemisphere. As the northern cap may be supposed, for simplicity of calculation, to increase at the same rate that the southern will diminish, the spherical form of the earth will

Fig. 5.

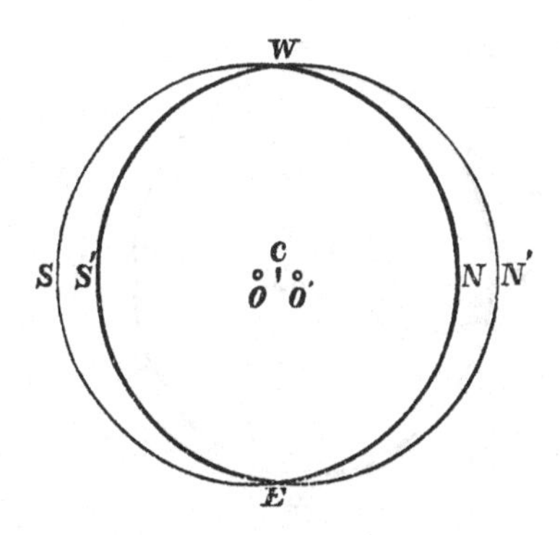

always be maintained. By the time that the northern cap has reached a maximum, the southern cap will have completely disappeared. The circle W N′ E S′ will now represent the earth with its cap on the northern hemisphere, and *o′* will be its centre of gravity; for *o′* is the centre of the circle W N′ E S′. And as the distance between the centres *o* and *o′* is equal to N N′, the thickness of the cap at the pole N N′ will therefore represent the extent to which the centre of gravity has been displaced. It will also represent the extent to which the ocean has risen at the north pole and sunk at the south. This is evident; for as the sphere W N′ E S′ is the same in all respects as the sphere W N E S, with the exception only that the cap is on the opposite side, the surface of the ocean at the poles will now be at the same distance from the centre *o′* as it was from the centre *o* when the cap covered the southern hemisphere. Hence the distance between *o* and *o′* must be equal to the extent of the submergence at the north pole and the emergence at the south. Neglect the attraction of the altering water on the water itself, which later on will come under our consideration.

We shall now consider the result when the earth is taken at its actual density, which is generally believed to be about 5·5. The density of ice being ·92, the density of the cap to that of the earth will therefore be as 1 to 6.

Fig. 6.

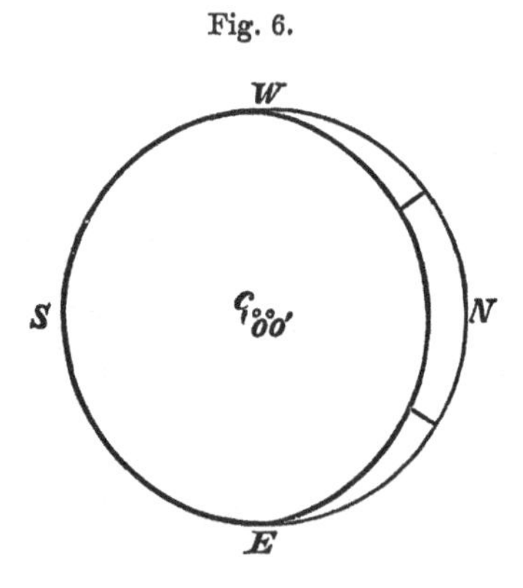

Let Fig. 6 represent the earth with an ice-cap on the northern hemisphere, whose thickness is, say, 6,000 feet at the pole. The centre of gravity of the earth without the cap is at *c*. When the cap is on, the centre of gravity is shifted to *o*, a point a little more than 500 feet to the north of *c*. Had the cap and the earth been of equal density, the centre of gravity would have been shifted to *o′* the centre of the figure, a point situated, of course, 3,000 feet to the north of *c*. Now it is very approximately true that the ocean will tend to adjust itself as a sphere around the centre of gravity, *o*. Thus it would of course sink at the south pole and rise to the same extent at the north, in any opening or channel in the ice allowing the water to enter.

Let the ice-cap be now transferred over to the southern

hemisphere, and the condition of things on the two hemispheres will in every particular be reversed. The centre of gravity will then lie to the south of *c*, or about 1,000 feet from its former position. Consequently the transferrence of the cap from the one hemisphere to the other will produce a total submergence of about 1,000 feet.

It is, of course, absurd to suppose that an ice-cap could ever actually reach down to the equator. It is probable that the great ice-cap of the glacial epoch nowhere reached even half-way to the equator. Our cap must therefore terminate at a moderately high latitude. Let it terminate somewhere about the latitude of the north of England, say at latitude 55°. All that we have to do now is simply to imagine our cap, up to that latitude, becoming converted into the fluid state. This would reduce the cap to less than one-half its former mass. But it would not diminish the submergence to anything like that extent. For although the cap would be reduced to less than one-half its former mass, yet its influence in displacing the centre of gravity would not be diminished to that extent. This is evident; for the cap now extending down to only latitude 55°, has its centre of gravity much farther removed from the earth's centre of gravity than it had when it extended down to the equator. Consequently it now possesses, in proportion to its mass, a much greater power in displacing the earth's centre of gravity.

There is another fact which must be taken into account. The common centre of gravity of the earth and cap is not exactly the point around which the ocean tends to adjust itself. It adjusts itself not in relation to the centre of gravity of the solid mass alone, but in relation to the common centre of gravity of the entire mass, solid and liquid. Now the water which is pulled over from the one hemisphere to the other by the attraction of the cap will also aid in displacing the centre of gravity. It will co-operate with the cap and carry the true centre of gravity to a point beyond that of the centre of gravity of the earth and cap, and thus increase the effect.

It is of course perfectly true that when the ice-cap does not extend down to the equator, as in the latter supposition, and is of less density than the globe, the ocean will not adjust itself uniformly around the centre of gravity; but the deviation from perfect uniformity is so trifling, as will be seen from the appended note of Sir William Thomson, that for all practical purposes it may be entirely left out of account.

In the *Reader* for January 13, 1866, I advanced an objection to the submergence theory on the grounds that the lowering of the ocean-level by the evaporation of the water to form the ice-cap, would exceed the submergence resulting from the displacement of the earth's centre of gravity. But, after my letter had gone to press, I found that I had overlooked some important considerations which seem to prove that the objection had no real foundation. For during a glacial period, say on the northern hemisphere, the entire mass of ice which presently exists on the southern hemisphere would be transferred to the northern, leaving the quantity of liquid water to a great extent unchanged.

Note on the preceding by Sir William Thomson, F.R.S.

"Mr. Croll's estimate of the influence of a cap of ice on the sea-level is very remarkable in its relation to Laplace's celebrated analysis, as being founded on that law of thickness which leads to expressions involving only the first term of the series of 'Laplace's functions,' or 'spherical harmonics.' The equation of the level surface, as altered by any given transferrence of solid matter, is expressed by equating the altered potential function to a constant. This function, when expanded in the series of spherical harmonics, has for its first term the potential due to the whole mass supposed collected at its altered centre of gravity. Hence a spherical surface round the altered centre of gravity is the *first* approximation in Laplace's method of solution for the altered level surface. Mr. Croll has with admirable tact chosen, of all the arbitrary suppositions that may be made foundations for rough estimates of the change of sea-level due to variations in the polar ice-crusts, *the* one which reduces to zero all terms after the first in the harmonic series, and renders that first approximation (which always expresses the *essence* of the result) the whole solution, undisturbed by terms irrelevant to the great physical question.

"Mr. Croll, in the preceding paper, has alluded with remarkable clearness to the effect of the change in the distribution of the water in increasing, by its own attraction, the deviation of the level surface above that which is due to the *given* change in the distribution of solid matter. The remark he makes, that it is round the centre of gravity of the altered solid and altered liquid that the altering liquid surface adjusts itself, expresses the essence of Laplace's celebrated demonstration of the stability of the ocean, and suggests the proper elementary solution of the problem to find the true alteration of sea-level produced by a given alteration of the solid. As an assumption leading to a simple calculation, let us suppose the solid earth to rise out of the water in a vast number of small flat-topped islands, each bounded by a perpendicular cliff, and let the proportion of water area to the whole be equal in all quarters. Let all of these islands in one hemisphere be covered with ice, of thickness according to the law assumed by Mr. Croll—that is, varying in simple proportion of the sine of the latitude. Let this ice be removed from the first hemisphere and similarly distributed over the islands of the second. By working out according to Mr. Croll's directions, it is easily found that the change of sea-level which this will produce will consist in a sinking in the first hemisphere and rising in the second, through heights varying according to the same law (that is, simple proportionality to sines of latitudes), and amounting at each pole to

$$\frac{(1-\omega)it}{1-\omega w},$$

where t denotes the thickness of the ice-crust at the pole; i the ratio of the density of ice, and w that of sea-water to the earth's mean density; and ω the ratio of the area of ocean to the whole surface.

"Thus, for instance, if we suppose $\omega = \frac{2}{3}$, and $t = 6{,}000$ feet, and take $\frac{1}{6}$ and $\frac{1}{5\frac{1}{2}}$ as the densities of ice and water respectively, we find for the rise of sea-level at one pole, and depression at the other,

$$\frac{\frac{1}{3} \times \frac{1}{6} \times 6000}{1 - \frac{2}{3} \times \frac{1}{5\frac{1}{2}}},$$

or approximately 380 feet.

"I shall now proceed to consider roughly what is the probable extent of submergence which, during the glacial epoch, may have resulted from the displacement of the earth's centre of gravity by means of the transferrence of the polar ice from the one hemisphere to the other."

Difference between Continental-ice and a Glacier.—An ordinary glacier descends in virtue of the slope of its bed, and, as a general rule, it is on this account thin at its commencement, and thickens as it descends into the lower valleys, where the slope is less and the resistance to motion greater. But in the case of continental ice matters are entirely different. The slope of the ground exercises little or no influence on the motion of the ice. In a continent of one or two thousand miles across, the general slope of the ground may be left out of account; for any slight elevation which the centre of such a continent may have will not compensate for the resistance offered to the flow of the ice by mountain ridges, hills, and other irregularities of its surface. The ice can move off such a surface only in consequence of pressure acting from the interior. In order to produce such a pressure, there must be a piling up of the ice in the interior; or, in other words, the ice-sheet must thicken from the edge inwards to the centre. We are necessarily led to the same conclusion, though we should not admit that the ice moves in consequence of pressure from behind, but should hold, on the contrary, that each particle of ice moves by gravity in virtue of its own weight; for in order to have such a motion there must be a slope, and as the slope is not on the ground, it must be on the ice itself: consequently we must conclude that the upper surface of the ice slopes upwards from the edge to the interior. What, then, is the least slope at which the ice will descend? Mr. Hopkins found that ice barely moves on a slope of one degree. We have therefore some data for arriving at least at a rough estimate of the probable thickness of an ice-sheet covering a continent, such, for example, as Greenland or the Antarctic Continent.

Probable Thickness of the Antarctic Ice-cap.—The antarctic continent is generally believed to extend, on an average, from the South Pole down to about, at least, lat. 70°. In round numbers, we may take the diameter of this continent at 2,800 miles. The distance from the edge of this ice-cap to its centre, the South Pole, will, therefore, be 1,400 miles. The whole of this continent, like Greenland, is undoubtedly covered with one continuous sheet of ice gradually thickening inwards from its edge to its centre. A slope of one degree continued for 1,400 miles will give twenty-four miles as the thickness of the ice at the pole. But suppose the slope of the upper surface of the cap to be only one-half this amount, viz., a half degree,—and we have no evidence that a slope so small would be sufficient to discharge the ice,—still we have twelve miles as the thickness of the cap at the pole. To those who have not been accustomed to reflect on the physical conditions of the problem, this estimate may doubtless be regarded as somewhat extravagant; but a slight consideration will show that it would be even more extravagant to assume that a slope of less than half a degree would be sufficient to produce the necessary outflow of the ice. In estimating the thickness of a sheet of continental ice of one or two thousand miles across, our imagination is apt to deceive us. We can easily form a pretty accurate sensuous representation of the thickness of the sheet; but we can form no adequate representation of its superficial area. We can represent to the mind with tolerable accuracy a thickness of a few miles, but we cannot do this in reference to the area of a surface 2,800 miles across. Consequently, in judging what proportion the thickness of the sheet should bear to its superficial area, we are apt to fall into the error of under-estimating the thickness. We have a striking example of this in regard to the ocean. The thing which impresses us most forcibly in regard to the ocean is its profound depth. A mean depth of, say, three miles produces a striking impression; but if we could represent to the mind the vast area of the ocean as correctly as we can do its depth, *shallowness* rather than *depth* would be the impression produced. A sheet of water 100 yards in diameter, and only one inch deep, would not be called a *deep* but a very *shallow* pool or thin layer of water. But such a layer would be a correct representation of the ocean in miniature. Were we in like manner to represent to the eye in miniature the antarctic ice-cap, we would call it a *thin crust of ice.* Taking the mean thickness of the ice at four miles, the antarctic ice-sheet would be represented by a carpet covering the floor of an ordinary-sized dining-room. Were those who consider the above estimate of the thickness of the antarctic ice-cap as extravagantly great called upon to sketch on paper a section of what they should deem a cap of moderate thickness, ninety-nine out of every hundred would draw one of much greater thickness than twelve miles at the centre.

The diagram on following page (Fig. 7) represents a section across the cap drawn to a natural scale; the upper surface of the sheet having a slope of half a degree. No one on looking at the section would pronounce it to be too thick at the centre, unless he were previously made aware that it represented a thickness of twelve miles at that place. It may be here mentioned that had the section been drawn upon a much larger scale—had it, for instance, been made seven feet long, instead of seven inches—it would have shown to the eye in a more striking manner the thinness of the cap.

But to avoid all objections on the score of over-estimating the thickness of the cap, I shall assume the angle of the upper surface to be only a quarter of a degree, and the thickness of the sheet one-half what it is represented in the section. The thickness at the pole will then be only six miles instead of twelve, and the mean thickness of the cap two instead of four miles.

Is there any well-grounded reason for concluding the above to be an over-estimate of the actual thickness of the antarctic ice? It is not so much in consequence of any *à priori* reason that can be urged against the probability of such a thickness of ice, but rather because it so far transcends our previous experience that we are reluctant to admit such an estimate. If we never had any experience of ice thicker than what is found in England, we should feel startled on learning for the first time that in the valleys of Switzerland the ice lay from 200 to 300 feet in depth. Again, if we had never heard of glaciers thicker than those of Switzerland, we could hardly credit the statement that in Greenland they are actually from 2,000 to 3,000 feet thick. We, in this country, have long been familiar with Greenland; but till very lately no one ever entertained the idea that that continent was buried under one continuous mass of ice, with scarcely a mountain top rising above the icy mantle. And had it not been that the geological phenomena of the glacial epoch have for so many years accustomed our minds to such an extraordinary condition of things, Dr. Rink's description of the Greenland ice would probably have been regarded as the extravagant picture of a wild imagination.

Let us now consider whether or not the facts of observation and experience, so far as they go,

Fig. 7.

S. Pole.

Section across Antarctic Ice-cap, drawn to a natural scale.
Length represented by section = 2,800 miles. Thickness at centre (South Pole) = 12 miles.
Slope of upper surface = half-degree.

bear out the conclusions to which physical considerations lead us in reference to the magnitude of continental ice; and more especially as regards the ice of the antarctic regions.

First. In so far as the antarctic ice-sheet is concerned, observation and experience to a great extent may be said to be a perfect blank. One or two voyagers have seen the outer edge of the sheet at a few places, and this is all. In fact, we judge of the present condition of the interior of the antarctic continent in a great measure from what we know of Greenland. But again, our experience of Greenland ice is almost wholly confined to the outskirts.

Few have penetrated into the interior, and, with the exception of Dr. Hayes and Professor Nordenskjöld, none, as far as I know, have passed to any considerable distance over the inland ice. Dr. Robert Brown in his interesting memoir on "Das Innere von Grönland," * gives an account of an excursion made in 1747 by a Danish officer of the name of Dalager, from Fredrikshaab, near the southern extremity of the continent, into the interior. After a journey of a day or two, he reached an eminence from which he saw the inland ice stretching in an unbroken mass as far as the eye could reach, but was unable to proceed further. Dr. Brown gives an account also of an excursion made in the beginning of March, 1830, by O. B. Kielsen, a Danish whale-fisher, from Holsteinborg (lat. 67° N.). After a most fatiguing journey of several days, he reached a high point from which he could see the ice of the interior. Next morning he got up early, and towards midday reached an extensive plain. From this the land sank inwards, and Kielsen now saw fully in view before him the enormous ice-sheet of the interior. He drove rapidly over all the little hills, lakes, and streams, till he reached a pretty large lake at the edge of the ice-sheet. This was the end of his journey, for after vainly attempting to climb up on the ice-sheet, he was compelled to retrace his steps, and had a somewhat difficult return. When he arrived at the fiord, he found the ice broken up, so that he had to go round

* Petermann's *Geog. Mittheilungen*, 1871, Heft. x., p. 377.

by the land way, by which he reached the depôt on the 9th of March. The distance which he traversed in a straight line from Holsteinborg into the interior measured eighty English miles.

Dr. Hayes's excursion was made, however, not upon the real inland ice, but upon a smaller ice-field connected with it; while Professor Nordenskjöld's excursion was made at a place too far south to afford an accurate idea of the actual condition of the interior of North Greenland, even though he had penetrated much farther than he actually did. However, the state of things as recorded by Hayes and by Nordenskjöld affords us a glimpse into the condition of things in the interior of the continent. They both found by observation, what follows as a necessary result from physical considerations, that the upper surface of the ice plain, under which hills and valleys are buried, gradually *slopes upwards towards the interior of the continent.* Professor Nordenskjöld states that when at the extreme point at which he reached, thirty geographical miles from the coast, he had attained an elevation of 2,200 feet, and that the inland ice *continued constantly to rise* towards the interior, so that the horizon towards the east, north, and south, was terminated by an ice-border almost as smooth as that of the ocean."*

Dr. Hayes and his party penetrated inwards to the distance of about seventy miles. On the first day they reached the foot of the great Mer de Glace; the second day's journey carried them to the upper surface of the ice-sheet. On the third day they travelled 30 miles, and the ascent, which had been about 6°, diminished gradually to about 2°. They advanced on the fourth day about 25 miles; the temperature being 30° below zero (Fah.). "Our station at the camp," he says, "was sublime as it was dangerous. We had attained an altitude of 5,000 feet above the sea-level, and were 70 miles from the coast, in the midst of a vast frozen Sahara immeasurable to the human eye. There was neither hill, mountain, nor gorge,

* Geol. Mag., 1872, vol. ix., p. 360.

anywhere in view. We had completely sunk the strip of land between the Mer de Glace and the sea, and no object met the eye but our feeble tent, which bent to the storm. Fitful clouds swept over the face of the full-orbed moon, which, descending towards the horizon, glimmered through the drifting snow that scudded over the icy plain—to the eye in undulating lines of downy softness, to the flesh in showers of piercing darts." *

Dr. Rink, referring to the inland ice, says that the elevation or height above the sea of this icy plain at its junction with the outskirts of the country, and where it begins to lower itself through the valleys to the friths, is, in the ramifications of the Bay of Omenak, found to be 2,000 feet, from which level *it gradually rises towards the interior.*†

Dr. Robert Brown, who, along with Mr. Whymper in 1867, attempted a journey to some distance over the inland ice, is of opinion that Greenland is not traversed by any ranges of mountains or high land, but that the entire continent, 1,200 miles in length and 400 miles in breadth, is covered with one continuous unbroken field of ice, the upper surface of which, he says, *rises by a gentle slope towards the interior.*‡

Suppose now the point reached by Hayes to be within 200 miles of the centre of dispersion of the ice, and the mean slope from that point to the centre, as in the case of the antarctic cap, to be only half a degree; this would give 10,000 feet as the elevation of the centre above the point reached. But the point reached was 5,000 feet above sea-level, consequently the surface of the ice at the centre of dispersion would be 15,000 feet above sea-level, which is about one-fourth what I have concluded to be the elevation of the surface of the antarctic ice-cap at its centre. And supposing we assume the general surface of the ground to have in the central region an elevation as great as 5,000 feet, which is not at all probable, still this would give 10,000 feet for the thickness of the ice at the centre

* "Open Polar Sea," p. 134.

† Journal of the Royal Geographical Society, 1853, vol. xxiii.

‡ "Physics of Arctic Ice," Quart. Journ. Geol. Soc. for February, 1871.

of the Greenland continent. But if we admit this conclusion in reference to the thickness of the Greenland ice, we must admit that the antarctic ice is far thicker, because the thickness, other things being equal, will depend upon the size, or, more properly, upon the diameter of the continent; for the larger the surface the greater is the thickness of ice required to produce the pressure requisite to make the rate of discharge of the ice equal to the rate of increase. Now the area of the antarctic continent must be at least a dozen of times greater than that of Greenland.

Second. That the antarctic ice must be far thicker than the arctic is further evident from the dimensions of the icebergs which have been met with in the Southern Ocean. No icebergs over three hundred feet in height have been found in the arctic regions, whereas in the antarctic regions, as we shall see, icebergs of twice and even thrice that height have been reported.

Third. We have no reason to believe that the thickness of the ice at present covering the antarctic continent is less than that which covered a continent of a similar area in temperate regions during the glacial epoch. Take, for example, the North American continent, or, more properly, that portion of it covered by ice during the glacial epoch. Professor Dana has proved that during that period the thickness of the ice on the American continent must in many places have been considerably over a mile. He has shown that over the northern border of New England the ice had a mean thickness of 6,500 feet, while its mean thickness over the Canada watershed, between St. Lawrence and Hudson's Bay, was not less than 12,000 feet, or upwards of two miles and a quarter (see *American Journal of Science and Art* for March, 1873).

Fourth. Some may object to the foregoing estimate of the amount of ice on the antarctic continent, on the grounds that the quantity of snowfall in that region cannot be much. But it must be borne in mind that, no matter however small the annual amount of snowfall may be, if more falls than is melted, the ice must continue to accumulate year by year till its thickness in the centre of the continent be sufficiently great to produce motion. The opinion that the snowfall of the antarctic regions is not great does not, however, appear to be borne out by the observation and experience of those who have visited those regions. Captain Wilkes, of the American Exploring Expedition, estimated it at 30 feet per annum ; and Sir James Ross says, that during a whole month they had only three days free from snow. The fact that perpetual snow is found at the sea-level at lat. 64° S. proves that the snowfall must be great. But there is another circumstance which must be taken into account, viz., that the currents carrying moisture move in from all directions towards the pole, consequently the area on which they deposit their snow becomes less and less as the pole is reached, and this must, to a corresponding extent, increase the quantity of snow falling on a given area. Let us assume, for example, that the clouds in passing from lat. 60° to lat. 80° deposit moisture sufficient to produce, say, 30 feet of snow per annum, and that by the time they reach lat. 80° they are in possession of only one-tenth part of their original store of moisture. As the area between lat. 80° and the pole is but one-eighth of that between lat. 60° and 80°, this would, notwithstanding, give 24 feet as the annual amount of snowfall between lat. 80° and the pole.*

Fifth. The enormous size and thickness of the icebergs which have been met with in the Southern Ocean testify to the thickness of the antarctic ice-cap.

We know from the size of some of the icebergs which have been met with in the southern hemisphere that the ice at the edge of the cap where the bergs break off must in some cases be considerably over a mile in thickness, for icebergs of more

* Some writers have objected to the conclusion that the antarctic ice-cap is thickest at the pole, on the ground that the snowfall there is probably less than at lower latitudes. The fact is, however, overlooked, that the greater thickness of an ice-cap at its centre is a physical necessity not depending on the rate of snowfall. Supposing the snowfall to be greater at, say, lat. 70° than at 80°, and greater at 80° than at the pole ; nevertheless, the ice will continue to accumulate till it is thicker at 80° than at 70°, and at the pole than it is at 80°.

than a mile in thickness have been found in the southern hemisphere. The following are the dimensions of a few of these enormous bergs taken from the Twelfth Number of the Meteorological Papers published by the Board of Trade, and from the excellent paper of Mr. Towson on the Icebergs of the Southern Ocean, published also by the Board of Trade.* With one or two exceptions, the heights of the bergs were accurately determined by angular measurement:—

Sept. 10th, 1856.—The *Lightning*, when in lat. 55° 33′ S., long. 140° W., met with an iceberg 420 feet high.

Nov., 1839.—In lat. 41° S., long. 87° 30′ E., numerous icebergs 400 feet high were met with.

Sept., 1840.—In lat. 37° S., long. 15° E., an iceberg 1,000 feet long and 400 feet high was met with.

Feb., 1860.—Captain Clark, of the *Lightning*, when in lat. 55° 20′ S., long. 122° 45′ W., found an iceberg 500 feet high and 3 miles long.

Dec. 1st, 1859.—An iceberg, 580 feet high, and from two and a half to three miles long, was seen by Captain Smithers, of the *Edmond*, in lat. 50° 52′ S., long. 43° 58′ W. So strongly did this iceberg resemble land, that Captain Smithers believed it to be an island, and reported it as such, but there is little or no doubt that it was in reality an iceberg. There were pieces of drift-ice under its lee.

Nov., 1856.—Three large icebergs, 500 feet high, were found in lat. 41° 0′ S., long. 42° 0′ E.

Jan., 1861.—Five icebergs, one 500 feet high, were met with in lat. 55° 46′ S., long. 155° 56′ W.

Jan., 1861.—In lat. 56° 10′ S., long. 160° 0′ W., an iceberg 500 feet high and half a mile long was found.

Jan., 1867.—The barque *Scout*, from the West Coast of

* It is a pity that at present no record is kept, either by the Board of Trade or by the Admiralty, of remarkable icebergs which may from time to time be met with. Such a record might be of little importance to navigation, but it would certainly be of great service to science.

America, on her way to Liverpool, passed some icebergs 600 feet in height, and of great length.

April, 1864.—The *Royal Standard* came in collision with an iceberg 600 feet in height.

Dec., 1856.—Four large icebergs, one of them 700 feet high, and another 500 feet, were met with in lat. 50° 14′ S., long. 42° 54′ E.

Dec. 25th, 1861.—The *Queen of Nations* fell in with an iceberg in lat. 53° 45′ S., long. 170° 0′ W., 720 feet high.

Dec., 1856.—Captain P. Wakem, ship *Ellen Radford,* found, in lat. 52° 31′ S., long. 43° 43′ W., two large icebergs, one at least 800 feet high.

Mr. Towson states that one of our most celebrated and talented naval surveyors informed him that he had seen icebergs in the southern regions 800 feet high.

March 23rd, 1855.—The *Agneta* passed an iceberg in lat. 53° 14′ S., long. 14° 41′ E., 960 feet in height.

Aug. 16th, 1840.—The Dutch ship, *General Baron von Geen,* passed an iceberg 1,000 feet high in lat. 37° 32′ S., long. 14° 10′ E.

May 15th, 1859.—The *Roseworth* found in lat. 53° 40′ S., long. 123° 17′ W., an iceberg as large as "Tristan d'Acunha."

In the regions where most of these icebergs were met with, the mean density of the sea is about 1·0256. The density of ice is ·92. The density of icebergs to that of the sea is therefore as 1 to 1·115; consequently every foot of ice above water indicates 8·7 feet below water. It therefore follows that those icebergs 400 feet high had 3,480 feet under water,—3,880 feet would consequently be the total thickness of the ice. The icebergs which were 500 feet high would be 4,850 feet thick, those 600 feet high would have a total thickness of 5,820 feet, and those 700 feet high would be no less than 6,790 feet thick, which is more than a mile and a quarter. The iceberg 960 feet high, sighted by the *Agneta,* would be actually 9,312 feet thick, which is upwards of a mile and three-quarters.

Although the mass of an iceberg below water compared to

that above may be taken to be about 8·7 to 1, yet it would not be always safe to conclude that the thickness of the ice below water bears the same proportion to its height above. If the berg, for example, be much broader at its base than at its top, the thickness of the ice below water would bear a less proportion to the height above water than as 8·7 to 1. But a berg such as that recorded by Captain Clark, 500 feet high and three miles long, must have had only 1-8·7 of its total thickness above water. The same remark applies also to the one seen by Captain Smithers, which was 580 feet high, and so large that it was taken for an island. This berg must have been 5,628 feet in thickness. The enormous berg which came in collision with the *Royal Standard* must have been 5,820 feet thick. It is not stated what length the icebergs 730, 960, and 1,000 feet high respectively were; but supposing that we make considerable allowance for the possibility that the proportionate thickness of ice below water to that above may have been less than as 8·7 to 1, still we can hardly avoid the conclusion that the icebergs were considerably above a mile in thickness. But if there are icebergs above a mile in thickness, then there must be land-ice somewhere on the southern hemisphere of that thickness. In short, the great antarctic ice-cap must in some places be over a mile in thickness at its edge.

Inadequate Conceptions regarding the Magnitude of Continental Ice.—Few things have tended more to mislead geologists in the interpretation of glacial phenomena than inadequate conceptions regarding the magnitude of continental ice. Without the conception of continental ice the known facts connected with glaciation would be perfectly inexplicable. It was only when it was found that the accumulated facts refused to be explained by any other conception, that belief in the very existence of such a thing as continental ice became common. But although most geologists now admit the existence of continental ice, yet, nevertheless, adequate conceptions of its real magnitude are by no means so common. Year by year, as the outstanding facts connected with glaciation accumulate, we are compelled to extend our conceptions of the magnitude of land-ice. Take the following as an example. It was found that the transport of the Wastdale Crag blocks, the direction of the striæ on the islands of the Baltic, on Caithness and on the Orkney, Shetland, and Faroe, islands, the boulder clay with broken shells in Caithness, Holderness, and other places, were inexplicable on the theory of land-ice. But it was so only in consequence of the inadequacy of our conceptions of the magnitude of the ice; for a slight extension of our ideas of its thickness has explained not only these phenomena,* but others of an equally remarkable character, such as the striation of the Long Island and the submerged rock-basins around our coasts described by Mr. James Geikie. In like manner, if we admit the theory of the glacial epoch propounded in former chapters, all that is really necessary to account for the submergence of the land is a slight extension of our hitherto pre-conceived estimate of the thickness of the ice on the antarctic continent. If we simply admit a conclusion to which all physical considerations, as we have seen, necessarily lead us, viz., that the antarctic continent is covered with a mantle of ice at least two miles in thickness, we have then a complete explanation of the cause of the submergence of the land during the glacial epoch.

Although of no great importance to the question under consideration, it may be remarked that, except during the severest part of the glacial epoch, we have no reason to believe that the total quantity of ice on the globe was much greater than at present, only it would then be all on one hemisphere. Remove two miles of ice from the antarctic continent, and place it on the northern hemisphere, and this, along with the ice that now exists on this hemisphere, would equal, in all probability, the quantity existing on our hemisphere during the glacial epoch; at least, before it reached its maximum severity.

* See Chapter XXVII., and also Geol. Mag. for May and June, 1870, and January, 1871.

CHAPTER XXIV.

THE PHYSICAL CAUSE OF THE SUBMERGENCE AND EMERGENCE OF THE LAND DURING THE GLACIAL EPOCH.—*Continued.*

Extent of Submergence from Displacement of Earth's Centre of Gravity.—Circumstances which show that the Glacial Submergence resulted from Displacement of the Earth's Centre of Gravity.—Agreement between Theory and observed Facts.—Sir Charles Lyell on submerged Areas during Tertiary Period.—Oscillations of Sea-level in Relation to Distribution.—Extent of Submergence on the Hypothesis that the Earth is fluid in the Interior.

Extent of Submergence from Displacement of Earth's Centre of Gravity.—How much, then, would the transference of the two miles of ice from the southern to the northern hemisphere raise the level of the ocean on the latter hemisphere? This mass, be it observed, is equal to only one-half that represented in our section. A considerable amount of discussion has arisen in regard to the method of determining this point. According to the method already detailed, which supposes the rise at the pole to be equal to the extent of the displacement of the earth's centre of gravity, the rise at the North Pole would be about 380 feet, taking into account the effect produced by the displaced water; and the rise in the latitude of Edinburgh would be 312 feet. The fall of level on the southern hemisphere would, of course, be equal to the rise of level on the northern. According to the method advanced by Mr. D. D. Heath,* the rise of level at the North Pole would be about 650 feet. Archdeacon Pratt's method† makes the rise still greater; while according to Rev. O. Fisher's method‡ the rise would be no

* Phil. Mag. for April, 1866, p. 323.

† Ibid., for March, 1866, p. 172.

‡ *Reader*, February 10, 1866.

less than 2,000 feet. There is, however, another circumstance which must be taken into account, which will give an additional rise of upwards of one hundred feet.

The greatest extent of the displacement of the earth's centre of gravity, and consequently the greatest rise of the ocean resulting from that displacement, would of course occur at the time of maximum glaciation, when the ice was all on one hemisphere. But owing to the following circumstance, a still greater rise than that resulting from the displacement of the earth's centre of gravity alone might take place at some considerable time, either before or after the period of mxaimum glaciation.

It is not at all probable that the ice would melt on the warm hemisphere at exactly the same rate as it would form on the cold hemisphere. It is probable that the ice would melt more rapidly on the warm hemisphere than it would form on the cold. Suppose that during the glacial epoch, at a time when the cold was gradually increasing on the northern and the warmth on the southern hemisphere, the ice should melt more rapidly off the antarctic continent than it was being formed on the arctic and sub-arctic regions; suppose also that, by the time a quantity of ice, equal to one-half what exists at present on the antarctic continent, had accumulated on the northern hemisphere, the whole of the antarctic ice had been melted away, the sea would then be fuller than at present by the amount of water resulting from the one mile of melted ice. The height to which this would raise the general level of the sea would be as follows:—

The antarctic ice-cap is equal in area to $\frac{1}{23.46}$ of that covered by the ocean. The density of ice to that of water being taken at 92 to 1, it follows that 25 feet 6 inches of ice melted off the cap would raise the general level of the ocean one foot, and the one mile of ice melted off would raise the level 200 feet. This 200 feet of rise resulting from the melted ice we must add to the rise resulting from the displacement of the earth's centre of gravity. The removal of the two miles of ice from the antarctic continent would displace the centre of gravity 190 feet, and the formation of a mass of ice equal to the one-half of this on the arctic regions would carry the centre of gravity 95 feet farther; giving in all a total displacement of 285 feet, thus producing a rise of sea-level at the North Pole of 285 feet, and in the latitude of Edinburgh of 234 feet. Add to this the rise of 200 feet resulting from the melted ice, and we have then 485 feet of submergence at the pole, and 434 feet in the latitude of Edinburgh. A rise to a similar extent might probably take place after the period of maximum glaciation, when the ice would be melting on the northern hemisphere more rapidly than it would be forming on the southern.

If we assume the antarctic ice-cap to be as thick as is represented in the diagram, the extent of the submergence would of course be double the above, and we might have in this case a rise of sea-level in the latitude of Edinburgh to the extent of from 800 to 1,000 feet. But be this as it may, it is evident that the quantity of ice on the antarctic continent is perfectly sufficient to account for the submergence of the glacial epoch, for we have little evidence to conclude that the *general* submergence much exceeded 400 or 500 feet.* We have evidence in England and other places of submergence to the extent of from 1,000 to 2,000 feet, but these may be quite local, resulting from subsidence of the land in those particular areas. Elevations and depressions of the land have taken place in all ages, and no doubt during the glacial epoch also.

Circumstances which show that the Glacial Submergence resulted from Displacement of the Earth's Centre of Gravity.—In favour of this view of the cause of the submergence of the glacial epoch, it is a circumstance of some significance, that in every part of the globe where glaciation has been found evidence of the submergence of the land has also been found along with it. The invariable occurrence of submergence along with glaciation

* In a former paper I considered the effects of another cause, viz., the melting of polar ice resulting from an increase of the Obliquity of the Earth's Orbit.—Trans. Glasgow Geol. Soc., vol. ii., p. 177. Phil. Mag., June, 1867. See also Chapter XXV.

points to some physical connection between the two. It would seem to imply, either that the two were the direct effects of a common cause, or that the one was the cause of the other; that is, the submergence the cause of the glaciation, or the glaciation the cause of the submergence. There is, I presume, no known cause to which the two can be directly related as effects. Nor do I think that there is any one who would suppose that the submergence of the land could have been the cause of its glaciation, even although he attributed all glacial effects to floating ice. The submergence of our country would, of course, have allowed floating ice to pass over it had there been any to pass over; but submergence would not have produced the ice, neither would it have brought the ice from the arctic regions where it already existed. But although submergence could not have been the cause of the glacial epoch, yet we can, as we have just seen, easily understand how the ice of the glacial epoch could have been the cause of the submergence. If the glacial epoch was brought about by an increase in the eccentricity of the earth's orbit, then a submergence of the land as the ice accumulated was a physical necessity.

There is another circumstance connected with glacial submergence which it is difficult to reconcile with the idea that it resulted from a subsidence of the land. It is well known that during the glacial epoch the land was not once under water only, but several times; and, besides, there were not merely several periods when the land stood at a lower level in relation to the sea than at present, but there were also several periods when it stood at a much higher level than now. And this holds true, not merely of our own country, but of every country on the northern hemisphere where glaciation has yet been found. All this follows as a necessary consequence from the theory that the oscillations of sea-level resulted from the transference of the ice from the one hemisphere to the other; but it is wholly inconsistent with the idea that they resulted from upheavals and subsidence of the land during a very recent period.

But this is not all, there is more still to be accounted for. It has been the prevailing opinion that at the time when the land was covered with ice, it stood at a much greater elevation than at present. It is, however, not maintained that the facts of geology establish such a conclusion. The greater elevation of the land is simply assumed as an hypothesis to account for the cold.* The facts of geology, however, are fast establishing the opposite conclusion, viz., that when the country was covered with ice, the land stood in relation to the sea at a lower level than at present, and that the continental periods or times when the land stood in relation to the sea at a higher level than now were the warm inter-glacial periods, when the country was free of snow and ice, and a mild and equable condition of climate prevailed. This is the conclusion towards which we are being led by the more recent revelations of surface geology, and also by certain facts connected with the geographical distribution of plants and animals during the glacial epoch.

The simple occurrence of a rise and fall of the land in relation to the sea-level in one or in two countries during the glacial epoch, would not necessarily imply any physical connection. The coincidence of these movements with the glaciation of the land might have been purely accidental; but when we find that a succession of such movements occurred, not merely in one or in two countries, but in every glaciated country where proper observations have been made, we are forced to the conclusion that the connection between the two is not accidental, but the result of some fixed cause.

If we admit that an increase in the eccentricity of the earth's orbit was the cause of the glacial epoch, then we must admit that all those results followed as necessary consequences. For if the glacial epoch lasted for upwards of one hundred thousand years or so, there would be a succession of cold and warm periods, and consequently a succession of elevations and depressions of sea-level. And the elevations of the sea-level would

* Phil. Mag. for November, 1868, p. 376.

take place during the cold periods, and the depressions during the warm periods.

But the agreement between theory and observed facts does not terminate here. It follows from theory that the greatest oscillations of sea-level would take place during the severest part of the glacial epoch, when the eccentricity of the earth's orbit would be at its highest value, and that the oscillations would gradually diminish in extent as the eccentricity diminished and the climate gradually became less severe. Now it is well known that this is actually what took place; the great submergence, as well as the great elevation or continental period, occurred during the earlier or more severe part of the glacial epoch, and as the climate grew less severe these changes became of less extent, till we find them terminating in our submerged forests and 25-foot raised beach.

It follows, therefore, according to the theory advanced, that the mere fact of an area having been under sea does not imply that there has been any subsidence or elevation of the land, and that consequently the inference which has been drawn from these submerged areas as to changes in physical geography may be in many cases not well founded.

Sir Charles Lyell, in his "Principles," publishes a map showing the extent of surface in Europe which has been covered by the sea since the earlier part of the Tertiary period. This map is intended to show the extraordinary amount of subsidence and elevation of the land which has taken place during that period. It is necessary for Sir Charles's theory of the cause of the glacial epoch that changes in the physical geography of the globe to an enormous extent should have taken place during a very recent period, in order to account for the great change of climate which occurred at that epoch. But if the foregoing results be anything like correct, it does not necessarily follow that there must have been great changes in the physical geography of Europe, simply because the sea covered those areas marked in the map, for this may have been produced by oscillations of sea-level, and not by changes in the land. In fact, the areas marked in Sir Charles's map as having been covered by the sea, are just those which would be covered were the sea-level raised a few hundred feet. No doubt there were elevations and subsidences in many of the areas marked in the map during the Tertiary period, and to this cause a considerable amount of the submergence might be due; but I have little doubt that by far the greater part must be attributed to oscillations of sea-level. It is no objection that the greater part of the shells and other organic remains found in the marine deposits of those areas are not indicative of a cold or glacial condition of climate, for, as we have seen, the greatest submergence would probably have taken place either before the more severe cold had set in or after it had to a great extent passed away. That the submergence of those areas probably resulted from elevations of sea-level rather than depressions of the land, is further evident from the following considerations. If we suppose that the climate of the glacial epoch was brought about mainly by changes in the physical geography of the globe, we must assume that these great changes took place, geologically speaking, at a very recent date. Then when we ask what ground is there for assuming that any such change in the relations of sea and land as is required actually took place, the submergence of those areas is adduced as the proof. Did it follow as a physical necessity that all submergence must be the result of subsidence of the land, and not of elevations of the sea, there would be some force in the reasons adduced. But such a conclusion by no means follows, and, *à priori*, it is just as likely that the appearance of the ice was the cause of the submergence as that the submergence was the cause of the appearance of the ice. Again, a subsidence of the land to the extent required would to a great extent have altered the configuration of the country, and the main river systems of Europe; but there is no evidence that any such change has taken place. All the main valleys are well known to have existed prior to the glacial epoch, and our rivers to have occupied the same channels then as they do now. In the case of some of the smaller streams, it is true, a slight deviation has resulted at some points from the filling up of their channels with drift during the glacial epoch; but as a general rule all the principal valleys and river systems are older than the glacial epoch. This, of course, could not be the case if a subsidence of the land sufficiently great to account for the submergence of the areas in question, or changes in the physical geography of Europe necessary to produce a glacial epoch, had actually taken place. The total absence of any geological evidence for the existence of any change which could explain either the submergence of the areas in question or the climate of the glacial epoch, is strong evidence that the submergence of the glacial epoch, as well as of the areas in question, was the result of a simple oscillation of sea-level resulting from the displacement of the earth's centre of gravity by the transferrence of the ice-cap from the southern to the northern hemisphere.

Oscillations of Sea-level in relation to Distribution.—The oscillations of sea-level resulting from the displacement of the earth's centre of gravity help to throw new light on some obscure points connected with the subject of the geographical distribution of plants and animals. At the time when the ice was on the southern hemisphere during the glacial epoch, and the northern hemisphere was enjoying a warm and equable climate, the sea-level would be several hundred feet lower than at present, the North Sea would probably be dry land, and Great Britain and Ireland joined to the continent, thus opening up a pathway from the continent to our island. As has been shown in former chapters, during the inter-glacial periods the climate would be much warmer and more equable than now, so that animals from the south, such as the hippopotamus, hyæna, lion, *Elephas antiquus* and *Rhinoceros megarhinus*, would migrate into this country, where at present they could not live in consequence of the cold. We have therefore an explanation, as was suggested on a former occasion,* of the fact that the bones of these animals are found mingled in the same grave with those of the musk ox, mammoth, reindeer, and other animals which lived in this

* Phil. Mag., November, 1868.

country during the cold periods of the glacial epoch; the animals from the north would cross over into this country upon the frozen sea during the cold periods, while those from the south would find the English Channel dry land during the warm periods.

The same reasoning will hold equally true in reference to the old and new world. The depth of Behring Straits is under 30 fathoms; consequently a lowering of the sea-level of less than 200 feet would connect Asia with America, and thus allow plants and animals, as Mr. Darwin believes, to pass from the one continent to the other.* During this period, when Behring Straits would be dry land, Greenland would be comparatively free from ice, and the arctic regions enjoying a comparatively mild climate. In this case plants and animals belonging to temperate regions could avail themselves of this passage, and thus we can explain how plants belonging to temperate regions may have, during the Miocene period, passed from the old to the new continent, and *vice versâ*.

As has already been noticed, during the time of the greatest extension of the ice, the quantity of ice on the southern hemisphere might be considerably greater than what exists on the entire globe at present. In that case there might, in addition to the lowering of the sea-level resulting from the displacement of the earth's centre of gravity, be a considerable lowering resulting from the draining of the ocean to form the additional ice. This decrease and increase in the total quantity of ice which we have considered would affect the level of the ocean as much at the equator as at the poles; consequently during the glacial epoch there might have been at the equator elevations and depressions of sea-level to the extent of a few hundred feet.

Extent of Submergence on the Hypothesis that the Earth is fluid in the Interior.—But we have been proceeding upon the supposition that the earth is solid to its centre. If we assume, however, what is the general opinion among geologists, that it

* "Origin of Species," chap. xi. Fifth Edition.

consists of a fluid interior surrounded by a thick and rigid crust or shell, then the extent of the submergence resulting from the displacement of the centre of gravity for a given thickness of ice must be much greater than I have estimated it to be. This is evident, because, if the interior of the globe be in a fluid state, it, in all probability, consists of materials differing in density. The densest materials will be at the centre, and the least dense at the outside or surface. Now the transferrence of an ice-cap from the one pole to the other will not merely displace the ocean—the fluid mass on the outside of the shell—but it will also displace the heavier fluid materials in the interior of the shell. In other words, the heavier materials will be attracted by the ice-cap more forcibly than the lighter, consequently they will approach towards the cap to a certain extent, sinking, as it were, into the lighter materials, and displacing them towards the opposite pole. This displacement will of course tend to shift the earth's centre of gravity in the direction of the ice-cap, because the heavier materials are shifted in this direction, and the lighter materials in the opposite direction. This process will perhaps be better understood from the following figures.

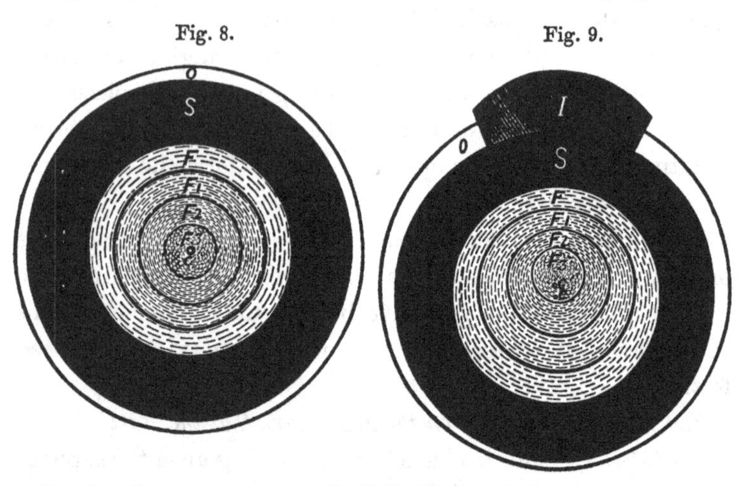

O. The Ocean. S. Solid Crust or Shell.
F, F^1, F^2, F^3. The various concentric layers of the fluid interior. The layers increase in density towards the centre.
I. The Ice-cap. C. Centre of gravity.
C^1. The displaced centre of gravity.

In Fig. 8, where there is no ice-cap, the centre of gravity of the earth coincides with the centre of the concentric layers of the fluid interior. In Fig. 9, where there is an ice-cap placed on one pole, the concentric layer F^1 being denser than layer F, is attracted towards the cap more forcibly than F, and consequently sinks to a certain depth in F. Again, F^2 being denser than F^1, it also sinks to a certain extent in F^1. And again F^3, the mass at the centre, being denser than F^2, it also sinks in F^2. All this being combined with the effects of the ice-cap, and the displaced ocean outside the shell, the centre of gravity of the entire globe will no longer be at C, but at C^1, a considerable distance nearer to the side of the shell on which the cap rests than C, and also a considerable distance nearer than it would have been had the interior of the globe been solid. There are here three causes tending to shift the centre of gravity, (1) the ice-cap, (2) the displaced ocean, and (3) the displaced materials in the interior. Two of the three causes mutually re-act on each other in such a way as to increase each other's effect. Thus the more the ocean is drawn in the direction of the ice-cap, the more effect it has in drawing the heavier materials in the interior in the same direction; and in turn the more the heavier materials in the interior are drawn towards the cap, the greater is the displacement of the earth's centre of gravity, and of course, as a consequence, the greater is the displacement of the ocean. It may be observed also that, other things being equal, the thinner the solid crust or shell is, and the greater the difference in the density of the fluid materials in the interior, the greater will be the extent of the displacement of the ocean, because the greater will be the displacement of the centre of gravity.

It follows that if we knew (1) the extent of the general submergence of the glacial epoch, and (2) the present amount of ice on the southern hemisphere, we could determine whether or not the earth is fluid in the interior.

CHAPTER XXV.

THE INFLUENCE OF THE OBLIQUITY OF THE ECLIPTIC ON CLIMATE AND ON THE LEVEL OF THE SEA.

The direct Effect of Change of Obliquity on Climate.—Mr. Stockwell on the maximum Change of Obliquity.—How Obliquity affects the Distribution of Heat over the Globe.—Increase of Obliquity diminishes the Heat at the Equator and increases that at the Poles.—Influence of Change of Obliquity on the Level of the Sea.—When the Obliquity was last at its superior Limit.—Probable Date of the 25-foot raised Beach.—Probable Extent of Rise of Sea-level resulting from Increase of Obliquity.—Lieutenant-Colonel Drayson's and Mr. Belt's Theories.—Sir Charles Lyell on Influence of Obliquity.

The direct Effect of Change in the Obliquity of the Ecliptic on Climate.—There is still another cause which, I feel convinced, must to a very considerable extent have affected climate during past geological ages. I refer to the change in the obliquity of the ecliptic. This cause has long engaged the attention of geologists and physicists, and the conclusion generally come to is that no great effect can be attributed to it. After giving special attention to the matter, I have been led to the very opposite conclusion. It is quite true, as has been urged, that the changes in the obliquity of the ecliptic cannot sensibly affect the climate of temperate regions; but it will produce a slight change on the climate of tropical latitudes, and a very considerable effect on that of the polar regions, especially at the poles themselves. We shall now consider the matter briefly.

It was found by Laplace that the obliquity of the ecliptic will oscillate to the extent of 1° 22′ 34″ on each side of 23° 28′, the obliquity in the year 1801.* This point has lately been

* Lieutenant-Colonel Drayson ("Last Glacial Epoch of Geology") and also Mr. Belt (Quart. Journ. of Science, October, 1874) state that Leverrier has lately investigated the question as to the extent of the variation of the plane of the

examined by Mr. Stockwell, and the results at which he has arrived are almost identical with those of Laplace. "The mean value of the obliquity," he says, "of both the apparent and fixed ecliptics to the equator is 23° 17′ 17″. The limits of the obliquity of the apparent ecliptic to the equator are 24° 35′ 58″ and 21° 58′ 36″; whence it follows that the greatest and least declinations of the sun at the solstices can never differ from each other to any greater extent than 2° 37′ 22″." *

This change will but slightly affect the climate of the temperate regions, but it will exercise a very considerable influence on the climate of the polar regions. According to Mr. Meech,† if 365·24 thermal days represent the present total annual quantity of heat received at the equator from the sun, 151·59 thermal days will represent the quantity received at the poles. Adopting his method of calculation, it turns out that when the obliquity of the ecliptic is at the maximum assigned by Laplace the quantity received at the equator would be 363·51 thermal days, and at the poles 160·04 thermal days. The equator would therefore receive 1·73 thermal days less heat, and the poles 8·45 thermal days more heat than at present.

ecliptic, and has arrived at results differing considerably from those of Laplace; viz., that the variation may amount to 4° 52′, whereas, according to Laplace, it amounts to only 1° 21′. I fear they are comparing things that are totally different; viz., the variation of the plane of the ecliptic in relation to its mean position with its variation in relation to the equator. Laplace estimated that the plane of the ecliptic would oscillate to the extent of 4° 53′ 33″ on each side of its mean position, a result almost identical with that of Leverrier, who makes it 4° 51′ 42″. But neither of these geometricians ever imagined that the ecliptic could change in relation to the equator to even one-third of that amount.

Laplace demonstrated that the change in the plane of the ecliptic affected the position of the equator, causing it to vary along with it, so that the equator could never possibly recede further than 1° 22′ 34″ from its mean position in relation to the ecliptic ("*Mécanique Céleste*," vol. ii., p. 856, Bowditch's Translation; see also Laplace's memoir, "Sur les Variations de l'Obliquité de l'Écliptique," *Connaissance des Temps* for 1827, p. 234), and I am not aware that Leverrier has arrived at a different conclusion.

* Memoir on the Secular Variations of the Elements of the Orbits of the Planets, "Smithsonian Contributions to Knowledge," vol. xvii.

† "Smithsonian Contributions to Knowledge," vol. ix.

ANNUAL AMOUNT OF SUN'S HEAT.

Amount in 1801. Obliquity 23° 28′.		Amount at maximum, 24° 50′ 34″.	Difference.
Latitude.	Thermal days.	Thermal days.	Thermal days.
0	365·24	363·51	− 1·73
40	288·55	288·32	− 0·23
70	173·04	179·14	+ 6·10
80	156·63	164·63	+ 8·00
90	151·59	160·04	+ 8·45

When the obliquity was at a maximum, the poles would therefore be receiving 19 rays for every 18 they are receiving at present. The poles would then be receiving nearly as much heat as latitude 76° is receiving at present.

The increase of obliquity would not sensibly affect the polar winter. It is true that it would slightly increase the breadth of the frigid zone, but the length of the winter at the poles would remain unaffected. After the sun disappears below the horizon his rays are completely cut off, so that a further descent of 1° 22′ 34″ would make no material difference in the climate. In the temperate regions the sun's altitude at the winter solstice would be 1° 22′ 34″ less than at present. This would slightly increase the cold of winter in those regions. But the increase in the amount of heat received by the polar regions would materially affect the condition of the polar summer. What, then, is the rise of temperature at the poles which would result from the increase of 8·45 thermal days in the total amount received from the sun?

An increase of 8·45 thermal days, or 1-18th of the total quantity received from the sun, according to the mode of calculation adopted in Chap. II. would produce, all other things being equal, a rise in the mean annual temperature equal to 14° or 15°.

According to Professor Dove* there is a difference of 7°·6 between the mean annual temperature of latitude 76° and the

* "Distribution of Heat on the Surface of the Globe," p. 14.

pole; the temperature of the former being 9°·8, and that of the latter 2°·2. Since it follows that when the obliquity of the ecliptic is at a maximum the poles would receive about as much heat per annum as latitude 76° receives at present, it may be supposed that the temperature of the poles at that period ought to be no higher than that of latitude 76° at the present time. A little consideration will, however, show that this by no means follows. Professor Dove's Tables represent correctly the mean annual temperature corresponding to every tenth degree of latitude from the equator to the pole. But it must be observed that the rate at which the temperature diminishes from the equator to the pole is not proportionate to the decrease in the total quantity of heat received from the sun as we pass from the equator to the pole. Were the mean annual temperature of the various latitudes proportionate to the amount of direct heat received, the equator would be much warmer than it actually is at present, and the poles much colder. The reason of this, as has been shown in Chapter II., is perfectly obvious. There is a constant transferrence of *heat* from the equator to the poles, and of *cold* from the poles to the equator. The warm water of the equator is constantly flowing towards the poles, and the cold water at the poles is constantly flowing to the equator. The same is the case in regard to the aërial currents. Consequently a great portion of the direct heat of the sun goes, not to raise the temperature of the equator, but to heat the poles. And, on the other hand, the cold materials at the poles are transferred to the equator, and thus lower the temperature of that part of the globe to a great extent. The present difference of temperature between lat. 76° and the pole, determined according to the rate at which the temperature is found to diminish between the equator and the pole, amounts to only about 7° or 8°. But were there no mutual transferrence of warm and cold materials between the equatorial and polar regions, and were the temperature of each latitude to depend solely upon the direct rays of the sun, the difference would far exceed that amount.

Now, when the obliquity of the ecliptic was at its superior limit, and the poles receiving about 1-18th more direct heat from the sun than at present, the increase of temperature due to this increase of heat would be far more than 7° or 8°. It would probably be nearly double that amount.

We may, therefore, conclude that when the obliquity of the ecliptic was at a maximum, and the poles were receiving 1-18th more heat than at present, the temperature of the poles ought to have been about 14° or 15° warmer than at the present day, *provided, of course, that this extra heat was employed wholly in raising the temperature.* Were the polar regions free from snow and ice, the greater portion of the extra heat would go to raise the temperature. But as those regions are covered with snow and ice, the extra heat would have no effect in raising the temperature, but would simply melt the snow and ice. The ice-covered surface upon which the rays fell could never rise above 32°. At the period under consideration, the total annual quantity of ice melted at the poles would be 1-18th more than at present.

The general effect which the change in the obliquity of the ecliptic would have upon the climate of the polar regions when combined with the effects resulting from the eccentricity of the earth's orbit would be this :—When the eccentricity was at a very high value, the hemisphere whose winter occurred in the aphelion (for physical reasons, which have already been discussed) * would be under a condition of glaciation, while the other hemisphere, having its winter in perihelion, would be enjoying a warm and equable climate. When the obliquity of the ecliptic was at a maximum, and 1-18th more heat falling at the poles than at present, the effect would be to modify to a great extent the rigour of the glaciation in the polar zone of the hemisphere under a glacial condition, and, on the other hand, to produce a more rapid melting of the ice on the other hemisphere enjoying the equable climate. The effects of eccentricity and obliquity thus combined would probably com-

* Chapter IV.

pletely remove the polar ice-cap from off the latter hemisphere, and forest-trees might then grow at the pole. Again, when the obliquity was at its minimum condition and less heat reaching the poles than at present, the glaciation of the former hemisphere would be increased and the warmth of the latter diminished.

The Influence of Change in the Obliquity of the Ecliptic on the Level of the Sea.—One very remarkable effect which seems to result indirectly from a variation of the obliquity under certain conditions, is an influence on the level of the sea. As this probably may have had something to do with those recent changes of sea-level with which the history of the submarine forests and raised beaches have made us all so familiar, it may be of interest to enter at some length into this part of this subject.

It appears almost certain that at the time when the northern winter solstice was in the aphelion last, a rise of the sea on the northern hemisphere to a considerable number of feet must have taken place from the combined effect of eccentricity and obliquity. About 11,700 years ago, the northern winter solstice was in the aphelion. The eccentricity at that time was 0187, being somewhat greater than it is now; but the winters occurring in aphelion instead of, as now, in perihelion, they would on that account be probably 10° or 15° colder than they are at the present day. It is probable, also, for reasons stated in a previous chapter, that the Gulf-stream at that time would be considerably less than now. This would tend to lower the temperature to a still greater extent. As snow instead of rain must have fallen during winter to a greater extent than at present, this no doubt must have produced a slight increase in the quantity of ice on the northern hemisphere had no other cause come into operation. But the condition of things, we have every reason to believe, must have been affected by the greater obliquity of the ecliptic at that period. We have no formula, except, perhaps, that given by Mr. Stockwell, from which to determine with perfect accuracy the extent of the obliquity at a period so remote as the one under consideration. If we adopt the formula given by Struve and Peters, which agrees pretty nearly with that obtained from Mr. Stockwell's formula, we have the obliquity at a maximum about the time that the solstice-point was in the aphelion. The formula given by Leverrier places the maximum somewhat later. At all events, we cannot be far from the truth in assuming that at the time the northern winter solstice was in the aphelion, the obliquity of the ecliptic would be about a maximum, and that since then it has been gradually diminishing. It is evident, then, that the annual amount of heat received by the arctic regions, and especially about the pole, would be considerably greater than at present. And as the heat received on those regions is chiefly employed in melting the ice, it is probable that the extra amount of ice which would then be melted in the arctic regions would prevent that slight increase of ice which would otherwise have resulted in consequence of the winter occurring in the aphelion. The winters at that period would be colder than they are at present, but the total quantity of ice on the northern hemisphere would not probably be greater.

Let us now turn to the southern hemisphere. As the southern winter would then occur in the perihelion, this would tend to produce a slight decrease in the quantity of ice on the southern hemisphere. But on this hemisphere the effects of eccentricity would not, as on the northern hemisphere, be compensated by those of obliquity; for both causes would here tend to produce the same effect; namely, a melting of the ice in the antarctic regions.

It is probable that at this time the quantity of warm water flowing from the equatorial regions into the Southern Ocean would be much greater than at present. This would tend to raise the temperature of the air of the antarctic regions, and thus assist in melting the ice. These causes, combined with the great increase of heat resulting from the change of obliquity, would tend to diminish to a considerable extent the quantity of ice on the southern hemisphere. I think we may assume that the slight increase of eccentricity at that period, the occurrence of the southern winter in perihelion, and the extra quantity of warm water flowing from the equatorial to the antarctic regions, would produce an effect on the south polar ice-cap equal to that produced by the increase in the obliquity of the ecliptic. It would, therefore, follow that for every eighteen pounds of ice melted annually at present at the south pole twenty pounds would then be melted.

Let us now consider the effect that this condition of things would have upon the level of the sea. It would evidently tend to produce an elevation of the sea-level on the northern hemisphere in two ways. 1st. The addition to the sea occasioned by the melting of the ice from off the antarctic land would tend to raise the general level of the sea. 2ndly. The removal of the ice would also tend to shift the earth's centre of gravity to the north of its present position—and as the sea must shift along with the centre, a rise of the sea on the northern hemisphere would necessarily take place.

The question naturally suggests itself, might not the last rise of the sea, relative to the land, have resulted from this cause? We know that during the period of the 25-foot beach, the time when the estuarine mud, which now forms the rich soil of the Carses of the Forth and Tay, was deposited, the sea, in relation to the land, stood at least 20 or 30 feet higher than at present. But immediately prior to this period we have the age of the submarine forests and peat-beds, when the sea relative to the land stood lower than it does now. We know also that these changes of level were not mere local affairs. There seems every reason to believe that our Carse clay, as Mr. Fisher states, is the equivalent of the marine mud, with *Scrobicularia*, which covers the submarine forests of England.* And on the other hand, those submarine forests are not confined to one locality. "They may be traced," says Mr. Jamieson, "round the whole of Britain and Ireland, from Orkney to Cornwall, from Mayo to the shores of Fife, and even, it would seem, along a great part of the western sea-board of Europe,

* Quart. Journ. Geol. Soc., June, 1866, p. 564.

as if they bore witness to a period of wide-spread elevation, when Ireland and Britain, with all its numerous islands, formed one mass of dry land, united to the continent, and stretching out into the Atlantic."* "These submarine forests," remarks De la Beche, also, "are to be found under the same general condition from the shores of Scandinavia to those of Spain and Portugal, and around the British islands."† Those buried forests are not confined to Europe, but are found in the valley of the Mississippi and in Nova Scotia, and other parts of North America. And again, the strata which underlie those forests and peat-beds bear witness to the fact of a previous elevation of the sea-level. In short, we have evidence of a number of oscillations of sea-level during post-tertiary times.‡

Had there been only one rise of the land relative to the sea-level, or one depression, it might quite reasonably, as already remarked, have been attributed to an upheaval or a sinking of the ground, occasioned by some volcanic, chemical, or other agency. But certainly those repeated oscillations of sea-level, extending as they do over so wide an area, look more like a rising and sinking of the sea than of the land. But, be this as it may, since it is now established, I presume, beyond controversy, that the old notion that the general level of the sea remains permanent, and that the changes must be all attributed to the land is wholly incorrect, and that the sea, as well as the land, is subject to changes of level, it is certainly quite legitimate to consider whether the last elevation of the sea-level relatively to the land may not have resulted from the rising of the sea rather than from the sinking of the land, in short, whether it may not be attributed to the cause we are now considering. The fact that those raised beaches and terraces are found at so many different heights, and also so discon-

* Quart. Journ. Geol. Soc., vol. xxi., p. 186.

† "Geological Observer," p. 446. See also Mr. James Geikie's valuable Memoir, "On the Buried Forests and Peat Mosses of Scotland." Trans. of the Royal Society of Edinburgh, vol. xxiv., and Chambers' "Ancient Sea-Margins."

‡ See Lyell's "Antiquity of Man," Second Edition, p. 282; "Elements," Sixth Edition, p. 162.

tinuously along our coasts, might be urged as an objection to the opinion that they were due to changes in the level of the sea itself. Space will not permit me to enter upon the discussion of this point at present; but it may be stated that this objection is more apparent than real. It by no means follows that beaches of the same age must be at the same level. This has been shown very clearly by Mr. W. Pengelly in a paper on "Raised Beaches," read before the British Association at Nottingham, 1866.

We have, as I think, evidence amounting to almost absolute certainty that 11,700 years ago the general sea-level on the northern hemisphere must have been higher than at present. And in order to determine the question of the 25-foot beach, we have merely to consider whether a rise to something like this extent probably took place at the period in question. We have at present no means of determining the exact extent of the rise which must have taken place at that period, for we cannot tell what quantity of ice was then melted off the antarctic regions. But we have the means of making a very rough estimate, which, at least, may enable us to determine whether a rise of some 20 or 30 feet may not possibly have taken place.

If we assume that the southern ice-cap extends on an average down to lat. 70°, we shall have an area equal to $\frac{1}{33 \cdot 163}$ of the entire surface of the globe. The proportion of land to that of water, taking into account the antarctic continent, is as 526 to 1272. The southern ice-cap will therefore be equal to $\frac{1}{23 \cdot 46}$ of the area covered by water. The density of ice to that of water being taken at ·92 to 1, it follows that 25 feet 6 inches of ice melted from off the face of the antarctic continent would raise the level of the ocean one foot. If 470 feet were melted off—and this is by no means an extravagant supposition, when we reflect that for every 18 pounds of ice presently melted an additional pound or two pounds, or perhaps more, would then be melted, and that for many ages in succession—the water thus produced from the melted ice would raise the level of the sea 18 feet 5 inches. The removal of the 470 feet of solid ice—

which must be but a very small fraction of the total quantity of ice lying upon the antarctic continent—would shift the earth's centre of gravity about 7 feet to the north of its present position. The shifting of the centre of gravity would cause the sea to sink on the southern hemisphere and rise on the northern. And the quantity of water thus transferred from the southern hemisphere to the northern would carry the centre of gravity about one foot further, and thus give a total displacement of the centre to the extent of about 8 feet. The sea would therefore rise about 8 feet at the North Pole, and in the latitude of Edinburgh about 6 feet 7 inches. This, added to the rise of 18 feet 5 inches, occasioned by the melting of the ice, would give 25 feet as the total rise in the latitude of Scotland 11,700 years ago.

Each square foot of surface at the poles 11,700 years ago would be receiving 18,223,100 foot-pounds more of heat annually than at present. If we deduct 22 per cent. as the amount absorbed in passing through the atmosphere, we have 14,214,000 foot-pounds. This would be sufficient to melt 2·26 feet of ice. But if 50, instead of 22, per cent. were cut off, 1·45 cubic feet would be melted. In this case the 470 feet of ice would be melted, independently of the effects of eccentricity, in about 320 years. And supposing that only one-fourth part of the extra heat reached the ground, 470 feet of ice would be removed in about 640 years.

As to the exact time that the obliquity was at a maximum, previous to that of 11,700 years ago, our uncertainty is still greater. If we are permitted to assume that the ecliptic passes from its maximum to its minimum state and back to its maximum again with anything like uniformity, at the rate assigned by Leverrier and others, the obliquity would not be far from a maximum about 60,000 years ago. Taking the rate of precession at $50''\cdot21129$, and assuming it to be uniform—which it probably is not—the winter solstice would be in the aphelion about 61,300 years ago.* In short, it seems not at all improbable that

* In order to determine the position of the solstice-point in relation to the aphelion, it will not do to assume, as is commonly done, that the point makes a

at the time the solstice-point was in the aphelion, the obliquity of the ecliptic would not be far from its maximum state. But at that time the value of the eccentricity was 0·023, instead of 0·0187, its value at the last period. Consequently the rise of the sea would probably be somewhat greater than it was 11,700 years ago. Might not this be the period of the 40-foot beach? In this case 11,000 or 12,000 years would be the age of the 25-foot beach, and 60,000 years the age of the 40-feet beach.

About 22,000 years ago the winter solstice was in the perihelion, and as the eccentricity was then somewhat greater than it is at present, the winters would be a little warmer and the climate more equable than it is at the present day. This perhaps might be the period of the submarine forests and lower peat-beds which underlie the Carse clays, *Scrobicularia* mud, and other deposits belonging to the age of the 25-feet beach. At any rate it is perfectly certain that a condition of climate at this period prevailed exceedingly favourable to the growth of peat. It follows also that at this time, owing to a greater accumulation of ice on the southern hemisphere, the sea-level would be a few feet lower than at present, and that forests and peat may have then grown on places which are now under the sea-level.

For a few thousand years before and after 11,700 years ago, when the winter solstice was evidently not far from the aphelion, and the sea standing considerably above its present level, would probably, as we have already stated, be the time when the Carse clays and other recent deposits lying above the present level of the river were formed. And it is also a singular fact that the condition of things at that period must

revolution from aphelion to aphelion in any regular given period, such as 21,000 years; for it is perfectly evident that owing to the great irregularity in the motion of the aphelion, no two revolutions will probably be performed in the same length of period. For example, the winter solstice was in the aphelion about the following dates: 11,700, 33,300, and 61,300 years ago. Here are two consecutive revolutions, the one performed in 21,600 years and the other in 28,000 years; the difference in the length of the two periods amounting to no fewer than 6,400 years.

have been exceedingly favourable to the formation of such estuarine deposits; for at that time the winter temperature of our island, as has been already shown, would be considerably lower than at present, and, consequently, during that season, snow, to a much larger extent than now, would fall instead of rain. The melting of the winter's accumulation of snow on the approach of summer would necessarily produce great floods, similar to what occur in the northern parts of Asia and America at the present day from this very same cause. The loose upper soil would be carried down by those floods and deposited in the estuaries of our rivers.

The foregoing is a rough and imperfect sketch of the history of the climate and the physical conditions of our globe for the past 60,000 years, in so far as physical and cosmical considerations seem to afford us information on the subject, and its striking agreement with that derived from geological sources is an additional evidence in favour of the opinion that geological and cosmical phenomena are physically related by a bond of causation.

Lieutenant-Colonel Drayson's Theory of the Cause of the Glacial Epoch.—In a paper read before the Geological Society by Lieutenant-Colonel Drayson, R.A., on the 22nd February, 1871,* that author states, that after calculating from the recorded positions of the pole of the heavens during the last 2,000 years, he finds the pole of the ecliptic is not the centre of the circle traced by the pole of the heavens. The pole of the heavens, he considers, describes a circle round a point 6° distant from the pole of the ecliptic and 29° 25′ 47″ from the pole of the heavens, and that about 13,700 years B.C. the angular distance of the two poles was 35° 25′ 47″. This would bring the Arctic Circle down to latitude 54° 34′ 13″ N. I fear that this is a conclusion that will not be generally accepted by those familiar with celestial mechanics. But, be this as it may, my present object is not to discuss the astronomical part of Colonel Drayson's theory, but

* Quart. Journ. Geol. Soc., vol. xxvii., p. 232. See also "The Last Glacial Epoch of Geology," by the same author.

to consider whether the conclusions which he deduces from his theory in regard to the cause of the glacial epoch be legitimate or not. Assuming for argument's sake that the obliquity of the ecliptic can possibly reach to 35° or 36°, so as to bring the Arctic Circle down to the centre of England, would this account for the glacial epoch? Colonel Drayson concludes that the shifting of the Arctic Circle down to the latitude of England would induce here a condition of climate similar to that which obtains in arctic regions. This seems to be the radical error of the theory. It is perfectly true that were the Arctic Circle brought down to latitude 54° 35′ part of our island would be in the arctic regions, but it does not on that account follow that our island would be subjected to an arctic climate.

The polar regions owe their cold not to the obliquity of the ecliptic, but to their distance from the equator. Indeed were it not for obliquity those regions would be much colder than they really are, and an increase of obliquity, instead of increasing their cold, would really make them warmer. The general effect of obliquity, as we have seen, is to diminish the amount of heat received in equatorial and tropical regions, and to increase it in the polar and temperate regions. The greater the obliquity, and, consequently, the farther the sun recedes from the equator, the smaller is the quantity of heat received by equatorial regions, and the greater the amount bestowed on polar and temperate regions. If, for example, we represent the present amount of heat received from the sun at the equator on a given surface at 100 parts, 42·47 parts will then represent the amount received at the poles on the same given surface. But were the obliquity increased to 35° the amount received at the equator would be reduced to 94·93 parts, and that at the poles increased to 59·81; being an increase at the poles of nearly one half. At latitude 60° the present quantity is equal to 57 parts; but about 63 parts would be received were the obliquity increased to 35°. It therefore follows that although the Arctic Circle were brought down to the latitude of London so that the British islands would become a part of the arctic regions, the mean temperature of these islands would not be lowered, but the reverse. The winters would no doubt be colder than they are at present, but the cold of winter would be far more than compensated for by the heat of summer. It is not a fair representation of the state of things, merely to say that an increase of obliquity tends to make the winters colder and the summers hotter, for it affects the summer heat far more than it does the winter cold. And the greater the obliquity the more does the increase of heat during summer exceed the decrease during winter. This is obvious because the greater the obliquity the greater the total annual amount of heat received.

If an increase of obliquity tended to produce an increase of ice in temperate and polar regions, and thus to lead to a glacial epoch, then the greater the obliquity the greater would be the tendency to produce such an effect. Conceive, then, the obliquity to go on increasing until it ultimately reached its absolute limit, 90°, and the earth's axis to coincide with the plane of the ecliptic. The Arctic Circle would then extend to the equator. Would this produce a glacial epoch? Certainly not. A square foot of surface at the poles would then be receiving as much heat per annum as a square foot at the equator at present, supposing the sun remained on the equator during the entire year. Less heat, however, would be reaching the equatorial regions than now. At present, as we have just seen, the annual quantity of heat received at either pole is to that received at the equator as 42·47 to 100; but at the period under consideration the poles would be actually obtaining one-half more heat than the equator. The amount received per square foot at the poles, to that received per square foot at the equator, would be in the ratio of half the circumference of a circle to its diameter, or as 1·5708 to 1. But merely to say that the poles would be receiving more heat per annum than the equator is at present, does not convey a correct idea of the excessive heat which the poles would then have to endure; for it must be borne in mind that the heat reaching the equator is spread over the whole year, whereas the poles would get their total amount during the six months of their summer. Consequently, for six months in the year the poles would be obtaining far more than double the quantity of heat received at present by the equator during the same length of time, and more than three times the quantity then received by the equator. The amount reaching the pole during the six months to that reaching the equator would be as 3·1416 to 1.

At the equator twelve hours' darkness alternates with twelve hours' sunshine, and this prevents the temperature from rising excessively high; but at the poles it would be continuous sunshine for six months without the ground having an opportunity of cooling for a single hour. At the summer solstice, when the sun would be in the zenith of the pole, the amount of heat received there every twenty-four hours would actually be nearly three-and-a-quarter times greater than that presently received at the equator. Now what holds true with regard to the poles would hold equally true, though to a lesser extent, of polar and temperate regions. We can form but a very inadequate idea of the condition of things which would result from such an enormous increase of heat. Nothing living on the face of the globe could exist in polar regions under so fearful a temperature as would then prevail during summer months. How absurd would it be to suppose that this condition of things would tend to produce a glacial epoch! Not only would every particle of ice in polar regions be dissipated, but the very seas around the pole would be, for several months in the year, at the boiling point.

If it could be shown from *physical principles*—which, to say the least, is highly improbable—that the obliquity of the ecliptic could ever have been as great as 35°, it would to a very considerable extent account for the comparative absence of ice in Greenland and other regions in high latitudes, such as we know was the case during the Carboniferous, Miocene, and other periods. But although a great increase of obliquity might cause a melting of the ice, yet it could not produce that mild condition of climate which we know prevailed in high latitudes during those periods; while no increase of obliquity, however great, could in any way tend to produce a glacial epoch.

Colonel Drayson, however, seems to admit that this great increase of obliquity would make our summers much warmer than they are at present. How, then, according to his theory, is the glacial epoch accounted for? The following is the author's explanation as stated in his own words:—

"At the date 13,700 B.C. the same conditions appear to have prevailed down to about 54° of latitude during winter as regards the sun being only a few degrees above the horizon. We are, then, warranted in concluding that the same climate prevailed down to 54° of latitude as now exists in winter down to 67° of latitude.

"Thus in the greater part of England and Wales, and in the whole of Scotland, icebergs of large size would be *formed each winter;* every river and stream would be frozen and blocked with ice, the whole country would be covered with a mantle of snow and ice, and those creatures which could neither migrate nor endure the cold of an arctic climate would be exterminated." —"The Last Glacial Epoch," p. 146.

"At the summer solstice the midday altitude of the sun for the latitude 54° would be about 71½°, an altitude equal to that which the sun now attains in the south of Italy, the south of Spain, and in all localities having a latitude of about 40°."

"There would, however, be this singular difference from present conditions, that in latitude 54° the sun at the period of the summer solstice would remain the whole twenty-four hours above the horizon; a fact which would give extreme heat to those very regions which, six months previously, had been subjected to an arctic cold. Not only would this greatly increased heat prevail in the latitude of 54°, but the sun's altitude would be 12° greater at midday in midsummer, and also 12° greater at midnight in high northern latitudes, than it ever attains now; consequently the heat would be far greater than at present, and high northern regions, even around the pole itself, would be subjected to a heat during summer far greater than any which now ever exists in those localities. The natural consequence would be, that the icebergs and ice which had during the severe winter accumulated in high latitudes would be rapidly thawed by this heat" (p. 148).

"Each winter the whole northern and southern hemispheres would be one mass of ice; each summer nearly the whole of the ice of each hemisphere would be melted and dispersed" (p. 150).

According to this theory, not only is the whole country covered each winter with a continuous mass of ice, but large icebergs are formed during that short season, and when the summer heat sets in all is melted away. Here we have a misapprehension not only as to the actual condition of things during the glacial epoch, but even as to the way in which icebergs and land-ice are formed. Icebergs are formed from land-ice, but land-ice is not formed during a single winter, much less a mass of sufficient thickness to produce icebergs. Land-ice of this thickness requires the accumulated snows of centuries for its production. All that we could really have, according to this theory, would be a thick covering of snow during winter, which would entirely disappear during summer, so that there could be no land-ice.

Mr. Thomas Belt's Theory.—The theory that the glacial epoch resulted from a great increase in the obliquity of the ecliptic has recently been advocated by Mr. Thomas Belt.* His conceptions on the subject, however, appear to me to be even more irreconcilable with physics than those we have been considering. Lieutenant-Colonel Drayson admits that the increase of heat to polar regions resulting from the great increase of obliquity would dissipate the ice there, but Mr. Belt does not even admit that an increase of obliquity would bring with it an increase of heat, far less that it would melt the polar ice.

* Quart. Journ. of Science, October, 1874.

On the contrary, he maintains that the tendency of obliquity is to increase the rigour of polar climate, and that this is the reason "that now around the poles some lands are being glaciated, for excepting for that obliquity snow and ice would not accumulate, excepting on mountain chains." "Thus," he says, "there exist glacial conditions at present around the poles, due primarily to the obliquity of the ecliptic." And he also maintains that if there were no obliquity and the earth's axis were perpendicular to the plane of its orbit, an eternal "spring would reign around the arctic circle," and that "under such circumstances the piling up of snow, or even its production at the sea-level, would be impossible, excepting perhaps in the immediate neighbourhood of the poles, where the rays of the sun would have but little heating power from its small altitude."

Mr. Belt has apparently been led to these strange conclusions by the following singular misapprehension of the effects of obliquity on the distribution of the sun's heat over the globe. "The obliquity of the ecliptic," he remarks, "*does not affect the mean amount of heat received at any one point from the sun,* but it causes the heat and the cold to predominate at different seasons of the year."

It is not necessary to dwell further on the absurdity of the supposition that an increase of obliquity can possibly account for the glacial epoch, but we may in a few words consider whether a decrease of obliquity would mitigate the climate and remove the snow from polar regions. Supposing obliquity to disappear and the earth's axis to become perpendicular to the plane of its orbit, it is perfectly true that day and night would be equal all over the globe, but then the quantity of heat received by the polar regions would be far less than at present. It is well known that at present at the equinoxes, when day and night are equal, snow and not rain prevails in the arctic regions, and can we suppose it could be otherwise in the case under consideration? How, we may well ask, could these regions, deprived of their summer, get rid of their snow and ice?

But even supposing it could be shown that a change in the obliquity of the ecliptic to the extent assumed by Mr. Belt and Lieutenant-Colonel Drayson would produce a glacial epoch, still the assumption of such a change is one which physical astronomy will not permit. Mr. Belt does not appear to dispute the accuracy of the methods by which it is proved that the variations of obliquity are confined within narrow limits; but he maintains that physical astronomers in making their calculations have left out of account some circumstances which materially affect the problem. These, according to Mr. Belt, are the following:—(1) Upheavals and subsidences of the land which may have taken place in past ages. (2) The unequal distribution of sea and land on the globe. (3) The fact that the equatorial protuberance is not a regular one, "but approaches in a general outline to an ellipse, of which the greater diameter is two miles longer than the other." (4) The heaping up of ice around the poles during the glacial period.

We may briefly consider whether any or all of these can sensibly affect the question at issue. In reference to the last-mentioned element, it is no doubt true that if an immense quantity of water were removed from the ocean and placed around the poles in the form of ice it would affect the obliquity of the ecliptic; but this is an element of change which is not available to Mr. Belt, because according to his theory the piling up of the ice is an effect which results from the change of obliquity.

In reference to the difference of two miles in the equatorial diameters of the earth, the fact must be borne in mind that the longer diameter passes through nearly the centre of the great depression of the Pacific Ocean,* whereas the shorter diameter passes through the opposite continents of Asia and America. Now, when we take into consideration the fact that these continents are not only two-and-a-half times denser than the ocean, but have a mean elevation of about 1,000 feet above the sea-level, it becomes perfectly obvious that the earth's mass must

* The longer diameter passes from long. 14° 23′ E. to long. 165° 37′ W.

be pretty evenly distributed around its axis of rotation, and that therefore the difference in the equatorial diameters can exercise no appreciable effect on the change of obliquity. It follows also that the present arrangement of sea and land is the best that could be chosen to prevent disturbance of motion.

That there ever were upheavals and depressions of the land of so enormous a magnitude as to lead to a change of obliquity to the extent assumed by Lieutenant-Colonel Drayson and Mr. Belt is what, I presume, few geologists would be willing to admit. Suppose the great table-land of Thibet, with the Himalaya Mountains, were to sink under the sea, it would hardly produce any sensible effect on the obliquity of the ecliptic. Nay more; supposing that all the land in the globe were sunk under the sea-level, or the ocean beds converted into dry land, still this would not materially affect obliquity. The reason is very obvious. The equatorial bulge is so immense that those upheavals and depressions would not to any great extent alter the oblate form of the earth. The only cause which could produce any sensible effect on obliquity, as has already been noticed, would be the removal of the water of the ocean and the piling of it up in the form of ice around the poles; but this is a cause which is not available to Mr. Belt.

Sir Charles Lyell's Theory.—I am also unable to agree with Sir Charles Lyell's conclusions in reference to the influence of the obliquity of the ecliptic on climate. Sir Charles says, "It may be remarked that if the obliquity of the ecliptic could ever be diminished to the extent of four degrees below its present inclination, such a deviation would be of geological interest, in so far as it would cause the sun's light to be disseminated over a broader zone inside of the arctic and anctartic circles. Indeed, if the date of its occurrence in past time could be ascertained, this greater spread of the solar rays, implying a shortening of the polar night, might help in some slight degree to account for a vegetation such as now characterizes lower latitudes, having had in the Miocene and Carboniferous periods a much wider range towards the pole."*

* "Principles," vol. i., p. 294. Eleventh Edition.

The effects, as we have seen, would be directly the reverse of what is here stated, viz., the more the obliquity was diminished the *less* would the sun's rays spread over the arctic and antarctic regions, and conversely the more the obliquity was increased the *greater* would be the amount of heat spread over polar latitudes. The farther the sun recedes from the equator, the greater becomes the amount of heat diffused over the polar regions; and if the obliquity could possibly attain its absolute limit (90°), it is obvious that the poles would then be receiving more heat than the equator is now.

CHAPTER XXVI.

COAL AN INTER-GLACIAL FORMATION.

Climate of Coal Period Inter-glacial in Character.—Coal Plants indicate an Equable, not a Tropical Climate.—Conditions necessary for Preservation of Coal Plants.—Oscillations of Sea-level necessarily implied.—Why our Coal-fields contain more than One Coal-seam.—Time required to form a Bed of Coal.—Why Coal Strata contain so little evidence of Ice-action.—Land Flat during Coal Period.—Leading Idea of the Theory.—Carboniferous Limestones.

An Inter-glacial Climate the one best suited for the Growth of the Coal Plants.—No assertion, perhaps, could appear more improbable, or is more opposed to all hitherto received theories, than the one that the plants which form our coal grew during a glacial epoch. But, nevertheless, if the theory of secular changes of climate, discussed in the foregoing chapters, be correct, we have in warm inter-glacial periods (as was pointed out several years ago)* the very condition of climate best suited for the growth of those kinds of trees and vegetation of which our coal is composed. It is the generally received opinion among both geologists and botanists that the flora of the Coal period does not indicate the existence of a tropical, but a moist, equable, and temperate climate. "It seems to have become," says Sir Charles Lyell, "a more and more received opinion that the coal plants do not on the whole indicate a climate resembling that now enjoyed in the equatorial zone. Tree-ferns range as far south as the southern parts of New Zealand, and Araucanian pines occur in Norfolk Island. A great preponderance of ferns and lycopodiums

* Phil. Mag. for August, 1864.

indicates moisture, equability of temperature, and freedom from frost, rather than intense heat."*

Mr. Robert Brown, the eminent botanist, considers that the rapid and great growth of many of the coal plants showed that they grew in swamps and shallow water of equable and genial temperature.

"Generally speaking," says Professor Page, "we find them resembling equisetums, marsh-grasses, reeds, club-mosses, tree-ferns, and coniferous trees; and these in existing nature attain their maximum development in warm, temperate, and sub-tropical, rather than in equatorial regions. The Wellingtonias of California and the pines of Norfolk Island are more gigantic than the largest coniferous tree yet discovered in the coal-measures."†

The Coal period was not only characterized by a great preponderance over the present in the quantity of ferns growing, but also in the number of different species. Our island possesses only about 50 species, while no fewer than 140 species have been enumerated as having inhabited those few isolated places in England over which the coal has been worked. And Humboldt has shown that it is not in the hot, but in the mountainous, humid, and shady parts of the equatorial regions that the family of ferns produces the greatest number of species.

"Dr. Hooker thinks that a climate warmer than ours now is, would probably be indicated by the presence of an increased number of flowering plants, which would doubtless have been fossilized with the ferns; whilst a lower temperature, *equal to the mean of the seasons now prevailing*, would assimilate our climate to that of such cooler countries as are characterized by a disproportionate amount of ferns." ‡

"The general opinion of the highest authorities," says Professor Hull, "appears to be that the climate did not resemble that of the equatorial regions, but was one in which

* "Elementary Geology," p. 399.

† "The Past and Present Life of the Globe," p. 102.

‡ "Memoirs of the Geological Survey," vol. ii., Part 2, p. 404.

the temperature was free from extremes; the atmosphere being warm and moist, somewhat resembling that of New Zealand and the surrounding islands, which we endeavour to imitate artificially in our hothouses."*

The enormous quantity of the carboniferous vegetation shows also that the climate under which it grew could not have been of a tropical character, or it must have been decomposed by the heat. Peat, so abundant in temperate regions, is not to be found in the tropics.

The condition most favourable to the preservation of vegetable remains, at least under the form of peat, is a cool, moist, and equable climate, such as prevails in the Falkland Islands at the present day. "In these islands," says Mr. Darwin, "almost every kind of plant, even the coarse grass which covers the whole surface of the land, becomes converted into this substance."†

From the evidence of geology we may reasonably infer that were the difference between our summer and winter temperature nearly annihilated, and were we to enjoy an equable climate equal to, or perhaps a little above, the present mean annual temperature of our island, we should then have a climate similar to what prevailed during the Carboniferous epoch.

But we have already seen that such must have been the character of our climate at the time that the eccentricity of the earth's orbit was at a maximum, and winter occurred when the earth was in the perihelion of its orbit. For, as we have already shown, the earth would in such a case be 14,212,700 miles nearer to the sun in winter than in summer. This enormous difference, along with other causes which have been discussed, would almost extinguish the difference between summer and winter temperature. The almost if not entire absence of ice and snow, resulting from this condition of things, would, as has already been shown, tend to raise the

* "Coal Fields of Great Britain," p. 45. Third Edition.
† "Journal of Researches," chap. xiii.

mean annual temperature of the climate higher than it is at present.

Conditions necessary for the Preservation of the Coal Plants.— But in order to the formation of coal, it is not simply necessary to have a condition of climate suitable for the growth, but also for the preservation, of a luxuriant vegetation. The very existence of coal is as much due to the latter circumstance as to the former; nay more, as we shall yet see, the fact that a greater amount of coal belongs to the Carboniferous period than to any other, was evidently due not so much to a more extensive vegetable growth during that age, suited to form coal, as to the fact that that flora has been better preserved. Now, as will be presently shown, we have not merely in the warm periods of a glacial epoch a condition of climate best suited for the growth of coal plants, but we have also in the cold periods of such an epoch the condition most favourable for the preservation of those plants.

One circumstance necessary for the preservation of plants is that they should have been covered over by a thick deposit of sand, mud, or clay, and for this end it is necessary that the area upon which the plants grew should have become submerged. It is evident that unless the area had become submerged, the plants could not have been covered over with a thick deposit; and, even supposing they had been covered over, they could not have escaped destruction from subaërial denudation unless the whole had been under water. Another condition favourable, if not essential, to the preservation of the plants, is that they should have been submerged in a cold and not in a warm sea. Assuming that the coal plants grew during a warm period of a glacial epoch, we have in the cold period which succeeded all the above conditions necessarily secured.

It is now generally admitted that the coal trees grew near broad estuaries and on immense flat plains but little elevated above sea-level. But that the *Lepidodendra*, *Sigillariæ*, and other trees, of which our coal is almost wholly composed, grew on dry ground, elevated above sea-level, and not in swamps and shallow water, as was at one time supposed, has been conclusively established by the researches of Principal Dawson and others. After the growth of many generations of trees, the plain is eventually submerged under the sea, and the whole, through course of time, becomes covered over with thick deposits of sand, gravel, and other sediments carried down by streams from the adjoining land. After this the submerged plain becomes again elevated above the sea-level, and forms the site of a second forest, which, after continuing to flourish for long centuries, is in turn destroyed by submergence, and, like the former, becomes covered over with deposits from the land. This alternate process of submergence and emergence goes on till we have a succession of buried forests one above another, with immense stratified deposits between. These buried forests ultimately become converted into beds of coal. This, I presume, is a fair representation of the generally admitted way in which our coal-beds had their origin. It is also worthy of notice that the stratified beds between the coal-seams are of marine and not of lacustrine origin. On this point I may quote the opinion of Professor Hull, a well-known authority on the subject: "Whilst admitting," he says, "the occasional presence of lacustrine strata associated with the coal-measures, I think we may conclude that the whole formation has been essentially of marine and estuarine origin."*

Coal-beds necessarily imply Oscillations of Sea-level.—It may also be observed that each coal-seam indicates both an elevation and a depression of the land. If, for example, there are six coal-seams, one above the other, this proves that the land must have been, at least, six times below and six times above sea-level. This repeated oscillation of the land has been regarded as a somewhat puzzling and singular circumstance. But if we assume coal to be an inter-glacial formation, this difficulty not only disappears, but all the various circumstances which we have been detailing are readily explained. We have to begin with a warm inter-glacial period, with a climate

* "Coal Fields of Great Britain," p. 67.

specially suited for the growth of the coal trees. During this period, as has been shown in the chapter on Submergence, the sea would be standing at a lower level than at present, laying bare large tracts of sea-bottom, on which would flourish the coal vegetation. This condition of things would continue for a period of 8,000 or 10,000 years, allowing the growth of many generations of trees. When the warm period came to a close, and the cold and glacial condition set in, the climate became unsuited for the growth of the coal plants. The sea would begin to rise, and the old sea-bottoms on which, during so long a period, the forests grew, would be submerged and become covered by sedimentary deposits brought down from the land. These forests becoming submerged in a cold sea, and buried under an immense mass of sediment, were then now protected from destruction, and in a position to become converted into coal. The cold continuing for a period of 10,000 years, or thereby, would be succeeded by another warm period, during which the submerged areas became again a land-surface, on which a second forest flourished for another 10,000 years, which in turn became submerged and buried under drift on the approach of the second cold period. This alternate process of submergence and emergence of the land, corresponding to the rise and fall of sea-level during the cold and warm periods, would continue so long as the eccentricity of the earth's orbit remained at a high value, till we might have, perhaps, five or six submerged forests, one above the other, and separated by great thicknesses of stratified deposits, these submerged forests being the coal-beds of the present day.

It is probable that the forests of the Coal period would extend inland over the country, but only such portions as were slightly elevated above sea-level would be submerged and covered over by sediment and thus be preserved, and ultimately become coal-seams. The process will be better understood from the following diagram. Let A B represent the surface of the ground prior to a glacial epoch, and to the formation of the beds of coal and stratified deposits represented in

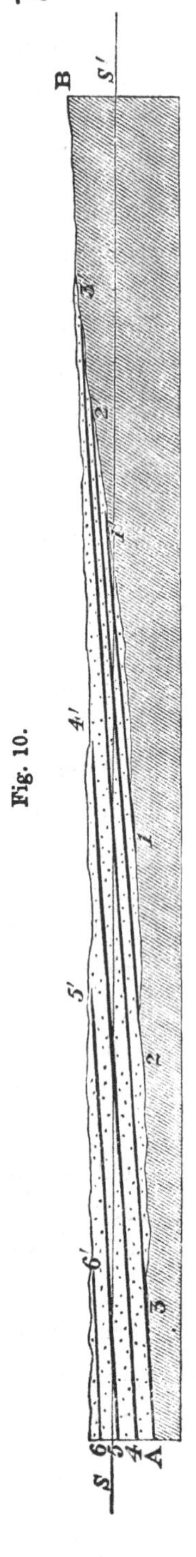

Fig. 10.

the diagram. Let S S′ be the normal sea-level. Suppose the eccentricity of the earth's orbit begins to increase, and the winter solstice approaches the perihelion, we have then a moderately warm period. The sea-level sinks to 1, and forests of sigillariæ and other coal trees cover the country from the sea-shore at 1, stretching away inland in the direction of B. In course of time the winter solstice moves round to aphelion and a cold period follows. The sea begins to rise and continues rising till it reaches 1′. Denudation and the severity of the climate destroy every vestige of the forest from 1′ backwards into the interior; but the portion 1 1′ being submerged and covered over by sediment brought down from the land is preserved. The eccentricity continuing to increase in extent, the second inter-glacial period is more warm and equable than the first, and the sea this time sinks to 2. A second forest now covers the country down to the sea-shore at 2. This second warm period is followed by the second cold period, more severe than the first, and the sea-level rises to 2′. Denudation and severity of climate now destroy every remnant of the forest, from 2′ inland, but of course the submerged portion of 2 2′, like the former portion 1 1′, is preserved. During the third warm period (the eccentricity being still on the increase) the sea-level sinks to 3, and the country for the third time is covered by forests, which extend down to 3. This third warm period is followed by a cold glacial period more severe than the preceding, and the sea-level rises to 3′, and the submerged portion of the forest from 3 to 3′ becomes covered with drift,—the rest as before being destroyed by denudation and the severity of the climate. We shall assume that the eccentricity has now reached a maximum, and that during the fourth inter-glacial period the sea-level sinks only to 4, the level to which it sank during the second inter-glacial period. The country is now covered for the fourth time by forests. The cold period which succeeds not being so severe as the last, the sea rises only to 4′, which, of course, marks the limit of the fourth forest. The eccentricity continuing to diminish, the fifth forest is only submerged up to 5′, and the sixth and last one up to 6′. The epoch of cold and warm periods being now at a close, the sea-level remains stationary at its old normal position S S′. Here we have six buried forests, the one above the other, which, through course of ages, become transformed into coal-beds.

It does not, however, necessarily follow that each separate coal-seam represents a warm period. It is quite possible that two or more seams separated from each other by thin partings or a few feet of sedimentary strata might have been formed during one warm period; for during a warm period minor oscillations of sea-level sufficient to submerge the land to some depth might quite readily have taken place from the melting of polar ice, as was shown in the chapter on Submergence.

It may be noticed that in order to make the section more distinct, its thickness has been greatly exaggerated. It will also be observed that beds 4, 5, and 6 extend considerably to the left of what is represented in the section.

But it is not to be supposed that the whole phenomena of the coal-fields can be explained without supposing a subsidence of the land. The great depth to which the coal-beds have been sunk, in many cases, must be attributed to a subsidence of the level. A series of beds formed during a glacial epoch, may, owing to a subsidence of the land, be sunk to a great depth, and become covered over with thousands of feet of sediment; and then on the occurrence of another glacial epoch, a new series of coal-beds may be formed on the surface. Thus the upper series may be separated from the lower by thousands of feet of sedimentary rock. There is another consequence resulting from the sinking of the land, which must be taken into account. Had there been no sinking of the land during the Carboniferous age, the quantity of coal-beds now remaining would be far less than it actually is, for it is in a great measure owing to their being sunk to a great depth that they have escaped destruction by the enormous amount of denudation which has taken place since that remote age. It therefore follows that only a very small fraction of the submerged forests of the Coal period do actually now exist in the form of coal. Generally it would only be those areas which happened to be sunk to a considerable depth, by a subsidence of the land, that would escape destruction from denudation. But no doubt the areas which would thus be preserved bear but a small proportion to those destroyed.

Length of Inter-glacial Period, as indicated by the Thickness of a Bed of Coal.—A fact favourable to the idea that the coal-seams were formed during inter-glacial periods is, that the length of those periods agrees pretty closely with the length of time supposed to be required to form a coal-seam of average thickness. Other things being equal, the thickness of a coal-seam would depend upon the length of the inter-glacial period. If the rate of precession and motion of the perihelion were always uniform the periods would all be of equal length. But although the rate of precession is not subject to much variation, such is not the case in regard to the motion of the perihelion, as will be seen from the tables of the longitude of the perihelion given in Chapter XIX. Sometimes the motion of the perihelion is rapid, at other times slow, while in some cases its motion is retrograde. In consequence of this, an inter-glacial period may not be more than some six or seven thousand years in length, while in other cases its length may be as much as fifteen or sixteen thousand years.

According to Boussingault, luxuriant vegetation at the present day takes from the atmosphere about a half ton of carbon per acre annually, or fifty tons per acre in a century. Fifty tons of carbon of the specific gravity of coal, about 1·5, spread evenly over the surface of an acre, would make a layer nearly one-third of an inch.* Humboldt makes the estimate a little higher, viz., one half-inch. Taking the latter estimate, it would require 7,200 years to form a bed of coal a yard thick. Dr. Heer, of Zurich, thinks that it would not require more than 1,400 years to form a bed of coal one yard thick;† while Mr. Maclaren thinks that a bed of coal one yard thick would be formed in 1,000 years.‡ Professor Phillip, calculating from the amount of carbon taken from the atmosphere, as determined by Liebig, considers that if it were converted into ordinary coal with about 75 per cent. of carbon, it would yield one inch in 127·5 years, or a yard in 4,600 years.§

There is here a considerable amount of difference in regard to the time required to form a yard of coal. The truth, however, may probably be somewhere between the two extremes, and we may assume 5,000 years to be about the time. In a warm period of 15,000 years we should then have deposited a seam of coal 9 feet thick, while during a warm period of 7,000 years we would have a seam of only 4 feet.

Reason why the Coal Strata present so little Evidence of Ice-action.—There are two objections which will, no doubt, present themselves to the reader's mind. (1.) If coal be an inter-glacial formation, why do the coal strata present so little evidence of ice-action? If the coal-seams represent warm inter-glacial periods, the intervening beds must represent cold or glacial periods, and if so, they ought to contain more abundant evidence of ice-action than they really do. (2.) In the case of the glacial epoch, almost every vestige of the vegetation of the warm periods was destroyed by the ice of the cold periods; why then did not the same thing take place during the glacial epoch of the Carboniferous period?

* See "Smithsonian Report for 1857," p. 138.

† Quart. Journ. Geol. Soc., May, 1865, p. civ.

‡ "Geology of Fife and the Lothians," p. 116.

§ "Life on the Earth," p. 133.

During the glacial epoch the face of the country was in all probability covered for ages with the most luxuriant vegetation; but scarcely a vestige of that vegetation now remains, indeed the very soil upon which it grew is not to be found. All that now remains is the wreck and desolation produced by the ice-sheet that covered the country during the cold periods of that epoch, consisting of transported blocks of stones, polished and grooved rocks, and a confused mass of boulder clay. Here we have in this epoch nothing tangible presenting itself but the destructive effects of the ice which swept over the land. Why, then, in reference to the glacial epochs of the Carboniferous age should we have such abundant evidence of the vegetation of the warm periods, and yet so little evidence of the effect of the ice of the cold periods? The answer to these two objections will go a great way to explain why we have so much coal belonging to the Carboniferous age, and so little belonging to any other age; and it will, I think, be found in the peculiar physical character of the country during the Carboniferous age. The areas on which the forests of the Coal period grew escaped the destructive power of glaciers and land-ice on account of the flat nature of the ground. There are few points on which geologists are more unanimous than in regard to the flat character of the country during the Coal period.

There does not seem to be any very satisfactory evidence that the interior of the country rose to any very great elevation. Mr. Godwin-Austen thinks that during the Coal period there must have been "a vast expanse of continuous horizontal surface at very slight elevations above the sea-level."* Of the widely spread terrestrial surface of the Coal-measure period, portions, he believes, attained a considerable elevation. But in contrast to this he states, "There is a feature which seems to distinguish this period physically from all subsequent periods, and which consists in the vast expanse of continuous horizontal surface which the land area presented, bordering on, and at very slight elevations above, the sea-level." † Hugh

* Quart. Journ. Geol. Soc., vol. xi., p. 535. † Ibid., vol. xii., p. 39.

Miller, describing in his usual graphic way the appearance of the country during the Coal period, says:—"It seems to have been a land consisting of immense flats, unvaried, mayhap, by a *single hill*, in which dreary swamps, inhabited by doleful creatures, spread out on every hand for hundreds and thousands of miles; and a gigantic and monstrous vegetation formed, as I have shown, the only prominent features of the scenery." *

Now, if this is in any way like a just representation of the general features of the country during the Coal period, it was physically impossible, no matter however severe the climate may have been, that there could have been in this country at that period anything approaching to continental ice, or perhaps even to glaciers of such dimensions as would reach down to near the sea-level, where the coal vegetation now preserved is supposed chiefly to have grown. The condition of things which would prevail would more probably resemble that of Siberia than that of Greenland.

The absence of all traces of ice-action in the strata of the coal-measures can in this case be easily explained. For as by supposition there were no glaciers, there could have been no scratching, grooving, or polishing of the rocks; neither could there have been any icebergs, for the large masses known as icebergs are the terminal portions of glaciers which have reached down to the sea. Again, there being no icebergs, there of course could have been no grinding or scratching of the rocks forming the floor of the ocean. True, during summer, when the frozen sea broke up, we should then have immense masses of floating ice, but these masses would not be of sufficient thickness to rub against the sea-bottom. But even supposing that they did occasionally touch the bottom here and there, we could not possibly find the evidence of this in any of the strata of the coal-measures. We could not expect to find any scratchings or markings on the sandstone or shale of those strata indicating the action of ice, for at that period there were no beds of sandstone or shale, but simply beds of sand and mud, which in

* Miller's "Sketch Book of Practical Geology," p. 192.

future ages became consolidated into sandstone and shale. A mass of ice might occasionally rub along the sea-bottom, and leave its markings on the loose sand or soft mud forming that bottom, but the next wave that passed over it would obliterate every mark, and leave the surface as smooth as before. Neither could we expect to find any large erratics or boulders in the coal strata, for these must come from the land, and as by supposition there were no glaciers or land-ice at that period, there was therefore no means of transporting them. In Greenland the icebergs sometimes carry large boulders, which are dropped into the sea as the icebergs melt away; but these blocks have all either been transported on the backs of glaciers from inland tracts, or have fallen on the field-ice along the shore from the face of crags and overhanging precipices. But as there were probably neither glaciers reaching to the sea, nor perhaps precipitous cliffs along the sea-shore, there could have been few or no blocks transported by ice and dropped into the sea of the Carboniferous period, and of course we need not expect to find them in the sandstone and shale which during that epoch formed the bed of the ocean. There would no doubt be coast-line ice and ground-ice in rivers, carrying away large quantities of gravel and stones; but these gravels and stones would of course be all water-worn, and although found in the strata of the coal-measures, as no doubt they actually are, they would not be regarded as indicating the action of ice. The simple absence of relics of ice-action in the coal-measures proves nothing whatever in regard to whether there were cold periods during their formation or not.

This comparative absence of continental ice might be one reason why the forests of the Carboniferous period have been preserved to a much greater extent than those of any other age.

It must be observed, however, that the conclusions at which we have arrived in reference to the comparative absence of continental ice applies only to the areas which now constitute our coal-fields. The accumulation of ice on the antarctic regions, and on some parts of the arctic regions, might have been as great during that age as it is at present. Had there been no continental ice there could have been no such oscillations of sea-level as is assumed in the foregoing theory. The leading idea of the theory, expressed in a few words, is, that the glacial epochs of the Carboniferous age were as severe, and the accumulation of ice as great, as during any other age, only there were large tracts of flat country, but little elevated above the sea-level, which were not covered by ice. These plains, during the warm inter-glacial periods, were covered with forests of sigillariæ and other coal trees. Portions of those forests were protected by the submergence which resulted from the rise of the sea-level during the cold or glacial periods and the subsequent subsidence of the land. Those portions now constitute our coal-beds.

But that coal may be an inter-glacial formation is no mere hypothesis, for we have in the well-known Dürnten beds —described in Chapter XV.—an actual example of such a formation.

Carboniferous Limestones.—As a general rule the limestones of the Carboniferous period, like the coal, are found in beds separated by masses of sandstone and other stratified deposits, which proves that the corals, crinoids, and other creatures, of the remains of which it is composed, did not live continuously on during the entire Limestone period. These limestones are a marine formation. If the land was repeatedly submerged the coal must of necessity have been produced in seams with stratified deposits between, but there is no reason why the same should have been the case with the limestones. If the climatic condition of the sea continued the same we should not have expected this alternate succession of life and death; but, according to the theory of alternate cold and warm periods, such a condition follows as a necessary consequence, for during the warm periods, when the land was covered with a luxuriant vegetation, the sea-bottom would be covered with mollusca, crinoids, corals, &c., fitted to live only in a moderately warm sea; but when the cold came on those creatures would die, and their remains, during the continuance of the cold period, would become slowly covered over with deposits of sand and clay. On the return of the warm period those deposits would soon become covered with life as before, forming another bed of limestone, and this alternation of life and death would go on as long as the glacial epoch continued.

It is true that in Derbyshire, and in the south of Ireland and some other places, the limestone is found in one mass of several hundred feet in thickness without any beds of sandstone or shale, but then it is nowhere found in one continuous mass from top to bottom without any lines of division. These breaks or divisions may as distinctly mark a cold period as though they had been occupied by beds of sandstone. The marine creatures ceased to exist, and when the rough surface left by their remains became smoothed down by the action of the waves into a flat plain, another bed would begin to form upon this floor so soon as life again appeared. Two agencies working together probably conspired to produce these enormous masses of limestone divided only by breaks marking different periods of elaboration. Corals grow in warm seas, and there only in water of a depth ranging from 20 to 30 fathoms. The cold of a period of glaciation would not only serve to destroy them, but they would be submerged so much beyond the depth proper for their existence that even were it possible that with the submergence a sufficient temperature was left, they would inevitably perish from the superincumbent mass of water. We are therefore, as it seems to me, warranted in concluding that the separate masses of Derbyshire limestone were formed during warm inter-glacial periods, and that the lines of division represent cold periods of glaciation during which the animals perished by the combined influence of cold and pressure of water. The submergence of the coral banks in deep water on a sea-bottom, which, like the land, was characteristically flat and even, implies its carrying away far into the bosom of the ocean, and consequently remote from any continent and the river-borne detritus thereof.

CHAPTER XXVII.

PATH OF THE ICE-SHEET IN NORTH-WESTERN EUROPE AND ITS RELATIONS TO THE BOULDER CLAY OF CAITHNESS.*

Character of Caithness Boulder Clay.—Theories of the Origin of the Caithness Clay.—Mr. Jamieson's Theory.—Mr. C. W. Peach's Theory.—The proposed Theory.—Thickness of Scottish Ice-sheet.—Pentlands striated on their Summits.—Scandinavian Ice-sheet.—North Sea filled with Land-ice.—Great Baltic Glacier.—Jutland and Denmark crossed by Ice.—Sir R. Murchison's Observations.—Orkney, Shetland, and Faroe Islands striated across — Loess accounted for.—Professor Geikie's Suggestion.—Professor Geikie and B. N. Peach's Observations on East Coast of Caithness.—Evidence from Chalk Flints and Oolitic Fossils in Boulder Clay.

The Nature of the Caithness Boulder Clay.—A considerable amount of difficulty has been felt by geologists in accounting for the origin of the boulder clay of Caithness. It is an unstratified clay, of a deep grey or slaty colour, resembling much that of the Caithness flags on which it rests. It is thus described by Mr. Jamieson (Quart. Jour. Geol. Soc., vol. xxii., p. 261):—

"The glacial drift of Caithness is particularly interesting as an example of a boulder clay which in its mode of accumulation and ice-scratched *débris* very much resembles that unstratified stony mud which occurs underneath glaciers—the '*moraine profonde*,' as some call it.

"The appearance of the drift along the Haster Burn, and in many other places in Caithness, is in fact precisely the same as that of the old boulder clay of the rest of Scotland, except that it is charged with remains of sea-shells and other marine organisms.

* From Geological Magazine, May and June, 1870; with a few verbal corrections, and a slight re-arrangement of the paragraphs.

"If want of stratification, hardness of texture, and abundance of well-glaciated stones and boulders are to be the tests for what we call genuine boulder clay, then much of the Caithness drift will stand the ordeal."

So far, therefore, as the mere appearance of the drift is concerned, it would at once be pronounced to be true Lower Till, the product of land-ice. But there are two circumstances connected with it which have been generally regarded as fatal to this conclusion.

(1) The striæ on the rocks show that the ice which formed the clay must have come from the sea, and not from the interior of the country; for their direction is almost at right angles to what it would have been had the ice come from the interior. Over the whole district, the direction of the grooves and scratches, not only of the rocks but even of the stones in the clay, is pretty nearly N.W. and S.E. "When examining the sections along the Haster Burn," says Mr. Jamieson, "in company with Mr. Joseph Anderson, I remarked that the striæ on the imbedded fragments generally agreed in direction with those of the rocks beneath. The scratches on the boulders, as usual, run lengthways along the stones when they are of an elongated form; and the position of these stones, as they lie imbedded in the drift, is, as a rule, such that their longer axes point in the same direction as do the scratches on the solid rock beneath; showing that the same agency that scored the rocks also ground and pushed along the drift."

Mr. C. W. Peach informs me that he seldom or never found a stone with two sets of striæ on it, a fact indicating, as Mr. Jamieson remarks, that the drift was produced by one great movement invariably in the same direction. Let it be borne in mind that the ice, which thus moved over Caithness in this invariable track, must either have come from the Atlantic to the N.W., or from the Moray Firth to the S.E.

(2) The boulder clay of Caithness is full of sea-shells and other marine remains. The shells are in a broken condition, and are interspersed like the stones through the entire mass of

the clay. Mr. Jamieson states that he nowhere observed any instance of shells being found in an undisturbed condition, "nor could I hear," he says, "of any such having been found; there seems to be no such thing as a bed of laminated silt with shells *in situ.*" The shell-fragments are scratched and ice-worn, the same as the stones found in the clay. Not only are the shells glaciated, but even the foraminifera, when seen through the microscope, have a rubbed and worn appearance. The shells have evidently been broken, striated, and pushed along by the ice at the time the boulder clay was being formed.

Theories regarding the Origin of the Caithness Clay.—Mr. Jamieson, as we have seen, freely admits that the boulder clay of Caithness has the appearance of true land-ice till, but from the N.W. and S.E. direction of the striæ on the rocks, and the presence of sea-shells in the clay, he has come to the conclusion that the glaciation of Caithness has been effected by floating ice at a time when the district was submerged. I have always felt convinced that Mr. Jamieson had not hit upon the true explanation of the phenomena.

(1) It is physically impossible that any deposit formed by icebergs could be wholly unstratified. Suppose a mass of the materials which would form boulder clay is dropped into the sea from, say an iceberg, the heavier parts, such as stones, will reach the bottom first. Then will follow lighter materials, such as sand, then clay, and last of all the mud will settle down over the whole in fine layers. The different masses dropped from the various icebergs, will, no doubt, lie in confusion one over the other, but each separate mass will show signs of stratification. A good deal of boulder clay evidently has been formed in the sea, but if the clay be unstratified, it must have been formed under glaciers moving along the sea-bottom as on dry ground. Whether *unstratified* boulder clay may happen to be formed under water or on dry land, it must in either case be the product of land-ice.* Those who imagine that materials,

* See Phil. Mag. for November, 1868, p. 374.

differing in specific gravity like those which compose boulder clay, dropped into water, can settle down without assuming the stratified form, should make the experiment, and they would soon satisfy themselves that the thing is physically impossible. The notion that unstratified boulder clay could be formed by deposits from floating ice, is not only erroneous, but positively pernicious, for it tends to lead those who entertain it astray in regard to the whole question of the origin of drift.

(2) It is also physically impossible that ice-markings, such as those everywhere found on the rocky face of the district, and on the pebbles and shells imbedded in the clay, could have been effected by any other agency than that of land-ice. I need not here enter into any discussion on this point, as this has been done at considerable length in another place.* In the present case, however, it is unnecessary, because if it can be shown that all the facts are accounted for in the most natural manner by the theory of land-ice, no one will contend for the floating-ice theory; for it is admitted that, with the exception of the direction of the striæ and the presence of the shells, all the facts agree better with the land-ice than with the floating-ice theory.

My first impression on the subject was that the glaciation of Caithness had been effected by the polar ice-cap, which, during the severer part of the glacial epoch, must have extended down to at least the latitude of the north of Scotland.

On a former occasion (see the *Reader* for 14th October, 1865) it was shown that all the northern seas, owing to their shallowness, must, at that period, have been blocked up with solid ice, which displaced the water and moved along the sea-bottoms the same as on dry land. In fact, the northern seas, including the German Ocean, being filled at the time with glacier-ice, might be regarded as dry land. Ice of this sort, moving along the bed of the German Ocean or North Sea, and over Caithness, could not fail to push before it the shells and other animal remains lying on the sea-bottom, and to mix

* See Phil. Mag. for November, 1868, pp. 366—374.

them up with the clay which now remains upon the land as evidence of its progress.

About two years ago I had a conversation with Mr. C. W. Peach on the subject. This gentleman, as is well known, has long been familiar with the boulder clay of Caithness. He felt convinced that the clay of that country is the true Lower Till, and not a more recent deposit, as Mr. Jamieson supposes. He expressed to me his opinion that the glaciation of Caithness had been effected by masses of land-ice crossing the Moray Firth from the mountain ranges to the south-east, and passing over Caithness in its course. The difficulty which seems to beset this theory is, that a glacier entering the Firth would not leave it and ascend over the Caithness coast. It would take the path of least resistance and move into the North Sea, where it would find a free passage into deeper water. Mr. Peach's theory is, however, an important step in the right direction. It is a part of the truth, but I believe not the whole truth. The following is submitted as a solution of the question.

The Proposed Theory.—It may now be regarded as an established fact that, during the severer part of the glacial period, Scotland was covered with one continuous mantle of ice, so thick as to bury under it the Ochil, Sidlaw, Pentland, Campsie, and other moderately high mountain ranges. For example, Mr. J. Geikie and Mr. B. N. Peach found that the great masses of the ice from the North-west Highlands, came straight over the Ochils of Perthshire and the Lomonds of Fife. In fact, these mountain ridges were not sufficiently high to deflect the icy stream either to the right hand or to the left; and the flattened and rounded tops of the Campsie, Pentland, and Lammermoor ranges bear ample testimony to the denuding power of ice.

Further, to quote from Mr. Jamieson, "the detached mountain of Schehallion in Perthshire, 3,500 feet high, is marked near the top as well as on its flanks, and this not by ice flowing down the sides of the hill itself, but by ice pressing over it from the north. On the top of another isolated hill, called

Morven, about 3,000 feet high, and situated a few miles to the north of the village of Ballater, in the county of Aberdeen, I found granite boulders unlike the rock of the hill, and apparently derived from the mountains to the west. Again, on the highest water-sheds of the Ochils, at altitudes of about 2,000 feet, I found this summer (1864) pieces of mica schist full of garnets, which seem to have come from the Grampian Hills to the north-west, showing that the transporting agent had overflowed even the highest parts of the Ochil ridge. And on the West Lomonds, in Fifeshire, at Clattering-well Quarry, 1,450 feet high, I found ice-worn pebbles of Red Sandstone and porphyry in the *débris* covering the Carboniferous Limestone of the top of the Bishop Hill. Facts like these meet us everywhere. Thus on the Perthshire Hills, between Blair Athol and Dunkeld, I found ice-worn surfaces of rocks on the tops of hills, at elevations of 2,200 feet, as if caused by ice pressing over them from the north-west, and transporting boulders at even greater heights." *

Facts still more important, however, in their bearing on the question before us were observed on the Pentland range by Mr. Bennie and myself during the summer of 1870. On ascending Allermuir, one of the hills forming the northern termination of the Pentland range, we were not a little surprised to find its summit ice-worn and striated. The top of the hill is composed of a compact porphyritic felstone, which is very much broken up; but wherever any remains of the original surface could be seen, it was found to be polished and striated in a most decided manner. These striæ are all in one uniform direction, nearly east and west; and on minutely examining them with a lens we had no difficulty whatever in determining that the ice which effected them came from the west and not from the east, a fact which clearly shows that they must have been made at the time when, as is well known, the entire Midland valley was filled with ice, coming from the North-west Highlands. On the summit of the hill we also found patches of boulder clay in

* Journ. Geol. Soc., vol. xxi., p. 165.

hollow basins of the rock. At one spot it was upwards of a foot in depth, and rested on the ice-polished surface. The clay was somewhat loose and sandy, as might be expected of a layer so thin, exposed to rain, frost, and snow, during the long course of ages which must have elapsed since it was deposited there. Of 100 pebbles collected from the clay, just as they turned up, every one, with the exception of three or four composed of hard quartz, presented a flattened and ice-worn surface; and forty-four were distinctly striated: in short, every stone which was capable of receiving and retaining scratches was striated. A number of these stones must have come from the Highlands to the north-west.*

The height of Allermuir is 1,617 feet, and, from its position, it is impossible that the ice could have gone over its summit, unless the entire Midland valley, at this place, had been filled with ice to the depth of more than 1,600 feet. The hill is situated about four or five miles to the south of Edinburgh, and forms, as has already been stated, the northern termination of the Pentland range. Immediately to the north lies the broad valley of the Firth of Forth, more than twelve miles across, offering a most free and unobstructed outlet for the great mass of ice coming along the Midland valley from the west. Now, when we reflect how easily ice can accommodate itself to the inequalities of the channel along which it moves, how it can turn to the right hand or to the left, so as to find for itself the path of least resistance, it becomes obvious that the ice never would have gone over Allermuir, unless not only the Midland valley at this point, but also the whole surrounding country had been covered with one continuous mass of ice to a depth of more than 1,600 feet. But it must not be supposed that the height of Allermuir represents the thickness of the ice; for on ascending Scald Law, a hill four miles to the south-west of Allermuir, and the highest of the Pentland range, we found, in the *débris* covering its summit, hundreds

* Specimens of the striated summit and boulder clay stones are to be seen in the Edinburgh Museum of Science and Art.

of transported stones of all sizes, from one to eighteen inches in diameter. We also dug up a Greenstone boulder about eighteen inches in diameter, which was finely polished and striated. As the height of this hill is 1,898 feet, the mass of ice covering the surrounding country must have been at least 1,900 feet deep. But this is not all. Directly to the north of the Pentlands, in a line nearly parallel with the east coast, and at right angles to the path of ice from the interior, there is not, with the exception of the solitary peak of East Lomond, and a low hill or two of the Sidlaw range, an eminence worthy of the name of a hill nearer than the Grampians in the north of Forfarshire, distant upwards of sixty miles. This broad plain, extending from almost the Southern to the Northern Highlands, was the great channel through which the ice of the interior of Scotland found an outlet into the North Sea. If the depth of the ice in the Firth of Forth, which forms the southern side of this broad hollow, was at least 1,900 feet, it is not at all probable that its depth in the northern side, formed by the Valley of Strathmore and the Firth of Tay, which lay more directly in the path of the ice from the North Highlands, could have been less. Here we have one vast glacier, more than sixty miles broad and 1,900 feet thick, coming from the interior of the country.

It is, therefore, evident that the great mass of ice entering the North Sea to the east of Scotland, especially about the Firths of Forth and Tay, could not have been less, and was probably much more, than from 1,000 to 2,000 feet in thickness. The grand question now to be considered is, What became of the huge sheet of ice after it entered the North Sea? Did it break up and float away as icebergs? This appears to have been hitherto taken for granted; but the shallowness of the North Sea shows such a process to have been utterly impossible. The depth of the sea in the English Channel is only about twenty fathoms, and although it gradually increases to about forty fathoms at the Moray Firth, yet we must go to the north and west of the Orkney and Shetland Islands ere we reach the 100 fathom line. Thus the average depth of the entire North Sea is not over forty fathoms, which is even insufficient to float an iceberg 300 feet thick.

No doubt the North Sea, for two reasons, is now much shallower than it was during the period in question. (1.) There would, at the time of the great extension of the ice on the northern hemisphere, be a considerable submergence, resulting from the displacement of the earth's centre of gravity.* (2.) The sea-bed is now probably filled up to a larger extent with drift deposits than it was at the ice period. But, after making the most extravagant allowance for the additional depth gained on this account, still there could not possibly have been water sufficiently deep to float a glacier of 1,000 or 2,000 feet in thickness. Indeed, the North Sea would have required to be nearly ten times deeper than it is at present to have floated the ice of the glacial period. We may, therefore, conclude with the most perfect certainty that the ice-sheet of Scotland could not possibly have broken up into icebergs in such a channel, but must have moved along on the bed of the sea in one unbroken mass, and must have found its way to the deep trough of the Atlantic, west of the Orkney and Shetland Islands, ere it broke up and floated away in the iceberg form.

It is hardly necessary to remark that the waters of the North Sea would have but little effect in melting the ice. A shallow sea like this, into which large masses of ice were entering, would be kept constantly about the freezing-point, and water of this temperature has but little melting power, for it takes 142 lbs. of water, at 33°, to melt one pound of ice. In fact, an icy sea tends rather to protect the ice entering it from being melted than otherwise. And besides, owing to fresh acquisitions of snow, the ice-sheet would be accumulating more rapidly upon its upper surface than it would be melting at its lower surface, supposing there were sea-water under that surface. The ice of Scotland during the glacial period must, of necessity, have found its way into warmer water than that of

* Phil. Mag. for April, 1866.

the North Sea before it could have been melted. But this it could not do without reaching the Atlantic, and in getting there it would have to pass round by the Orkney Islands, along the bed of the North Sea, as land-ice.

This will explain how the Orkney Islands may have been glaciated by land-ice; but it does not, however, explain how Caithness should have been glaciated by that means. These islands lay in the very track of the ice on its way to the Atlantic, and could hardly escape being overridden; but Caithness lay considerably to the left of the path which we should expect the ice to have taken. The ice would not leave its channel, turn to the left, and ascend upon Caithness, unless it were forced to do so. What, then, compelled the ice to pass over Caithness?

Path of the Scandinavian Ice.—We must consider that the ice from Scotland and England was but a fraction of that which entered the North Sea. The greater part of the ice of Scandinavia must have gone into this sea, and if the ice of our island could not find water sufficiently deep in which to float, far less would the much thicker ice of Scandinavia do so. The Scandinavian ice, before it could break up, would thus, like the Scottish ice, have to cross the bed of the North Sea and pass into the Atlantic. It could not pass to the north, or to the north-west, for the ocean in these directions would be blocked up by the polar ice. It is true that along the southern shore of Norway there extends a comparatively deep trough of from one to two hundred fathoms. But this is evidently not deep enough to have floated the Scandinavian ice-sheet; and even supposing it had been sufficiently deep, the floating ice must have found its way to the Atlantic, and this it could not have done without passing along the coast. Now, its passage would not only be obstructed by the mass of ice continually protruding into the sea directly at right angles to its course, but it would be met by the still more enormous masses of ice coming off the entire Norwegian coast-line. And, besides this, the ice entering the Arctic Ocean from Lapland and the northern parts of Siberia, except the very small portion which might find an outlet into the Pacific through Behring's Straits, would have to pass along the Scandinavian coast in its way to the Atlantic. No matter, then, what the depth of this trough may have been, if the ice from the land, after entering it, could not make its escape, it would continue to accumulate till the trough became blocked up; and after this, the great mass from the land would move forward as though the trough had no existence. Thus, the only path for the ice would be by the Orkney and Shetland Islands. Its more direct and natural path would, no doubt, be to the south-west, in the direction of our shores; and in all probability, had Scotland been a low flat island, instead of being a high and mountainous one, the ice would have passed completely over it. But its mountainous character, and the enormous masses of ice at the time proceeding from its interior, would effectually prevent this, so that the ice of Scandinavia would be compelled to move round by the Orkney Islands. Consequently, these two huge masses of moving ice—the one from Scotland and the much greater one from Scandinavia—would meet in the North Sea, probably not far from our shores, and would move, as represented in the diagram, side by side northwards into the Atlantic as one gigantic glacier.

Nor can this be regarded as an anomalous state of things; for in Greenland and the antarctic continent the ice does not break up into icebergs on reaching the sea, but moves along the sea-bottom in a continuous mass until it reaches water sufficiently deep to float it. It is quite possible that the ice at the present day may nowhere traverse a distance of three or four hundred miles of sea-bottom, but this is wholly owing to the fact that it finds water sufficiently deep to float it before having travelled so far. Were Baffin's Bay and Davis's Straits, for example, as shallow as the North Sea, the ice of Greenland would not break up into icebergs in these seas, but cross in one continuous mass to and over the American continent.

The median line of the Scandinavian and Scottish ice-sheets

would be situated not far from the east coast of Scotland. The Scandinavian ice would press up as near to our coast as the resistance of the ice from this side permitted. The enormous mass of ice from Scotland, pressing out into the North Sea, would compel the Scandinavian ice to move round by the Orkneys, and would also keep it at some little distance from Scotland. Where, on the other hand, there was but little resistance offered by ice from the interior of this country (and this might be the case along many parts of the English coast), the Scandinavian ice might reach the shores, and even overrun the country for some distance inland.

We have hitherto confined our attention to the action of ice proceeding from Norway; but if we now consider what took place in Sweden and the Baltic, we shall find more conclusive proof of the downward pressure of Scandinavian ice on our own shores. The western half of Gothland is striated in the direction of N.E. and S.W., and that this has been effected by a huge mass of ice covering the country, and not by local glaciers, is apparent from the fact observed by Robert Chambers,* and officers of the Swedish Geological Survey, that the general direction of the groovings and striæ on the rocks bears little or no relation to the conformation of the surface, showing that the ice was of sufficient thickness to move straight forward, regardless of the inequalities of the ground.

At Gottenburg, on the shores of the Cattegat, and all around Lake Wener and Lake Wetter, the ice-markings are of the most remarkable character, indicating, in the most decided manner, that the ice came from the interior of the country to the north-east in one vast mass. All this mass of ice must have gone into the shallow Cattegat, a sea not sufficiently deep to float even an ordinary glacier. The ice coming off Gothland would therefore cross the Cattegat, and thence pass over Jutland into the North Sea. After entering the North Sea, it would be obliged to keep between our shores and the ice coming direct from the western side of Scandinavia.

* "Tracings of the North of Europe," 1850, pp. 48—51.

But this is not all. A very large proportion of the Scandinavian ice would pass into the Gulf of Bothnia, where it could not possibly float. It would then move south into the Baltic as land-ice. After passing down the Baltic, a portion of the ice would probably move south into the flat plains in the north of Germany, but the greater portion would keep in the bed of the Baltic, and of course turn to the right round the south end of Gothland, and thence cross over Denmark into the North Sea. That this must have been the path of the ice is, I think, obvious from the observations of Murchison, Chambers, Hörbye, and other geologists. Sir Roderick Murchison found—though he does not attribute it to land-ice—that the Aland Islands, which lie between the Gulf of Bothnia and the Baltic, are all striated in a north and south direction.*

Upsala and Stockholm, a tract of flat country projecting for some distance into the Baltic, is also grooved and striated, not in the direction that would be effected by ice coming from the interior of Scandinavia, but north and south, in a direction parallel to what must have been the course of the ice moving down the Baltic. † This part of the country must have been striated by a mass of ice coming from the direction of the Gulf of Bothnia. And that this mass must have been great is apparent from the fact that Lake Malar, which crosses the country from east to west, at right angles to the path of the ice, does not seem to have had any influence in deflecting the icy stream. That the ice came from the north and not from the south is also evident from the fact that the northern sides of rocky eminences are polished, rounded, and ice-worn, while the southern sides are comparatively rough. The northern banks of Lake Malar, for example, which, of course, face the south, are rough, while the southern banks, which must have offered opposition to the advance of the ice, are smoothed and rounded in a most singular manner.

* Quart. Journ. Geol. Soc., vol. ii., p. 364.

† "Tracings of the North of Europe," by Robert Chambers, pp. 259, 285. "Observations sur les Phénomènes d'Erosion en Norvège," by M. Hörbye, 1857. See also Professor Erdmann's "Formations Quaternaires de la Suède."

Again, that the ice, after passing down the Baltic, turned to the right along the southern end of Gothland, is shown by the direction of the striæ and ice-groovings observed on such islands as Gothland, Öland, and Bornholm. Sir R. Murchison found that the island of Gothland is grooved and striated in one uniform direction from N.E. to S.W. "These groovings," says Sir Roderick, "so perfectly resemble the flutings and striæ produced in the Alps by the actual movement of glaciers, that neither M. Agassiz nor any one of his supporters could detect a difference." He concludes, however, that the markings could not have been made by land-ice, because Gothland is not only a low, flat island in the middle of the Baltic, but is "at least 400 miles distant from any elevation to which the term of mountain can be applied." This, of course, is conclusive against the hypothesis that Gothland and the other islands of the Baltic could have been glaciated by ordinary glaciers; but it is quite in harmony with the theory that the Gulf of Bothnia and the entire Baltic were filled with one continuous mass of land-ice, derived from the drainage of the greater part of Sweden, Lapland, and Finland. In fact, the whole glacial phenomena of Scandinavia are inexplicable on the hypothesis of local glaciers.

That the Baltic was completely filled by a mass of ice moving from the north is further evidenced by the fact that the mainland, not only at Upsala, but at several places along the coast of Gothland, is grooved and striated parallel to the shore, and often at right angles to the markings of the ice from the interior, showing that the present bed of the Baltic was not large enough to contain the icy stream. For example, along the shores between Kalmar and Karlskrona, as described by Sir Roderick Murchison and by M. Hörbye, the striations are parallel to the shore. Perhaps the slight obstruction offered by the island of Öland, situated so close to the shore, would deflect the edge of the stream at this point over on the land. The icy stream, after passing Karlskrona, bent round to the west along the present entrance to the Baltic, and again

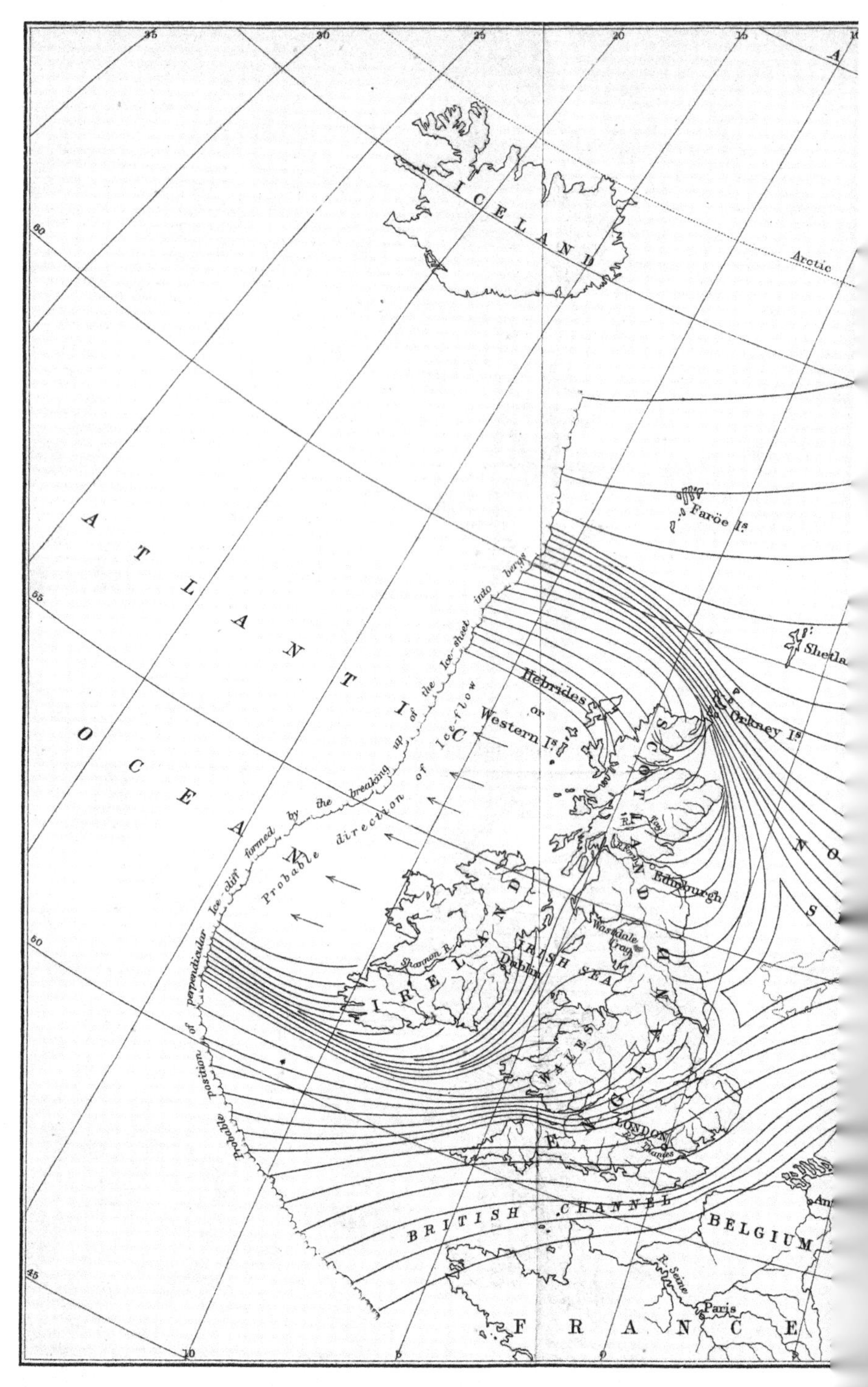

CHART SHOWING THE PROBABLE PATH OF THE ICE IN NORT

The lines also represent the act

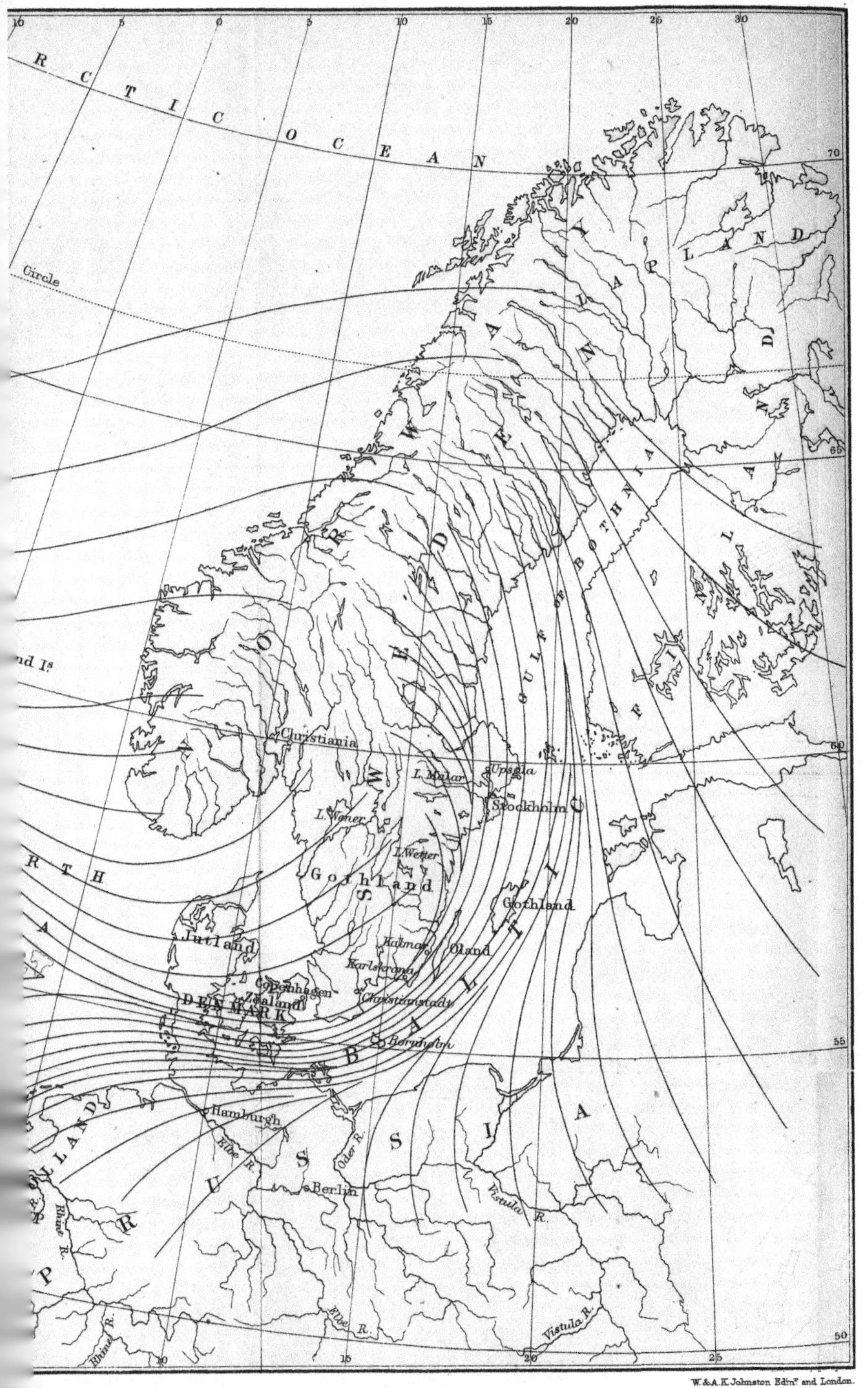

ESTERN EUROPE DURING THE PERIOD OF MAXIMUM GLACIATION.

direction of the striae on the rocks.

invaded the mainland, and crossed over the low headland of Christianstadt, and thence passed westward in the direction of Zealand.

This immense Baltic glacier would in all probability pass over Denmark, and enter the North Sea somewhere to the north of the River Elbe, and would then have to find an outlet to the Atlantic through the English Channel, or pass in between our eastern shores and the mass from Gothland and the north-western shores of Europe. The entire probable path of the ice may be seen by a reference to the accompanying chart (Plate V.) That the ice crossed over Denmark is evident from the fact that the surface of that country is strewn with *débris* derived from the Scandinavian peninsula.

Taking all these various considerations into account, the conclusion is inevitable that the great masses of ice from Scotland would be obliged to turn abruptly to the north, as represented in the diagram, and pass round into the Atlantic in the direction of Caithness and the Orkney Islands.

If the foregoing be a fair representation of the state of matters, it is physically impossible that Caithness could have escaped being overridden by the land-ice of the North Sea. Caithness, as is well known, is not only a low, flat tract of land, little elevated above the sea-level, and consequently incapable of supporting large glaciers; but, in addition, it projects in the form of a headland across the very path of the ice. Unless Caithness could have protected itself by pushing into the sea glaciers of one or two thousand feet in thickness, it could not possibly have escaped the inroads of the ice of the North Sea. But Caithness itself could not have supported glaciers of this magnitude, neither could it have derived them from the adjoining mountainous regions of Sutherland, for the ice of this county found a more direct outlet than along the flat plains of Caithness.

The shells which the boulder clay of Caithness contains have thus evidently been pushed out of the bed of the North Sea by the land-ice, which formed the clay itself.

The fact that these shells are not so intensely arctic as those found in some other quarters of Scotland, is no evidence that the clay was not formed during the most severe part of the glacial epoch, for the shells did not live in the North Sea at the time that it was filled with land-ice. The shells must have belonged to a period prior to the invasion of the ice, and consequently before the cold had reached its greatest intensity. Neither is there any necessity for supposing the shells to be pre-glacial, for these shells may have belonged to an inter-glacial period. In so far as Scotland is concerned, it would be hazardous to conclude that a plant or an animal is either pre-glacial or post-glacial simply because it may happen not to be of an arctic or of a boreal type.

The same remarks which apply to Caithness apply to a certain extent to the headland at Fraserburgh. It, too, lay in the path of the ice, and from the direction of the striæ on the rocks, and the presence of shells in the clay, as described by Mr. Jamieson, it bears evidence also of having been overridden by the land-ice of the North Sea. In fact, we have, in the invasion of Caithness and the headland at Fraserburgh by the land-ice of the North Sea, a repetition of what we have seen took place at Upsala, Kalmar, Christianstadt, and other flat tracts along the sides of the Baltic.

The scarcity, or perhaps entire absence of Scandinavian boulders in the Caithness clay is not in any way unfavourable to the theory, for it would only be the left edge of the North Sea glacier that could possibly pass over Caithness; and this edge, as we have seen, was composed of the land-ice from Scotland. We might expect, however, to find Scandinavian blocks on the Shetland and Faroe Islands, for, as we shall presently see, there is pretty good evidence to prove that the Scandinavian ice passed over these islands.

The Shetland and Faroe Islands glaciated by Land-ice.—It is also worthy of notice that the striæ on the rocks in the Orkney, Shetland, and Faroe Islands, all point in the direction of Scandinavia, and are what would be effected by land-ice moving in the paths indicated in the diagram. And it is a fact of some significance, that when we proceed north to Iceland, the striæ, according to the observations of Robert Chambers, seem to point towards North Greenland. Is it possible that the entire Atlantic, from Scandinavia to Greenland, was filled with land-ice? Astounding as this may at first appear, there are several considerations which render such a conclusion probable. The observations of Chambers, Peach, Hibbert, Allan, and others, show that the rocky face of the Shetland and Faroe Islands has been ground, polished, and striated in a most remarkable manner. That this could not have been done by ice belonging to the islands themselves is obvious, for these islands are much too small to have supported glaciers of any size, and the smallest of them is striated as well as the largest. Besides, the uniform direction of the striæ on the rocks shows that it must have been effected by ice passing over the islands. That the striations could not have been effected by floating icebergs at a time when the islands were submerged is, I think, equally obvious, from the fact that not only are the tops of the highest eminences ice-worn, but the entire surface down to the present sea-level is smoothed and striated; and these striations conform to all the irregularities of the surface. This last fact Professor Geikie has clearly shown is wholly irreconcilable with the floating-ice theory.* Mr. Peach † found vertical precipices in the Shetlands grooved and striated, and the same thing was observed by Mr. Thomas Allan on the Faroe Islands.‡ That the whole of these islands have been glaciated by a continuous sheet of ice passing over them was the impression left on the mind of Robert Chambers after visiting them. § This is the theory which alone explains all the facts. The only difficulty which besets it is the enormous thickness of the ice demanded by the theory. But this difficulty is very much diminished

* "Glacial Drift of Scotland," p. 29.

† Geological Magazine, vol. ii., p. 343. Brit. Assoc. Rep., 1864 (sections), p. 59.

‡ Trans. Roy. Soc. Edin., vol. vii., p. 265.

§ "Tracings of Iceland and the Faroe Islands," p. 49.

when we reflect that we have good evidence, from the thickness of icebergs which have been met with in the Southern Ocean,* that the ice moving off the antarctic continent must be in some places considerably over a mile in thickness. It is then not so surprising that the ice of the glacial epoch, coming off Greenland and Northern Europe, should not have been able to float in the North Atlantic.

Why the Ice of Scotland was of such enormous Thickness.—The enormous thickness of the ice in Scotland, during the glacial epoch, has been a matter of no little surprise. It is remarkable how an island, not more than 100 miles across, should have been covered with a sheet of ice so thick as to bury mountain ranges more than 1,000 feet in height, situated almost at the sea-shore. But all our difficulties disappear when we reflect that the seas around Scotland, owing to their shallowness, were, during the glacial period, blocked up with solid ice. Scotland, Scandinavia, and the North Sea, would form one immense table-land of ice, from 1,000 to 2,000 feet above the sea-level. This table-land would terminate in the deep waters of the Atlantic by a perpendicular wall of ice, extending probably from the west of Ireland away in the direction of Iceland. From this barrier icebergs would be continually breaking off, rivalling in magnitude those which are now to be met with in the antarctic seas.

The great Extension of the Loess accounted for.—An effect which would result from the blocking up of the North Sea with land-ice, would be that the waters of the Rhine, Elbe, and Thames would have to find an outlet into the Atlantic through the English Channel. Professor Geikie has suggested to me that if the Straits of Dover were not then open—quite a possible thing—or were they blocked up with land-ice, say by the great Baltic glacier crossing over from Denmark, the consequence would be that the waters of the Rhine and Elbe would be dammed back, and would inundate all the low-lying tracts of country to the south; and this might account for the extra-

* See Chap. XXIII.

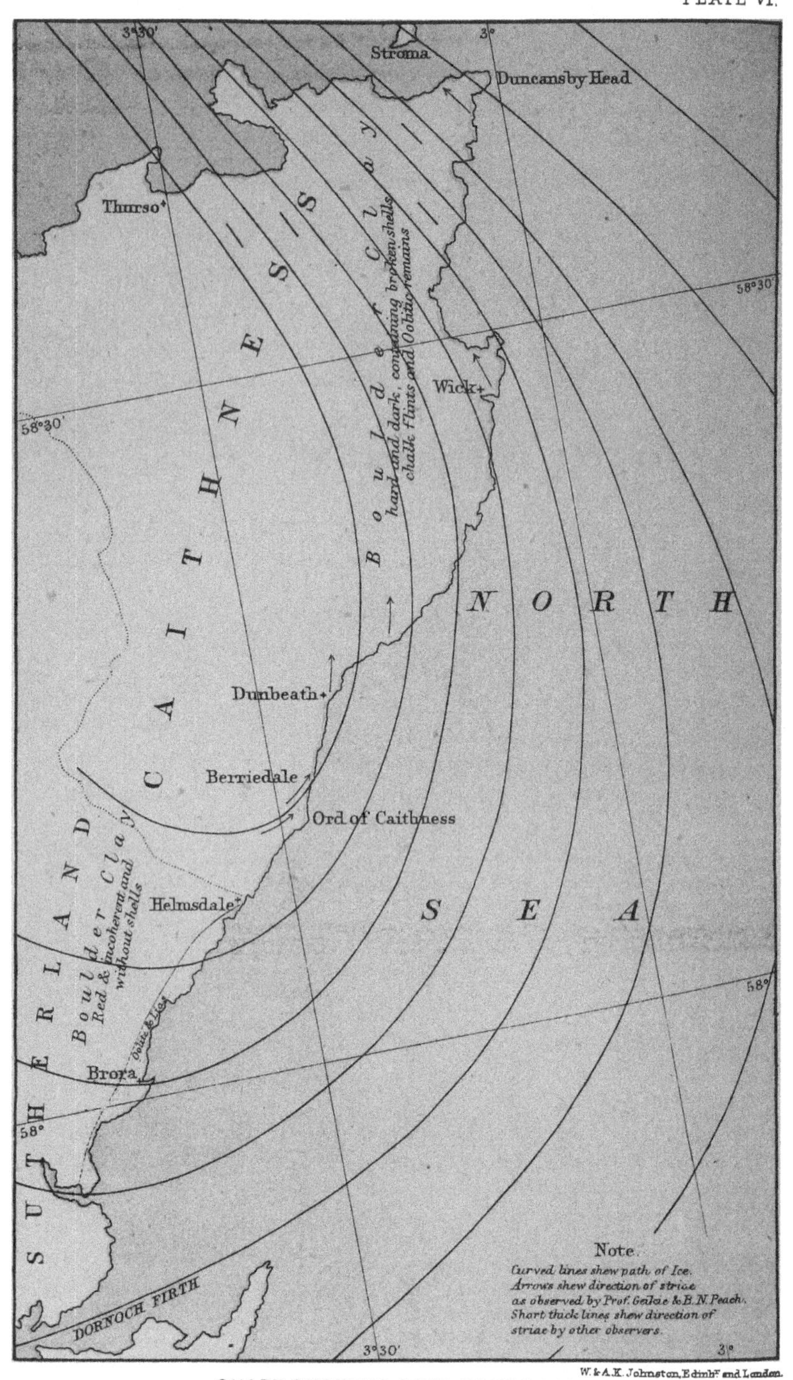
3°30'
3°
Stroma
Duncansby Head
Thurso
CAITHNESS
Boulder Clay
hard and dark, containing broken shells chalk flints and Oolitic remains
58°30'
Wick
NORTH SEA
Dunbeath
Berriedale
Ord of Caithness
SUTHERLAND
Boulder Clay
Red & incoherent and without shells
Helmsdale
Oolitic & Lias
Brora
58°
DORNOCH FIRTH
Note.
Curved lines shew path of Ice.
Arrows shew direction of striae as observed by Prof. Geikie & B. N. Peach.
Short thick lines shew direction of striae by other observers.
W. & A. K. Johnston, Edinb^r and London.

CHART SHOWING PATH OF THE ICE

ordinary extension of the Loess in the basin of the Rhine, and in Belgium and the north of France.*

Note on the Glaciation of Caithness.

I have very lately received a remarkable confirmation of the path of the Caithness ice in observations communicated to me by Professor Geikie and Mr. B. N. Peach. The latter geologist says, "Near the Ord of Caithness and on to Berriedale the striæ pass off the land and out to sea; but near Dunbeath, 6 miles north-east of Berriedale, they begin to creep up out of the sea on to the land and range from about 15° to 10° east of north. *Where the striæ pass out to the sea* the boulder clay is made up of the materials from inland and contains no shells, but *immediately the striæ begin to creep up on to the land* then shells begin to make their appearance; and there is a difference, moreover, in the colour of the clay, for in the former case it is red and incoherent, and in the latter hard and dark-coloured." The accompanying chart (Plate VI.) shows the outline of the Caithness coast and the direction of the striæ as observed by Professor Geikie and Mr. Peach, and no demonstration could be more conclusive as to the path of the ice and the obstacles it met than these observations, supplemented and confirmed as they are by other recorded facts to which I shall presently allude. Had the ice-current as it entered the North Sea off the Sutherland coast met with no obstacle it would have ploughed its way outwards till it broke off in glaciers and floated away. But it is clear that the great press of Scandinavian ice and the smaller mass of land-ice from the Morayshire coast converging in the North Sea filled up its entire bed, and these, meeting the opposing current from the Sutherland coast, turned it back upon itself, and forced it over the north-

* Mr. Thomas Belt has subsequently advanced (Quart. Jour. Geol. Soc., vol. xxx., p. 490), a similar explanation of the steppes of Siberia. He supposes that an overflow of ice from the polar basin dammed back all the rivers flowing northward, and formed an immense lake which extended over the lowlands of Siberia, and deposited the great beds of sand and silt with occasional fresh-water shells and elephant remains, of which the steppes consist.

east part of Caithness. The farther south on the Sutherland coast that the ice entered the sea the deeper would it be able to penetrate into the ocean-bed before it met an opposition sufficiently strong to turn its course, and the wider would be its sweep; but when we come to the Sutherland coast we reach a point where the land-ice—as, for example, near Dunbeath—is forced to bend round before it even reaches the sea-shore, as will be seen from the accompanying diagram.

We are led to the same conclusions regarding the path of the ice in the North Sea from the presence of oolitic fossils and chalk flints found likewise in the boulder clay of Caithness, for these, as we shall see, evidently must have come from the sea. At the meeting of the British Association, Edinburgh, 1850, Hugh Miller exhibited a collection of boreal shells with fragments of oolitic fossils, chalk, and chalk flints from the boulder clay of Caithness collected by Mr. Dick, of Thurso. My friend, Mr. C. W. Peach, found that the chalk flints in the boulder clay of Caithness become more abundant as we proceed northward, while the island of Stroma in the Pentland Firth he found to be completely strewn with them. This same observer found, also, in the Caithness clay stones belonging to the Oolitic and Lias formations, with their characteristic fossils, while ammonites, belemnites, fossil wood, &c., &c., were also found loose in the clay.* The explanation evidently is, that these remains were derived from an outcrop of oolitic and cretaceous beds in the North Sea. It is well known that the eastern coast of Sutherlandshire is fringed with a narrow strip of oolite, which passes under the sea, but to what distance is not yet ascertained. Outside the Oolitic formation the chalk beds in all probability crop out. It will be seen from a glance at the accompanying chart (Plate VI.) that the ice which passed over the north-eastern part of Caithness must have crossed the out-cropping chalk beds.

As has already been stated in the foregoing chapter, the headland of Fraserburgh, north-eastern corner of Aberdeen-

* Proc. Roy. Phys. Soc., Edin., vols. ii. and iii.

shire, bears evidence, both from the direction of the striæ and broken shells in the boulder clay, of having been overridden also by land-ice from the North Sea. This conclusion is strengthened by the fact that chalk flints and oolitic fossils have also been abundantly met with in the clay by Dr. Knight, Mr. James Christie, Mr. W. Ferguson, Mr. T. F. Jamieson, and others.

CHAPTER XXVIII.

NORTH OF ENGLAND ICE-SHEET, AND TRANSPORT OF WASTDALE CRAG BLOCKS.*

Transport of Blocks; Theories of.—Evidence of Continental Ice.—Pennine Range probably striated on Summit.—Glacial Drift in Centre of England.—Mr. Lacy on Drift of Cotteswold Hills.—England probably crossed by Land-ice.—Mr. Jack's Suggestion.—Shedding of Ice North and South.—South of England Ice-sheet.—Glaciation of West Somerset.—Why Ice-markings are so rare in South of England.—Form of Contortion produced by Land-ice.

CONSIDERABLE difficulty has been felt in accounting for the transport of the Wastdale granite boulders across the Pennine chain to the east. Professors Harkness,† and Phillips,‡ Messrs. Searles Wood, jun.,§ Mackintosh,|| and I presume all who have written on the subject, agree that these blocks could not have been transported by land-ice. The agency of floating ice under some form or other is assumed by all.

We have in Scotland phenomena of an exactly similar nature. The summits of the Ochils, the Pentlands, and other mountain ranges in the east of Scotland, at elevations of from 1,500 to 2,000 feet, are not only ice-marked, but strewn over with boulders derived from rocks to the west and north-west. Many of them must have come from the Highlands distant some 50 or 60 miles. It is impossible that these stones could have been transported, or the summits of the hills striated, by means of ordinary glaciers. Neither can the phenomena be attributed to the agency of icebergs carried along by currents. For we should require to assume not merely a submergence of the land

* From Geol. Mag. for January, 1871.

† Quart. Journ. Geol. Soc., xxvi., p. 517.

‡ British Assoc. Report for 1864 (sections), p. 65.

§ Quart. Journ. Geol. Soc., xxvi., p. 90.

|| Geol. Mag., vii., p. 349.

to the extent of 2,000 feet or so,—an assumption which might be permitted,—but also that the currents bearing the icebergs took their rise in the elevated mountains of the Highlands (a most unlikely place), and that these currents radiated in all directions from that place as a centre.

In short, the glacial phenomena of Scotland are wholly inexplicable upon any other theory than that, during at least a part of the glacial epoch, the entire island from sea to sea was covered with one continuous mass of ice of not less than 2,000 feet in thickness.

In my paper on the Boulder Clay of Caithness (see preceding chapter), I have shown that if the ice was 2,000 feet or so in thickness, it must, in its motion seawards, have followed the paths indicated by the curved lines in the chart accompanying that paper (See Plate V.). In so far as Scotland is concerned [and Scandinavia also], these lines represent pretty accurately not only the paths actually taken by the boulders, but also the general direction of the ice-markings on all the elevated mountain ridges. But if Scotland was covered to such an extent with ice, it is not at all probable that Westmoreland and the other mountainous districts of the North of England could have escaped being enveloped in a somewhat similar manner. Now if we admit the supposition of a continuous mass of ice covering the North of England, all our difficulties regarding the transport of the Wastdale blocks across the Pennine chain disappear. An inspection of the chart above referred to will show that these blocks followed the paths which they ought to have done upon the supposition that they were conveyed by continental ice.

That Wastdale Crag itself suffered abrasion by ice moving over it, in the direction indicated by the lines in the diagram, is obvious from what has been recorded by Dr. Nicholson and Mr. Mackintosh. They both found the Crag itself beautifully *moutonnée* up to its summit, and striated in a W.S.W. and E.N.E. direction. Mr. Mackintosh states that these scorings run obliquely up the sloping face of the crag. Ice scratches crossing valleys and running up the sloping faces of hills and over their summits are the sure marks of continental ice, which meet the eye everywhere in Scotland. Dr. Nicholson found in the drift covering the lower part of the crag, pebbles of the Coniston flags and grits from the west.*

The fact that in Westmoreland the direction of the ice-markings, as a general rule, corresponds with the direction of the main valleys, is no evidence whatever that the country was not at one period covered with a continuous sheet of ice; because, for long ages after the period of continental ice, the valleys would be occupied by glaciers, and these, of course, would necessarily leave the marks of their presence behind. This is just what we have everywhere in Scotland. It is on the summits of the hills and elevated ridges, where no glacier could possibly reach, that we find the sure evidence of continental ice. But that land-ice should have passed over the tops of hills 1,000 or 2,000 feet in height is a thing hitherto regarded by geologists as so unlikely that few of them ever think of searching in such places for ice-markings, or for transported stones. Although little has been recorded on this point, I hardly think it likely that there is in Scotland a hill under 2,000 feet wholly destitute of evidence that ice has gone over it. If there were hills in Scotland that should have escaped being overridden by ice, they were surely the Pentland Hills; but these, as was shown on a former occasion,† were completely buried under the mass of ice covering the flat surrounding country. I have no doubt whatever that if the summits of the Pennine range were carefully examined, say under the turf, evidence of ice-action, in the form of transported stones or scratches on the rock, would be found.‡

* Trans. Edin. Geol. Soc., vol. i., p. 136.

† Geol. Mag. for June, 1870. See Chap. XXVII.

‡ This was done by Mr. R. H. Tiddeman of the Geological Survey of England (Quart. Journ. Geol. Soc. for November, 1872), and the result established the correctness of the above opinion as to the existence of a North of England ice-sheet. Additional confirmation has been derived from the important observations of Mr. D. Mackintosh, and also of Mr. Goodchild, of the Geological Survey of England.

Nor is the fact that the Wastdale boulders are not rounded and ice-marked, or found in the boulder clay, but lie on the surface, any evidence that they were not transported by land-ice. For it would not be the stones *under* the ice, but those falling on the upper surface of the sheet, that would stand the best chance of being carried over mountain ridges. But such blocks would not be crushed and ice-worn; and it is on the surface of the clay, and not imbedded in it, that we should expect to find them.

It is quite possible that the dispersion of the Wastdale boulders took place at various periods. During the period of local glaciers the blocks would be carried along the line of the valleys.

All I wish to maintain is that the transport of the blocks across the Pennine chain is easily accounted for if we admit, what is very probable, that the great ice-covering of Scotland overlapped the high grounds of the North of England. The phenomenon is the same in both places, and why not attribute it to the same cause?

There is another curious circumstance connected with the drift of England which seems to indicate the agency of an ice-covering.

As far back as 1819, Dr. Buckland, in his Memoir on the Quartz Rock of Lickey Hill,* directed attention to the fact, that on the Cotteswold Hills there are found pebbles of hard red chalk which must have come from the Wolds of Yorkshire and Lincolnshire. He pointed out also that the slaty and porphyritic pebbles probably came from Charnwood Forest, near Leicester. Professor Hull, of the Geological Survey, considers that "almost all the Northern Drift of this part of the country had been derived from the *débris* of the rocks of the Midland Counties."† He came also to the conclusion that the slate fragments may have been derived from Charnwood Forest. In

* Trans. Geol. Soc., vol. v., p. 516 (first series).

† Quart. Journ. Geol. Soc., vol. xi., p. 492. "Memoir of the Country around Cheltenham," 1857. "Geology of the Country around Woodstock," 1859.

the Vale of Moreton he found erratic boulders from two feet to three feet in diameter. The same northern character of the drift of this district is remarked by Professor Ramsay and Mr. Aveline, in their Memoir of the Geology of parts of Gloucestershire. In Leicestershire and Northamptonshire the officers of the Geological Survey found in abundance drift which must have come from Lincolnshire and Yorkshire to the north-east.

Mr. Lucy, who has also lately directed attention to the fact that the Cotteswold Hills are sprinkled over with boulders from Charnwood Forest, states also that, on visiting the latter place, he found that many of the stones contained in it had come from Yorkshire, still further to the north-east.*

Mr. Searles Wood, jun., in his interesting paper on the Boulder Clay of the North of England,† states that enormous quantities of the chalk *débris* from the Yorkshire Wold are found in Leicester, Rutland, Warwick, Northampton, and other places to the south and south-west. Mr. Wood justly concludes that this chalk *débris* could not have been transported by water. "If we consider," he says, "the soluble nature of chalk, it must be evident that none of this débris can have been detached from the parent mass, either by water-action, or by any other atmospheric agency then moving ice. The action of the sea, of rivers, or of the atmosphere, upon chalk, would take the form of dissolution, the degraded chalk being taken up in minute quantities by the water, and held in suspension by it, and in that form carried away; so that it seems obvious that this great volume of rolled chalk can have been produced in no other way than by the agency of moving ice; and for that agency to have operated to an extent adequate to produce a quantity that I estimate as exceeding a layer 200 feet thick over the entire Wold, nothing less than the complete envelopment of a large part of the Wold by ice for a long period would suffice."

* Geol. Mag., vol. vii., p. 497.
† Quart. Journ. Geol. Soc., vol. xxvi., p. 90.

I have already assigned my reasons for disbelieving the opinion that such masses of drift could have been transported by floating ice; but if we refer it to land-ice, it is obvious that the ice could not have been in the form of local glaciers, but must have existed as a sheet moving in a south and south-west direction, from Yorkshire, across the central part of England. But how is this to harmonize with the theory of glaciation, which is advanced to explain the transport of the Shap boulders?

The explanation has, I think, been pointed out by a writer in the *Glasgow Herald*,* of the 26th November, 1870, in a review of Mr. Lucy's paper.

In my paper on the Boulder Clay of Caithness, I had represented the ice entering the North Sea from the east coast of Scotland and England, as all passing round the north of Scotland. But the reviewer suggests that the ice entering at places to the south of, say, Flamborough Head, would be deflected southwards instead of northwards, and thus pass over England. "It is improbable, however," says the writer, "that this joint ice-sheet would, as Mr. Croll supposes, all find its way round the north of Scotland into the deep sea. The southern uplands of Scotland, and probably also the mountains of Northumberland, propelled, during the coldest part of the glacial period, a land ice-sheet in an eastward direction. This sheet would be met by another streaming outward from the south-western part of Norway—in a diametrically opposite direction. In other words, an imaginary line might be drawn representing the course of some particular boulder in the *moraine profonde* from England met by a boulder from Norway, in the same straight line. With a dense ice-sheet to the north of this line, and an open plain to the south, it is clear that all the ice travelling east or west from points to the south of the starting-points of our two boulders would be 'shed' off to the south. There would be a point somewhere along the line, at which the

* My colleague, Mr. R. L. Jack.

ice would turn as on a pivot—this point being nearer England or Scandinavia, as the degree of pressure exercised by the respective ice-sheets should determine. There is very little doubt that the point in question would be nearer England. Further, the direction of the joint ice-sheet could not be *due* south unless the pressure of the component ice-sheets should be exactly equal. In the event of that from Scandinavia pressing with greater force, the direction would be to the south-west. This is the direction in which the drifts described by Mr. Lucy have travelled."

I can perceive no physical objection to this modification of the theory. What the ice seeks is the path of least resistance, and along this path it will move, whether it may lie to the south or to the north. And it is not at all improbable that an outlet to the ice would be found along the natural hollow formed by the valleys of the Trent, Avon, and Severn. Ice moving in this direction would no doubt pass down the Bristol Channel and thence into the Atlantic.

Might not the shedding of the north of England ice-sheet to the north and south, somewhere not far from Stainmoor, account for the remarkable fact pointed out by Mr. Searles Wood, that the boulder clay, with Shap boulders, to the north of the Wold is destitute of chalk; while, on the other hand, the chalky boulder clay to the south of the Wold is destitute of Shap boulders? The ice which passed over Wastdale Crag moved to the E.N.E., and did not cross the chalk of the Wold; while the ice which bent round to the south by the Wold came from the district lying to the south of Wastdale Crag, and consequently did not carry with it any of the granite from that Crag. In fact, Mr. Searles Wood has himself represented on the map accompanying his Memoir this shedding of the ice north and south.

These theoretical considerations are, of course, advanced for what they are worth. Hitherto geologists have been proceeding upon the supposition of an ice-sheet and an open North Sea;

but the latter is an impossibility. But if we suppose the seas around our island to have been filled with land-ice during the glacial epoch, the entire glacial problem is changed, and it does not then appear so surprising that ice should have passed over England.

Note on the South of England Ice-sheet.

If what has already been stated regarding the north of England be anything like correct, it is evident that the south of England could not possibly have escaped glaciation. If the North Sea was so completely blocked up by Scandinavian ice, that the great mass of ice from the Cumberland mountains entering the sea on the east coast was compelled to bend round and find a way of escape across the centre of England in the direction of the Bristol Channel, it is scarcely possible that the immense mass of ice filling the Baltic Sea and crossing over Denmark could help passing across at least a portion of the south of England. The North Sea being blocked up, its natural outlet into the Atlantic would be through the English Channel; and it is not likely that it could pass through without impinging to some extent upon the land. Already geologists are beginning to recognise the evidence of ice in this region.

Mr. W. C. Lucy, in the *Geological Magazine* for June, 1874, records the finding by himself of evidences of glaciation in West Somerset, in the form of "rounded rocky knolls," near Minehead, like those of glaciated districts; of a bed of gravel and clay 70 feet deep, which he considered to be boulder clay. He also mentions the occurrence near Portlock of a large mass of sandstone well striated, only partially detached from the parent rock. In the same magazine for the following month Mr. H. B. Woodward records the discovery by Mr. Usher of some "rum stuff" near Yarcombe, in the Black Down Hills of Devonshire, which, on investigation, proved to be boulder clay; and further, that it was not a mere isolated patch, but occurred in several other places in the same district. Mr. C. W Peach informs me that on the Cornwall coast, near Dodman Point, at an elevation of about 60 feet above sea-level, he found the rock surface well striated and ice-polished. In a paper on the Drift Deposits of the Bath district, read before the Bath Natural History and Antiquarian Field Club, March 10th, 1874, Mr. C. Moore describes the rock surfaces as grooved, with deep and long-continued furrows similar to those usually found on glaciated rocks, and concludes that during the glacial period they were subjected to ice-action. This conclusion is confirmed by the fact of there being found, immediately overlying these glaciated rocks, beds of gravel with intercalated clay-beds, having a thickness of 30 feet, in which mammalian remains of arctic types are abundant. The most characteristic of which are *Elephas primigenius*, *E. antiquus*, *Rhinoceros tichorhinus*, *Bubalus moschatus*, and *Cervus tarandus*.

There is little doubt that when the ground is better examined many other examples will be found. One reason, probably, why so little evidence of glaciation in the south of England has been recorded, is the comparative absence of rock surfaces suitable for retaining ice-markings. There is, however, one class of evidence which might determine the question of the glaciation of the south of England as satisfactorily as markings on the rock. The evidence to which I refer is that of contorted beds of sand or clay. In England contortions from the sinking of the beds are, of course, quite common, but a thoughtful observer, who has had a little experience of ice-formed contortions, can easily, without much trouble, distinguish the latter from the former. Contortions resulting from the lateral pressure of the ice assume a different form from those produced by the sinking of the beds. In Scotland, for example, there is one well-marked form of contortion, which not only proves the existence of land-ice, but also the direction in which it moved. The form of contortion to which I refer is the bending back of the stratified beds upon themselves, somewhat in the form of a fishing-hook. This form of contortion will be better understood from the accompanying figure.

Fig. 11.

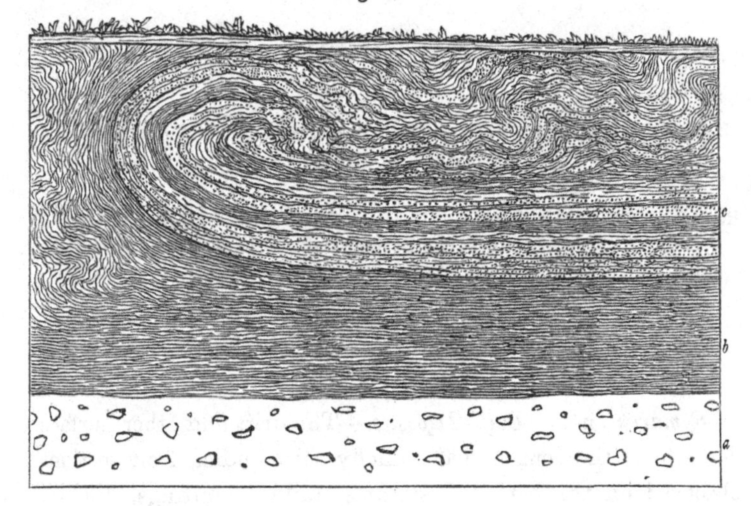

SECTION OF CONTORTED DRIFT NEAR MUSSELBURGH.

a Boulder Clay; *b* Laminated Clay; *c* Sand, Gravel, and Clay, contorted. Depth of Section, twenty-two feet.—H. SKAE.

CHAPTER XXIX.

EVIDENCE FROM BURIED RIVER CHANNELS OF A CONTINENTAL PERIOD IN BRITAIN.*

Remarks on the Drift Deposits.—Examination of Drift by Borings.—Buried River Channel from Kilsyth to Grangemouth.—Channels not excavated by Sea nor by Ice.—Section of buried Channel at Grangemouth.—Mr. Milne Home's Theory.—German Ocean dry Land.—Buried River Channel from Kilsyth to the Clyde.—Journal of Borings.—Marine Origin of the Drift Deposits.—Evidence of Inter-glacial Periods.—Oscillations of Sea-level.—Other buried River Channels.

Remarks on the Drift Deposits.—The drift and other surface deposits of the country have chiefly been studied from sections observed on the banks of streams, railway cuttings, ditches, foundations of buildings, and other excavations. The great defect of such sections is that they do not lay open a sufficient depth of surface. They may, no doubt, represent pretty accurately the character and order of the more recent deposits which overlie the boulder clay, but we are hardly warranted in concluding that the succession of deposits belonging to the earlier part of the glacial epoch, the period of the true till, is fully exhibited in such limited sections.

Suppose, for example, the glacial epoch proper—the time of the lower boulder clay—to have consisted of a succession of alternate cold and warm periods, there would, in such a case, be a series of separate formations of boulder clay; but we could hardly expect to find on the flat and open face of the country, where the surface deposits are generally not of great depth, those various formations of till lying the one superimposed upon the other. For it is obvious that the till formed during one ice-period would, as a general rule, be either swept away

* The greater portion of this chapter is from the Trans. of Geol. Soc. of Edinburgh, for 1869.

or re-ground and laid down by the ice of the succeeding period. If the very hardest rocks could not withstand the abrading power of the enormous masses of ice which passed over the surface of the country during the glacial epoch, it is hardly to be expected that the comparatively soft boulder clay would be able to do so. It is probable that the boulder clay of one period would be used as grinding materials by the ice of the succeeding periods. The boulder clay which we find in one continuous mass may, therefore, in many cases, have been ground off the rocks underneath at widely different periods.

If we wish to find the boulder clays belonging to each of the successive cold periods lying, the one superimposed on the other in the order of time in which they were formed, we must go and search in some deep gorge or valley, where the clay has not only accumulated in enormous masses, but has been partially protected from the destructive power of the ice. But it is seldom that the geologist has an opportunity of seeing a complete section down to the rock-head in such a place. In fact, excepting by bores for minerals, or by shafts of pits, the surface, to a depth of one or two hundred feet, is never passed through or laid open.

Examination of Drift by Borings.—With the view of ascertaining if additional light would be cast on the sequence of events, during the formation of the boulder clay, by an examination of the journals of bores made through a great depth of surface deposits, a collection of about 250 bores, put down in all parts of the mining districts of Scotland, was made. An examination of these bores shows most conclusively that the opinion that the boulder clay, or lower till, is one great undivided formation, is wholly erroneous.

These 250 bores, as already stated,* represent a total thickness of 21,348 feet, giving 86 feet as the mean thickness of the deposits passed through. Twenty of these bores have one boulder clay, with beds of stratified sand or gravel beneath the clay; 25 have 2 boulder clays, with stratified beds of sand and

* Chapter XV., p. 253.

gravel between; 10 have 3 boulder clays; one has 4 boulder clays; 2 have 5 boulder clays; and one has no fewer than 6 separate masses of boulder clay, with stratified beds of sand and gravel between; 16 have two or three separate boulder clays, differing altogether in colour and hardness, without any stratified beds between. We have, therefore, out of 250 bores, 75 of them representing a condition of things wholly different from that exhibited to the geologist in ordinary sections.

These bores bear testimony to the conclusion that the glacial epoch consisted of a succession of cold and warm periods, and not of one continuous and unbroken period of ice, as was at one time generally supposed.

The full details of the character of the deposits passed through by these bores, and their bearing on the history of the glacial epoch, have been given by Mr. James Bennie, in an interesting paper read before the Glasgow Geological Society,* to which I would refer all those interested in the subject of surface geology. But it is not to the mere contents of the bores that I wish at present to direct attention, but to a new and important result, to which they have unexpectedly led.

Buried River Channel, Kilsyth to Grangemouth, Firth of Forth.—These borings reveal the existence of a deep pre-glacial, or perhaps inter-glacial, trough or hollow, extending from the Clyde above Bowling across the country by Kilsyth, along the valley of the Forth and Clyde Canal, to the Firth of Forth at Grangemouth. This trough is filled up with immense deposits of mud, sand, gravel, and boulder clay. These deposits not only fill it up, but they cover it over to such an extent that it is absolutely impossible to find on the surface a single trace of it; and had it not been for borings, and other mining operations, its existence would probably never have been known. In places where the bottom of the trough is perhaps 200 feet below the sea-level, we find on the surface not a hollow, but often an immense ridge or elliptical knoll of sand, gravel, or boulder clay, rising sometimes to 150 or 200 feet above the present sea-level.

* Trans. of the Geol. Soc. of Glasgow, vol. iii., part i., page 133.

I need not here enter into any minute details regarding the form, depth, and general outline of this trough, or of the character of the deposits covering it, these having already been described by Mr. Bennie, but shall proceed to the consideration of circumstances which seem to throw light on the physical origin of this curious hollow, and to the proof which it unexpectedly affords that Scotland, during probably an early part of the glacial epoch, stood higher in relation to the sea-level than it does at present; or rather, as I would be disposed to express it, the sea stood much lower than at present.

From the fact that all along the line of this trough the surface of the country is covered with enormous beds of stratified sands and gravels of marine origin, which proves that the sea must have at a recent period occupied the valley, my first impression was that this hollow had been scooped out by the sea. This conclusion appeared at first sight quite natural, for at the time that the sea filled the valley, owing to the Gulf-stream impinging on our western shores, a strong current would probably then pass through from the Atlantic on the west to the German Ocean on the east. However, considerations soon began to suggest themselves wholly irreconcilable with this hypothesis.

The question immediately arose, if the tendency of the sea occupying the valley is to deepen it, by wearing down its rocky bottom, and removing the abraded materials, then why is the valley filled up to such a prodigious extent with marine deposits? Does not the fact of the whole valley being filled up from sea to sea with marine deposits to a depth of from 100 to 200 feet, and in some places, to even 400 feet, show that the tendency of the sea filling this valley is to silt it up rather than to deepen it? What conceivable change of conditions could account for operations so diverse?

That the sea could not have cut out this trough, is, however, susceptible of direct proof. The height of the surface of the valley at the watershed or highest part, about a mile to the east of Kilsyth, where the Kelvin and the Bonny Water, running in opposite directions,—the one west into the Clyde, and the other east into the Carron,—take their rise, is 160 feet above the sea-level. Consequently, before the sea could pass through the valley at present, the sea-level would require to be raised 160 feet.

But in discussing the question as to the origin of this pre-glacial hollow, we must suppose the surface deposits of the valley all removed, for this hollow was formed before these deposits were laid down. Let us take the average depth of these deposits at the watershed to be 50 feet. It follows that, assuming the hollow in question to have been formed by the sea, the sea-level at the time must have been at least 110 feet higher than at present.

Were the surface deposits of the country entirely removed, the district to the west and north-west of Glasgow would be occupied by a sea which would stretch from the Kilpatrick Hills, north of Duntocher, to Paisley, a distance of about five miles, and from near Houston to within a short way of Kirkintilloch, a distance of more than twelve miles. This basin would contain a few small islands and sunken rocks, but its mean depth, as determined from a great number of surface bores obtained over its whole area, would be not much under 70 or 80 feet. But we shall, however, take the depth at only 50 feet. Now, if we raise the sea-level so as to allow the water just barely to flow over the water-shed of the valley, the sea in this basin would therefore be 160 feet deep. Let us now see what would be the condition of things on the east end of the valley. The valley, for several miles to the east of Kilsyth, continues very narrow, but on reaching Larbert it suddenly opens into the broad and flat carse lands through which the Forth and Carron wind. The average depth at which the sea would stand at present in this tract of country, were the surface removed, as ascertained from bores, would be at least 100 feet, or about double that in the western basin. Consequently, when the sea was sufficiently high to pass over the water-shed, the water would be here 210 feet in depth, and several miles in breadth.

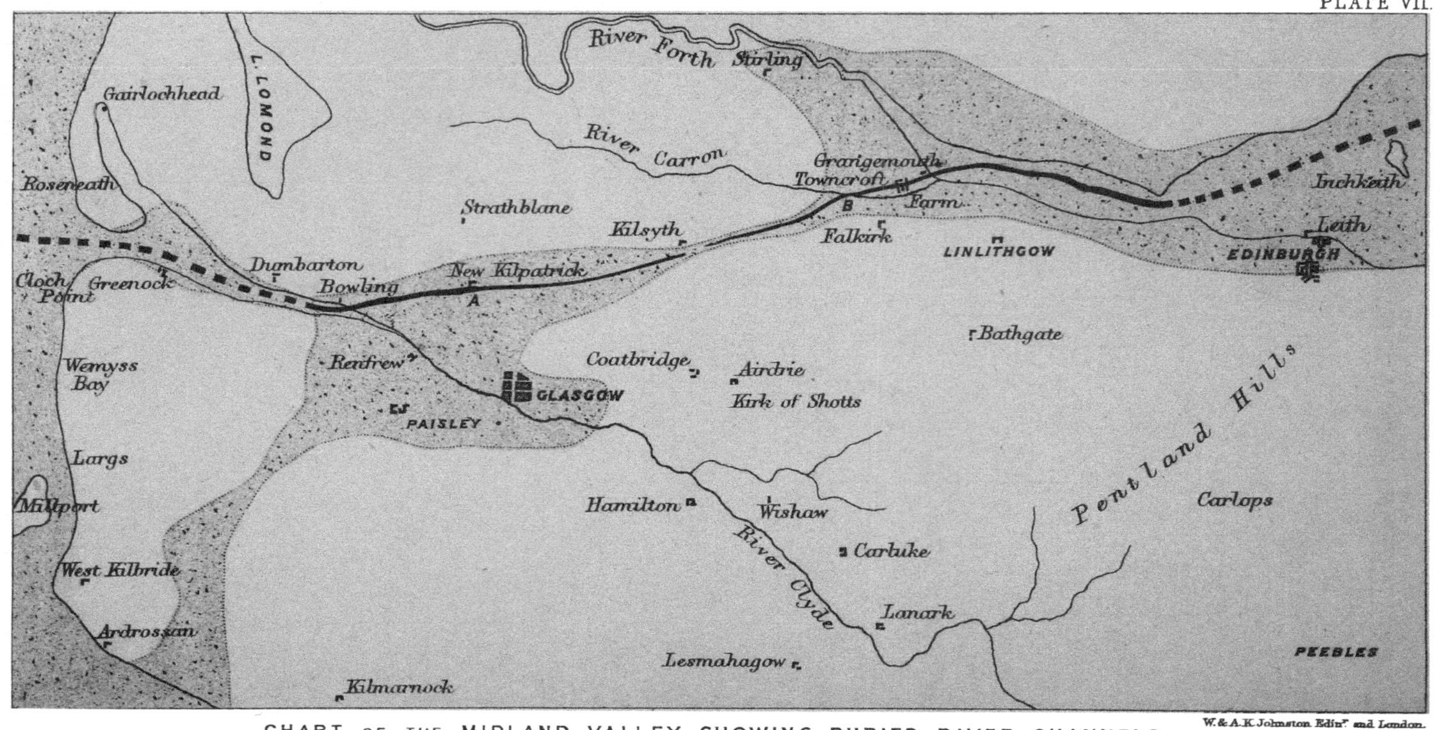

W. & A. K. Johnston Edin^r and London.

CHART OF THE MIDLAND VALLEY, SHOWING BURIED RIVER CHANNELS.

The blue parts represent the area which would be covered by sea were the land submerged to the extent of 200 feet. The heavy black lines A and B represent the buried River Channels.

But in order to have a current of some strength passing through the valley, let us suppose the sea at the time to have stood 150 feet higher in relation to the land than at present. This would give 40 feet as the depth of the sea on the water-shed, and 200 feet as the depth in the western basin, and 250 feet as the depth in the eastern.

An examination of the Ordnance Survey map of the district will show that the 200 feet contour lines which run along each side of the valley from Kilsyth to Castlecary come, in several places, to within one-third of a mile of each other. From an inspection of the ground, I found that, even though the surface deposits were removed off the valley, it would not sensibly affect the contours at those places. It is therefore evident that though the sea may have stood even 200 feet higher than at present, the breadth of the strait at the water-shed and several other points could not have exceeded one-third of a mile. It is also evident that at those places the current would be flowing with the greatest velocity, for here it was not only narrowest, but also shallowest. A reference to Plate VII. will show the form of the basins. The stippled portion, coloured blue, represents the area which would be covered by the sea were the land submerged to the extent of 200 feet.

Let us take the breadth of the current in the western basin at, say, three miles. This is two miles less than the breadth of the basin itself. Suppose the current at the narrow parts between Kilsyth and Castlecary to have had a velocity of, say, five miles an hour. Now, as the mean velocity of the current at the various parts of its course would be inversely proportionate to the sectional areas of those parts, it therefore follows that the mean velocity of the current in the western basin would be only 1-45th of what it was in the narrow pass between Kilsyth and Castlecary. This would give a mile in nine hours as the velocity of the water in the western basin. In the eastern basin the mean velocity of the current, assuming its breadth to be the same as in the western, would be only a mile in eleven hours. In the central part of the current the velocity at the surface would probably be considerably above the mean, but at the sides and bottom it would, no doubt, be under the mean. In fact, in these two basins the current would be almost insensible.

The effect of such a current would simply be to widen and deepen the valley all along that part between Kilsyth and Castlecary where the current would be flowing with considerable rapidity. But it would have little or no effect in deepening the basins at each end, but the reverse. It would tend rather to silt them up. If the current flowed from west to east, the materials removed from the narrow part between Kilsyth and Castlecary, where the velocity of the water was great, would be deposited when the current almost disappeared in the eastern part of the valley. Sediment carried by a current flowing at the rate of five miles an hour, would not remain in suspension when the velocity became reduced to less than five miles a day.

But even supposing it were shown that the sea under such conditions could have deepened the valley along the whole distance from the Clyde to the Forth, still this would not explain the origin of the trough in question. What we are in search of is not the origin of the valley itself, but the origin of a deep and narrow hollow running along the bottom of it. A sea filling the whole valley, and flowing with considerable velocity, would, under certain conditions, no doubt deepen and widen it, but it would not cut out along its bottom a deep, narrow trough, with sides often steep, and in some places perpendicular and even overhanging.

This hollow is evidently an old river-bed scooped out of the rocky valley by a stream, flowing probably during an early part of the glacial period.

During the latter part of the summer of 1868, I spent two or three weeks of my holidays in tracing the course of this buried trough from Kilsyth to the river Forth at Grangemouth, and I found unmistakable evidence that the eastern portion of it, stretching from the watershed to the Forth, had been cut out, not by the sea, but by a stream which must have followed almost the present course of the Bonny Water.

I found that this deep hollow enters the Forth a few hundred yards to the north of Grangemouth Harbour, at the extraordinary depth of 260 feet below the present sea-level. At the period when the sea occupied the valley of the Forth ard Clyde Canal, the bottom of the trough at this spot would therefore be upwards of 400 feet below the level of the sea.

A short distance to the west of Grangemouth, and also at Carron, several bores were put down in lines almost at right angles across the trough, and by this means we have been enabled to form a pretty accurate estimate of its depth, breadth, and shape at those places. I shall give the details of one of those sections.

Between Towncroft Farm and the river Carron, a bore was put down to the depth of 273 feet before the rock was reached. About 150 yards to the north of this there is another bore, giving 234 feet as the depth to the rock; 150 yards still further north the depth of the surface deposits, as determined by a third bore, is 155 feet. This last bore is evidently outside of the hollow, for one about 150 yards north of it gives the same depth of surface, which seems to be about its average depth for a mile or two around. About half a mile to the south of the hollow at this place the surface deposits are 150 feet deep. From a number of bores obtained at various points within a circuit of 1½ miles, the surface appears to have a pretty uniform depth of 150 feet or thereby. For the particulars of these "bores" I am indebted to the kindness of Mr. Mackay, of Grangemouth.

To the south of the trough (see Fig. 12) there is a fault running nearly parallel to it, having a down-throw to the north, and cutting off the coal and accompanying strata to the south. But an inspection of the section will show that the hollow in question is no way due to the fault, but has been scooped out of the solid strata.

The main coal wrought extensively here is cut off by the

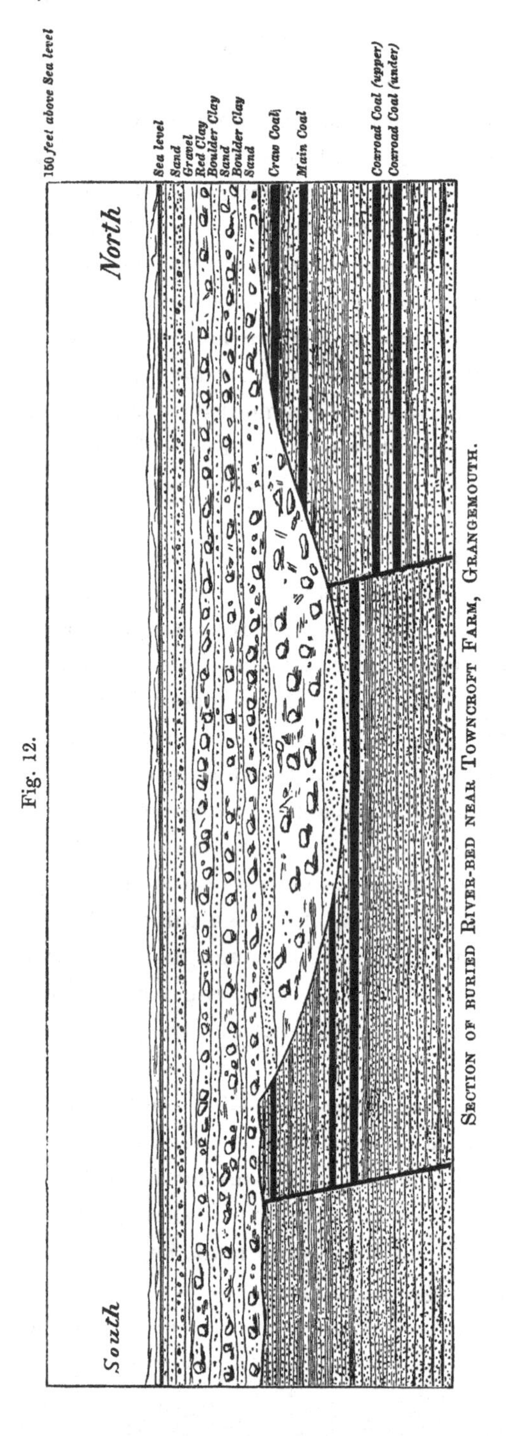

Fig. 12.

SECTION OF BURIED RIVER-BED NEAR TOWNCROFT FARM, GRANGEMOUTH.

trough, as will be seen from the section. Mr. Dawson, of Carron Iron Works, informs me that at Carronshore pit, about a mile and a quarter above where this section is taken, the coal was found to be completely cut off by this trough. In one of the workings of this pit, about forty years ago, the miners cut into the trough at 40 fathoms below the surface, when the sand rushed in with irresistible pressure, and filled the working. Again, about a mile below where the section is taken, or about two miles below Carronshore, and just at the spot where the trough enters the Firth, it was also cut into in one of the workings of the Heuck pit at a depth of 40 fathoms from the surface. Fortunately, however, at this point the trough is filled with boulder clay instead of sand, and no damage was sustained. Here, for a distance of two miles, the Main coal and "Upper Coxroad" are cut off by this hollow; or rather, I should say this hollow has been cut through the coal-seams. The "Under Coxroad," lying about 14 fathoms below the position of the "Main" coal, as will be seen in the descriptive section (Fig. 12), is not reached by the trough, and passes undisturbed under it.

This hollow would seem to narrow considerably as it recedes westwards, for at Carronshore pit-shaft the surface is 138 feet deep; but not much over 150 yards to the south of this is the spot where the coal was cut off by the trough at a depth of 40 fathoms or 240 feet. Here it deepens upwards of 100 feet in little more than 150 yards. That it is narrow at this place is proved by the fact, that a bore put down near Carronbank, a little to the south, shows the surface to be only 156 feet deep.

In the section (Fig. 12) the line described as "150 *feet above sea-level*" registers the height of the sea-level at the time when the central valley was occupied by sea 40 feet deep at the watershed. Now, if this hollow, which extends right along the whole length of the valley, had been cut out by the sea, the surface of the rock 150 feet below the present surface of the ground would be the sea-bottom at the time, and the line marked "150 *feet above sea-level*" would be the surface of the sea. The sea would therefore be here 300 feet deep for several miles around. It cannot be supposed that the sea acting on a broad flat plain of several miles in extent should cut out a deep, narrow hollow, like the one exhibited in the section, and leave the rest of the plain a flat sea-bottom.

And it must be observed, that this is not a hollow cut merely in a sea-beach, but one extending westward to Kilsyth. Now, if this hollow was cut out by the sea, it must have been done, not by the waves beating on the beach, but by a current flowing through the valley. The strongest current that could possibly pass through the narrow part between Kilsyth and Castlecary would be wholly insensible when it reached Grangemouth, where the water was 300 feet deep, and several miles broad. Consequently, it is impossible that the current could have scooped out the hollow represented in the section.

Again, if this hollow had been scooped out by the sea, it ought to be deepest between Kilsyth and Castlecary, where the current was narrowest; but the reverse is actually the case. It is shallowest at the place where the current was narrowest, and deepest at the two ends where the current was broadest. In the case of a trough cut by a sea current, we must estimate its depth from the level of the sea. Its depth is the depth of the water in it while it was being scooped out. The bottom of the trough in the highest and narrowest part of the valley east of Kilsyth is 40 feet above the present sea-level. Consequently, its depth at this point at the period in question, when the sea-level was 150 feet higher than at present, would be 110 feet. The bottom of the trough at Grangemouth is 260 feet below the present sea-level; add to this 150 feet, and we have 410 feet as its depth here at the time in question. If this hollow was scooped out by the sea, how then does it thus happen that at the place where the current was strongest and confined to a narrow channel by hills on each side, it cut its channel to a depth of only 110 feet, whereas at the place where it had scarcely any motion it has cut, on a flat and open plain several miles broad, a channel to a depth of 410 feet?

But, suppose we estimate the relative amount of work performed by the sea at Kilsyth and Grangemouth, not by the actual depth of the bottom of the trough at these two places below the sea-level at the time that the work was performed, but by the present actual depth of the bottom of the trough below the rocky surface of the valley, this will still not help us out of the difficulty. Taking, as before, the height of the rocky bed of the valley at the watershed at 110 feet above the present sea-level, and the bottom of the trough at 40 feet, this gives 70 feet as the depth scooped out of the rock at that place. The depth of the trough at Grangemouth below the rocky surface is 118 feet. Here we have only 70 feet cut out at the only place where there was any resistance to the current, as well as the place where it possessed any strength; whereas at Grangemouth, where there was no resistance, and no strength of current, 118 feet has been scooped out. Such a result as this is diametrically opposed to all that we know of the dynamics of running water.

We may, therefore, conclude that it is physically impossible that this hollow could have been cut out by the sea.

Owing to the present tendency among geologists to attribute effects of this kind to ocean-currents, I have been induced to enter thus at much greater length than would otherwise have been necessary into the facts and arguments against the possibility of the hollow having been excavated by the sea. In the present case the discussion is specially necessary, for here we have positive evidence of the sea having occupied the valley for ages, along which this channel has been cut. Consequently, unless it is proved that the sea could not possibly have scooped out the channel, most geologists would be inclined to attribute it to the sea-current which is known to have passed through the valley rather than to any other cause.

But that it is a hollow of denudation, and has been scooped out by some agent, is perfectly certain. By what agent, then, has the erosion been made? The only other cause to which it can possibly be attributed is either land-ice or river-action.

The supposition that this hollow was scooped out by ice is not more tenable than the supposition that the work has been done by the sea. A glacier filling up the entire valley and descending into the German Ocean would unquestionably not only deepen the valley, but would grind down the surface over which it passed all along its course. But such a glacier would not cut a deep and narrow channel along the bottom of the valley. A glacier that could do this would be a small and narrow one, just sufficiently large to fill this narrow trough;

for if it were much broader than the trough, it would grind away its edges, and make a broad trough instead of a narrow one. But a glacier so small and narrow as only to fill the trough, descending from the hills at Kilsyth to the sea at Grangemouth, a distance of fifteen miles, is very improbable indeed. The resistance to the advance of the ice along such a slope would cause the ice to accumulate till probably the whole valley would be filled.*

There is no other way of explaining the origin of this hollow, but upon the supposition of its being an old river-bed. But there is certainly nothing surprising in the fact of finding an old watercourse under the boulder clay and other deposits. Unless the present contour of the country be very different from what it was at the earlier part of the glacial epoch, there must have then been watercourses corresponding to the Bonny Water and the river Carron of the present day; and that the remains of these should be found under the present surface deposits is not surprising, seeing that these deposits are of such enormous thickness. When water began to flow down our valleys, on the disappearance of the ice at the close of the glacial epoch, the Carron and the Bonny Water would not be able to regain their old rocky channels, but would be obliged to cut, as they have done, new courses for themselves on the surface of the deposits under which their old ones lay buried.

Although an old pre-glacial or inter-glacial river-bed is in

* Mr. Milne Home has advanced, in his "Estuary of the Firth of Forth," p. 91, the theory that this trough had been scooped out during the glacial epoch by icebergs floating through the Midland valley from west to east when it was submerged. The bottom of the trough, be it observed, at the water-shed at Kilsyth, is 300 feet above the level of its bottom at Grangemouth; and this Mr. Milne Home freely admits. But he has not explained how an iceberg, which could float across the shallow water at Kilsyth, say, 100 feet deep, could manage to grind the rocky bottom at Grangemouth, where it was not less than 400 feet deep. "The impetus acquired in the Kyle at Kilsyth," says Mr. Milne Home, "would keep them moving on, and the prevailing westerly winds would also aid, so that when *grating* on the subjacent carboniferous rocks they would not have much difficulty in scooping out a channel both wider and deeper than at Kilsyth." But how could they "*grate* on the subjacent carboniferous rocks" at Grangemouth, if they managed to *float* at Kilsyth? Surely an iceberg that could "*grate*" at Grangemouth would "*ground*" at Kilsyth.

itself an object of much interest and curiosity, still, it is not on that account that I have been induced to enter so minutely into the details of this buried hollow. There is something of far more importance attached to this hollow than the mere fact of its being an old watercourse. For the fact that it enters the Firth of Forth at a depth of 260 feet below the present sea-level, proves incontestably that at the time this hollow was occupied by a stream, *the land must have stood at least between* 200 *and* 300 *feet higher in relation to the sea-level than at present.*

We have seen that the old surface of the country in the neighbourhood of Grangemouth, out of which this ancient stream cut its channel, is at least 150 feet below the present sea-level. Now, unless this surface had been above the sea-level at that time, the stream would not have cut a channel in it. But it has not merely cut a channel, but cut one to a depth of 120 feet. It is impossible that this channel could have been occupied by a river of sufficient volume to fill it. It is not at all likely that the river which scooped it out could have been much larger than the Carron of the present day, for the area of drainage, from the very formation of the country, could not have been much greater above Grangemouth than at present. An elevation of the land would, no doubt, increase the area of the drainage of the stream measured from its source to where it might then enter the sea, because it would increase the length of the stream; but it would neither increase the area of drainage, nor the length of the stream above Grangemouth. Kilsyth would be the watershed then as it is now.

What we have here is not the mere channel which had been occupied by the ancient Carron, but the valley in which the channel lay. It may, perhaps, be more properly termed a buried river valley; formed, no doubt, like other river valleys by the denuding action of rain and river.

The river Carron at present is only a few feet deep. Suppose the ancient Carron, which flowed in this old channel, to have been say 10 feet deep. This would show that the land in relation to the sea at that time must have stood at least 250 feet higher than at present. If 10 feet was the depth of this old river, and Grangemouth the place where it entered the sea, then 250 feet would be the extent of the elevation. But it is probable that Grangemouth was not the mouth of the river; it would likely be merely the place where it joined the river Forth of that period. We have every reason to believe that the bed of the German Ocean was then dry land, and that the Forth, Tay, Tyne, and other British rivers flowing eastward, as Mr. Godwin-Austin supposes, were tributaries to the Rhine, which at that time was a huge river passing down the bed of the German Ocean, and entering the Atlantic to the west of the Orkney Islands. That the German Ocean, as well as the sea-bed of the Western Hebrides, was dry land at a very recent geological period, is so well known, that, on this point, I need not enter into details. We may, therefore, conclude that the river Forth, after passing Grangemouth, would continue to descend until it reached the Rhine. If, by means of borings, we could trace the old bed of the Forth and the Rhine up to the point where the latter entered the Atlantic, in the same way as we have done the Bonny Water and the Carron, we should no doubt obtain a pretty accurate estimate as to the height at which the land stood at that remote period. Nothing whatever, I presume, is known as to the depth of the deposits covering the bed of the German Ocean along what was then the course of the Rhine. It must, no doubt, be something enormous. We are also in ignorance as to the thickness of the deposits covering the ancient bed of the Forth. A considerable number of bores have been put down at various parts of the Firth of Forth in connection with the contemplated railway bridge across the Firth, but in none of those bores has the rock been reached. Bores to a depth of 175 feet have been made without even passing through the deposits of silt which probably overlie an enormous thickness of sand and boulder clay. Even in places where the water is 40 fathoms deep and quite narrow, the bottom is not rock but silt.

It is, however, satisfactory to find on the land a confirmation of what has long been believed from evidence found in the seas around our island, that at a very recent period the sea-level in relation to the land must have been some hundreds of feet lower than at the present day, and that our island must have at that time formed a part of the great eastern continent.

A curious fact was related to me by Mr. Stirling, the manager of the Grangemouth collieries, which seems to imply a great elevation of the land at a period long posterior to the time when this channel was scooped out.

In sinking a pit at Orchardhead, about a mile to the north of Grangemouth, the workmen came upon the boulder clay after passing through about 110 feet of sand, clay, and gravel. On the upper surface of the boulder clay they found cut out what Mr. Stirling believes to have been an old watercourse. It was 17 feet deep, and not much broader. The sides of the channel appear to have been smooth and waterworn, and the whole was filled with a fine sharp sand beautifully stratified. As this channel lay about 100 feet below the present sea-level, it shows that if it actually be an old watercourse, it must have been scooped out at a time when the land in relation to the sea stood at least 100 feet higher than at present.

Buried River Channel from Kilsyth to the Clyde.—In all probability the western half of this great hollow, extending from the watershed at Kilsyth to the Clyde, is also an old river channel, probably the ancient bed of the Kelvin. This point cannot, however, be satisfactorily settled until a sufficient number of bores have been made along the direct line of the hollow, so as to determine with certainty its width and general form and extent. That the western channel is as narrow as the eastern is very probable. It has been found that its sides at some places, as, for example, at Garscadden, are very steep. At one place the north side is actually an overhanging buried precipice, the bottom of which is about 200 feet below the sea-level. We know also that the coal and ironstone in that quarter are cut through by the trough, and the miners there have to exercise great caution in driving their workings, in case they might cut into it. The trough along this district is filled with sand, and is known to the miners of the locality as the "sand-dyke." To cut into running sand at a depth of 40 or 50 fathoms is a very dangerous proceeding, as will be seen from the details given in Mr. Bennie's paper * of a disaster which occurred about twenty years ago to a pit near Duntocher, where this trough was cut into at a depth of 51 fathoms from the surface.

The depth of this hollow, below the present sea-level at Drumry, as ascertained by a bore put down, is 230 feet. For several miles to the east the depth is nearly as great. Consequently, if this hollow be an old river-bed, the ancient river that flowed in it must have entered the Clyde at a depth of more than 200 feet below the present sea-level; and if so, then it follows that the rocky bed of the ancient Clyde must lie buried under more than 200 feet of surface deposits from Bowling downwards to the sea. Whether this is the case or not we have no means at present of determining. The manager to the Clyde Trustees informs me, however, that in none of the borings or excavations which have been made has the rock ever been reached from Bowling downwards. The probability is, that this deep hollow passes downwards continuously to the sea on the western side of the island as on the eastern.†

The following journals of a few of the borings will give the reader an idea of the character of the deposits filling the

* Trans. of the Geol. Soc. of Glasgow, vol. iii., p. 141.

† Mr. John Young and Mr. Milne Home advanced the objection, that several trap dykes cross the valley of the Clyde near Bowling, and come to so near the present surface of the land, that the Clyde at present flows across them with a depth not exceeding 20 feet. I fear that Mr. Young and Mr. Milne Home have been misinformed in regard to the existence of these dykes. About a mile *above* Bowling there are one or two dykes which approach to the river-bank, and may probably cross, but these could not possibly cut off a channel entering the Clyde at Bowling. In none of the borings or excavations which have been made by the Clyde Trustees has the rock been reached from Bowling downwards. I may also state that the whole Midland valley, from the Forth of Clyde to the Firth of Forth, has been surveyed by the officers of the Geological Survey, and only a single dyke has been found to cross the buried channels, viz., one (Basalt rock) running eastward from Kilsyth to the canal bridge near Dullatur. But as this is not far from the water-shed between the two channels it cannot affect the question at issue. See sheet 31 of Geological Survey Map of Scotland.

channels. The beds which are believed to be boulder clay are printed in italics :—

Borings made through the Deposits filling the Western Channel.

Bore, Drumry Farm, on Lands of Garscadden.

	ft.	ins.
Surface soil	2	6
Sand and gravel	3	6
Dry sand	11	0
Blue mud	8	6
Light mud and sand beds	13	0
Sand	31	6
Sand and mud	8	0
Sand and gravel	19	6
Sand	8	6
Gravel	24	4
Sand	5	0
Gravel	9	6
Sand	71	6
Sand (coaly)	1	0
Sand	9	0
Sand (coaly)	1	0
Sand	10	3
Red clay and gravel	4	8
Sand	1	5
Gravel	2	0
Sand	2	8
Gravel	10	6
Sand	1	6
Gravel	8	10
Clay stones and gravel	33	3
	297	10

Bore on Mains of Garscadden, one mile north-east of Drumry.

	ft.	ins.
Surface soil	1	0
Blue clay and stones	60	0
Red clay and stones	18	0
Soft clay and sand beds	7	0
Gravel	6	0
Large gravel	9	0
Sand and gravel	7	0
Hard gravel	1	6
Sand and gravel	16	6
Dry sand	30	0
Black sand	2	0
Dry sand	33	0
Wet sand	8	0
Light mud	5	0
Sand	3	0
Gravel	5	6
Sandstone, black	0	6
Blue clay and stones	1	4
Whin block	0	10
Sandy clay	4	6
	219	8

Bore nearly half a mile south-west of Millichen.

	ft.	ins.
Sandy clay	5	0
Brown clay and stones	17	0
Mud	15	0
Sandy mud	31	0
Sand and gravel with water	28	0
Sandy clay and gravel	17	0
Sand	5	0
Mud	6	0
Sand	14	0
Gravel	30	0
Brown sandy clay and stones	30	0
Hard red gravel	4	6
Light mud and sand	1	8
Light clay and stones	6	6
Light clay and whin block	26	0
Fine sandy mud	36	0
Brown clay and gravel and stones	14	4
Dark clay and stones	68	0
	355	0

Bore at West Millichen, about 100 yards east of farm-house.

	ft.	ins.
Soil	1	6
Muddy sand and stones	4	6
Soft mud	4	0
Sand and gravel	45	0
Sandy mud and stones	20	6
Coarse gravel	11	6
Clay and gravel	1	4
Fine mud	7	0
Sand and gravel	2	0
Sandy mud	30	6
Brown sandy clay and stones	25	0
Sand and gravel	6	0
Brown sandy clay and stones	12	0
Sand	2	0
Brown sandy clay and stones	4	0
Mud	5	0
Mud and sand	10	9
Sand and stones	2	9
Blue clay and stones	5	0
	200	4

BORINGS MADE THROUGH THE DEPOSITS FILLING THE EASTERN CHANNEL.

No. 1. Between Towncroft Farm and Carron River—200 yards from river. Height of surface, 12 feet above sea-level.

	Feet.		Feet.
Surface sand	6	*Soft blue till*	17
Blue mud	4	*Hard blue till*	140
Sand	4	Sand	20
Gravel	3		
Sand	33		273
Red clay	46		

No. 2. About 150 yards north of No. 1. Height of surface, 12 feet above sea-level.

	Feet.		Feet.
Surface sand	6	Sand	20
Blue mud	3	*Hard blue till and stones*	24
Shell bed	1	Sand	2
Gravel	2	*Hard blue till and stones*	40
Blue mud	8	Sand	7
Gravel	3	*Hard blue till*	24
Blue muddy sand	15		
Red clay	49		234
Blue till and stones	20		

No. 3. About 150 yards north of No. 2. Height of surface, 12 feet above sea-level.

	Feet.		Feet.
Surface sand	6	*Soft till*	36
Soft mud with shells	11	Sand (pure)	2
Blue mud and sand (hard)	3	*Soft till and sand*	17
Channel (rough gravel)	3	Gravel	8
Fine sand	8	*Hard blue till*	14
Running sand (red and fine)	17		
Red clay	30		155

No. 4. About 100 yards from No. 1.

	Feet.		Feet.
Surface	5	*Soft blue till*	32
Blue mud	5	*Hard blue till and stones*	104
Black sand	3	Grey sand not passed through	22
Gravel	3		
Red clay and stones	34		252
Red clay	44	Rock-head not reached.	

No. 5. About 50 yards north of No. 4.

	Feet.		Feet.
Surface	6	*Blue till and stones*	20
Blue mud	3	Sand	20
Shell bed	1	*Hard blue till and stones*	24
Channel	2	Sand	2
Blue mud	8	*Hard blue till and stones*	40
Channel	3	Sand	7
Blue mud and sand	15	*Hard blue till*	24
Red clay and sand	10		
Red clay	49		211

No. 6. Between Heuck and Carron River.

	Feet.		Feet.
Sandy clay	7	Brown clay	39
Mud	16	*Blue till and stones*	54
Brown sandy clay and stones	3		
Mud	36		155

The question arises as to what is the origin of the stratified sands and gravels filling up the buried river channels. Are they of marine or of freshwater origin? Mr. Dugald Bell* and Mr. James Geikie† are inclined to believe that as far as regards those filling the western channel they are of lacustrine origin; that they were formed in lakes, produced by the damming back of the water resulting from the melting of the ice. I am, however, for the following reasons, inclined to agree with Mr. Bennie's opinion that they are of marine origin. It will be seen, by a comparison of the journals of the borings made through the deposits in the eastern channel with those in the western, that they are of a similar character; so that, if we suppose those in the western channel to be of freshwater origin, we may from analogy infer the same in reference to the origin of those in the eastern channel. But, as we have already seen, the deposits extend to the Firth of Forth at Grangemouth, where they are met with at a depth of 260 feet below sea-level. Consequently, if we conclude them to be of freshwater origin, we are forced to the assumption, not that the water formed by the melted ice was dammed back, but that the sea itself was dammed back, and that by a wall extending to a depth of not less than two or three hundred feet, so as to allow of a lake being formed in which the deposits might accumulate; assuming, of course, that the absolute level of the land was the same then as it is now.

But as regards the stratified deposits of Grangemouth, we have direct evidence of their marine origin down to the bottom of the Red Clay that immediately overlies the till and its intercalated beds, which on an average is no less than 85 feet, and in some cases 100 feet, below the present surface. From this deposit, Foraminifera, indicating an arctic condition of sea, were determined by Mr. David Robertson. Marine shells were also found in this bed, and along with them the remains of a seal, which was determined by Professor Turner to be of an exceed-

* Trans. Geol. Soc. Glasgow, vol. iv., p. 166.

† "Great Ice Age," chap. xiii.

ingly arctic type, thus proving that these deposits were not only marine but glacial.

Direct fossil evidence as to the character of the deposits occupying the western basin, is, however, not so abundant, but this may be owing to the fact that during the sinking of pits, no special attention has been paid to the matter. At Blairdardie, in sinking a pit-shaft through these deposits, shells were found in a bed of sand between two immense masses of boulder clay. The position of this bed will be better understood from the following section of the pit-shaft :—

	Feet.		Feet.
Surface soil	4½	Mud and running sand	11
Blue clay	9	Hard clay, boulders, and broken rock	27
Hard stony clay	69		
Sand with a few *shells*	3		
Stony clay and boulders	46½		170

But as the shells were not preserved, we have, of course, no means of determining whether they were of marine or of freshwater origin.

In another pit, at a short distance from the above, *Cyprina Islandica* was found in a bed at the depth of 54 feet below the surface.*

In a paper read by Mr. James Smith, of Jordanhill, to the Geological Society, April 24th, 1850,† the discovery is recorded of a stratified bed containing *Tellina proxima* intercalated between two distinct boulder clays. The bed was discovered by Mr. James Russell in sinking a well at Chapelhall, near Airdrie. Its height above sea-level was 510 feet. The character of the shell not only proves the marine origin of the bed, but also the existence of a submergence to that extent during an inter-glacial period.

On the other hand, the difficulty besetting the theory of the marine origin of the deposits is this. The intercalated boulder clays bear no marks of stratification, and are evidently the true unstratified till formed when the country was covered

* See further particulars in Mr. Bennie's paper on the Surface Geology of the district around Glasgow, Trans. Geol. Soc. of Glasgow, vol. iii.

† See also Smith's "Newer Pliocene Geology," p. 139.

by ice. But the fact that these beds are both underlaid and overlaid by stratified deposits would, on the marine theory, imply not merely the repeated appearance and disappearance of the ice, but also the repeated submergence and emergence of the land. If the opinion be correct that the submergences and emergences of the glacial epoch were due to depressions and elevations of the land, and not to oscillations of sea-level, then the difficulty in question is, indeed, a formidable one. But, on the other hand, if the theory of submergences propounded in Chapters XXIII. and XXIV. be the true one, the difficulty entirely disappears. The explanation is as follows, viz., during a cold period of the glacial epoch, when the winter solstice was in aphelion, the low grounds would be covered with ice, under which a mass of till would be formed. After the cold began to decrease, and the ice to disappear from the plains, the greatest rise of the ocean, for reasons already stated, would take place. The till covering the low grounds would be submerged to a considerable depth and would soon be covered over by mud, sand, and gravel, carried down by streams from the high ground, which, at the time, would still be covered with snow and ice. In course of time the sea would begin to sink and a warm and continental period of, perhaps, from 6,000 to 10,000 years, would follow, when the sea would be standing at a much lower level than at present. The warm period would be succeeded by a second cold period, and the ice would again cover the land and form a second mass of till, which, in some places, would rest directly on the former till, while in other places it would be laid down upon the surface of the sands and gravels overlying the first mass. Again, on the disappearance of the ice the second mass of till would be covered over in like manner by mud, sand, and gravel, and so on, while the eccentricity of the earth's orbit continued at a high value. In this way we might have three, four, five, or more masses of till separated by beds of sand and gravel.

It will be seen from Table IV. of the eccentricity of the

earth's orbit, given in Chapter XIX., that the former half of that long succession of cold and warm periods, known as the glacial epoch, was much more severe than the latter half. That is to say, in the former half the accumulation of ice during the cold periods, and its disappearance in polar regions during the warm periods, would be greater than in the latter half. It was probable that it was during the warm periods of the earlier part of the glacial epoch that the two buried channels of the Midland valley were occupied by rivers, and that it was during the latter and less severe part of the glacial epoch that these channels became filled up with that remarkable series of deposits which we have been considering.

Other buried River Channels.—A good many examples of buried river channels have been found both in Scotland and in England, though none of them of so remarkable a character as the two occupying the valley of the Forth and Clyde Canal which have been just described. I may, however, briefly refer to one or two localities where some of these occur.

(1.) An ancient buried river channel, similar to the one extending from Kilsyth to Grangemouth, exists in the coal-fields of Durham, and is known to miners in the district as the "Wash." Its course was traced by Mr. Nicholas Wood, F.G.S., and Mr. E. F. Boyd, from Durham to Newcastle, a distance of fourteen miles.* It traverses, after passing the city of Durham, a portion of the valley of the Wear, passes Chester-le-Street, and then follows the valley of the river Team, and terminates at the river Tyne. And what is remarkable, it enters the Tyne at a depth of 140 feet below the present level of the sea. This curious hollow lies buried, like the Scottish one just alluded to, under an enormous mass of drift, and it is only through means of boring and other mining operations that its character has been revealed. The bottom and sides of this channel everywhere bear evidence of long exposure to the abrading influence of water in motion; the rocky bottom being smoothed, furrowed, and water-worn. The river Wear of the present day flows to

* British Association Report for 1863, p. 89. *Geologist* for 1863, p. 384.

the sea over the surface of the drift at an elevation of more than 100 feet above this buried river-bed. At the time that this channel was occupied by running water the sea-level must have been at least 140 feet lower than at present. This old river evidently belongs to the same continental period as those of Scotland.

(2.) From extensive borings and excavations, made at the docks of Hull and Grimsby, it is found that the ancient bed of the Humber is buried under more than 100 feet of silt, clay, and gravel. At Hull the bottom of this buried trough was found to be 110 feet below the sea-level. And what is most interesting at both these places, the remains of a submerged forest was found at a depth of from thirty to fifty feet below the sea-level. In some places two forests were found divided by a bed of leafy clay from five to fifteen feet thick.

(3.) In the valleys of Norfolk we also find the same conditions exhibited. The ancient bed of the Yare and other rivers of this district enter the sea at a depth of more than 100 feet below the present sea-level. At Yarmouth the surface was found 170 feet thick, and the deep surface extends along the Yare to beyond Norwich. Buried forests are also found here similar to those on the Humber.

It is probable that all our British rivers flow into the sea over their old buried channels, except in cases where they may have changed their courses since the beginning of the glacial epoch.

(4.) In the Sanquhar Coal Basin, at the foot of the Kello Water, an old buried river course was found by Mr. B. N. Peach. It ran at right angles to the Kello, and was filled with boulder clay which cut off the coal; but, on driving the mine through the clay, the coal was found in position on the other side.

(5.) An old river course, under the boulder clay, is described by Mr. Milne Home in his memoir on the Mid-Lothian coal-fields. It has been traced out from Niddry away in a N.E. direction by New Craighall. At Niddry, the hollow is about 100 yards wide and between 60 and 70 feet deep. It seems to deepen and widen as it approaches towards the sea, for at New Craighall it is about 200 yards wide and 97 feet deep. This old channel will probably enter the sea about Musselburgh. Like the channels in the Midland Valley of Scotland already described, it is so completely filled up by drift that not a trace of it is to be seen on the surface. And like these, also, it must have belonged to a period when the sea-level stood much lower than at present.

(6.) At Hailes' Quarry, near Edinburgh, there is to be seen a portion of an ancient water-course under the boulder drift. A short account of it was given by Dr. Page in a paper read before the Edinburgh Geological Society.* The superincumbent sandstone, he says, has been cut to a depth of 60 feet. The width of the channel at the surface varies from 12 to 14 feet, but gradually narrows to 2 or 3 feet at the bottom. The sides and bottom are smoothed and polished, and the whole is now filled with till and boulders.

(7.) One of the most remarkable buried channels is that along the Valley of Strathmore, supposed to be the ancient bed of the Tay. It extends from Dunkeld, the south of Blairgowrie, Ruthven, and Forfar, and enters the German Ocean at Lunan Bay. Its length is about 34 miles.

"No great river," says Sir Charles Lyell, "follows this course, but it is marked everywhere by lakes or ponds, which afford shell-marl, swamps, and peat moss, commonly surrounded by ridges of detritus from 50 to 70 feet high, consisting in the lower part of till and boulders, and in the upper of stratified gravels, sand, loam, and clay, in some instances curved or contorted." †

"It evidently marks an ancient line, by which, first, a great glacier descended from the mountains to the sea, and by which, secondly, at a later period, the principal water drainage of this country was effected." ‡

* See Geological Magazine, vol. ii., p. 38.
† Proc. Geol. Soc., vol. iii., 1840, p. 342.
‡ "Antiquity of Man" (Third Edition), p. 249.

(8.) A number of examples of ancient river courses, underneath the boulder clay, are detailed by Professor Geikie in his glacial drift of Scotland. Some of the cases described by him have acquired additional interest from the fact of their bearing decided testimony to the existence of inter-glacial warm periods. I shall briefly refer to a few of the cases described by him.

In driving a trial mine in a pit at Chapelhall, near Airdrie, the workmen came upon what they believed to be an old river course. At the end of the trial mine the ironstone, with its accompanying coal and fire-clay, were cut off at an angle of about 20° by a stiff, dark-coloured earth, stuck full of angular pieces of white sandstone, coal, and shale, with rounded pebbles of greenstone, basalt, quartz, &c. Above this lay a fine series of sand and clay beds. Above these stratified beds lay a depth of 50 or 60 feet of true boulder clay. The channel ran in the direction of north-east and south-west. Mr. Russell, of Chapelhall, informs Professor Geikie that another of the same kind, a mile farther to the north-west, had been traced in some of the pit workings.

"It is clear," says Professor Geikie, "that whatever may be the true explanation of these channels and basins, they unquestionably belong to the period of the boulder clay. The Chapelhall basin lies, indeed, in a hollow of the carboniferous rocks, but its stratified sands and clays rest on an irregular floor of true till. The old channel near the banks of the Calder is likewise scooped out of sandstones and shales; but it has a coating of boulder clay, on which its finely-laminated sands and clays repose, *as if the channel itself had once been filled with boulder clay, which was re-excavated to allow of the deposition of the stratified deposits. In all cases, a thick mantle of coarse, tumultuous boulder clay buries the whole.*"*

Professor Geikie found between the mouth of the Pease Burn and St. Abb's Head, Berwickshire, several ancient buried channels. One at the Menzie Cleuch, near Redheugh Shore, was

* "Glacial Drift of Scotland," p. 65. Trans. Geol. Soc. Glas., vol. i., part 2.

filled to the brim with boulder clay. Another, the Lumsden Dean, half a mile to the east of Fast Castle, on the bank of the Carmichael Burn, near the parish church of Carmichael,—an old watercourse of the boulder-clay period—is to be seen. The valley of the Mouse Water he instances as a remarkable example.

One or two he found in Ayrshire, and also one on the banks of the Lyne Water, a tributary of the Tweed.

(9.) In the valley of the Clyde, above Hamilton, several buried river channels have been observed. They are thus described by Mr. James Geikie:—*

"In the Wishaw district, two deep, winding troughs, filled with sand and fine gravel, have been traced over a considerable area in the coal workings.† These troughs form no feature at the surface, but are entirely concealed below a thick covering of boulder clay. They appear to be old stream-courses, and are in all probability the pre-glacial ravines of the Calder Water and the Tillon Burn. The 'sand-dyke' that represents the pre-glacial course of the Calder Water runs for some distance parallel to the present course of the stream down to Wishaw House, where it is intersected by the Calder, and the deposits which choke it up are well seen in the steep wooded banks below the house and in the cliff on the opposite side. It next strikes to south-east, and is again well exposed on the road-side leading down from Wishaw to the Calder Water. From this point it has been traced underground, more or less continuously, as far as Wishaw Ironworks. Beyond this place the coal-seams sink to a greater depth, and therefore cease to be intersected by the ancient ravine, the course of which, however, may still be inferred from the evidence obtained during the sinking of shafts and trial borings. In all probability it runs south, and enters the old course of the Clyde a little below Cambusnethan House. Only a portion of the old ravine

* "Memoir, Geological Survey of Scotland," Sheet 23, p. 42.

† Mr. Robert Dick had previously described, in the Trans. Geol. Soc. Edinburgh, vol. i., p. 345, portions of these buried channels. He seems, however, to have thought that they formed part of one and the same channel.

of the Tillon Burn is shown upon the Map. It is first met with in the coal-workings of Cleland Townhead (Sheet 31). From this place it winds underground in a southerly direction until it is intersected by the present Tillon Burn, a little north of Glencleland (Sheet 31). It now runs to south-west, keeping parallel to the burn, and crosses the valley of the Calder just immediately above the mouth of the Tillon. From this point it can be traced in pit-shafts, open-air sections, borings, and coal-workings, by Ravenscraig, Nether Johnstone, and Robberhall Belting, on to the Calder Water below Coursington Bridge (Sheet 31). It would thus appear that in pre-glacial times the Calder and the Tillon were independent streams, and that since glacial times the Calder Water, forsaking its pre-glacial course, has cut its way across the intervening ground, ploughing out deep ravines in the solid rocks, until eventually it united with the Tillon. Similar buried stream courses occur at other places. Thus, at Fairholme, near Larkhall, as already mentioned (par. 94), the pre-glacial course of the Avon has been traced in pit-shafts and borings for some distance to the north. Another old course, filled up with boulder clay, is exposed in a burn near Plotcock, a mile south-west from Millheugh; and a similar pre-glacial ravine was met with in the cement-stone workings at Calderwood.* Indeed, it might be said with truth that nearly all the rocky ravines through which the waters flow, especially in the carboniferous areas, are of post-glacial age—the pre-glacial courses lying concealed under masses of drift. Most frequently, however, the present courses of the streams are partly pre-glacial and partly post-glacial. In the pre-glacial portions the streams flow through boulder clay, in the post-glacial reaches their course, as just mentioned, is usually in rocky ravines. The Avon and the Calder, with their tributaries, afford numerous illustrations of these phenomena."

* A description of this channel was read to the Natural History Society of Glasgow by Mr. James Coutts, the particulars of which will appear in the Transactions of the Society.

The question naturally arises, When were those channels scooped out? To what geological period must those ancient rivers be referred? It will not do to conclude that those channels must be pre-glacial simply because they contain boulder clay. Had the glacial epoch been one unbroken period of cold, and the boulder clay one continuous formation, then the fact of finding boulder clay in those channels would show that they were pre-glacial. But when we find undoubted geological evidence of a warm condition of climate of long continuance, during the severest part of the glacial epoch,. when the ice, to a great extent, must have disappeared, and water began to flow as usual down our valleys, all that can reasonably be inferred from the fact of finding till in those channels, is that they must be older than the till they contain. We cannot infer that they are older than all the till lying on the face of the country. The probability, however, is, that some of them are of pre-glacial and others of inter-glacial origin. That many of these channels have been used as watercourses during the glacial epoch, or rather during warm periods of that epoch, is certain, from the fact that they have been filled with boulder clay, then re-excavated, and finally filled up again with the clay.

CHAPTER XXX.

THE PHYSICAL CAUSE OF THE MOTION OF GLACIERS.—THEORIES OF GLACIER-MOTION.

Why the Question of Glacier-motion has been found to be so difficult.—The Regelation Theory.—It accounts for the Continuity of a Glacier, but not for its Motion.—Gravitation proved by Canon Moseley insufficient to shear the Ice of a Glacier.—Mr. Mathew's Experiment.—No Parallel between the bending of an Ice Plank and the shearing of a Glacier.—Mr. Ball's Objection to Canon Moseley's Experiment.—Canon Moseley's Method of determining the Unit of Shear.—Defect of Method.—Motion of a Glacier in some Way dependent on Heat.—Canon Moseley's Theory.—Objections to his Theory.—Professor James Thomson's Theory.—This Theory fails to explain Glacier-motion.—De Saussure and Hopkins's "Sliding" Theories.—M. Charpentier's "Dilatation" Theory.—Important Element in the Theory.

THE cause of the motion of glaciers has proved to be one of the most difficult and perplexing questions within the whole domain of physics. The main difficulty lies in reconciling the motion of the glacier with the physical properties of the ice. A glacier moves down a valley very much in the same way as a river, the motion being least at the sides and greatest at the centre, and greater at the surface than at the bottom. In a cross section scarcely two particles will be moving with the same velocity. Again, a glacier accommodates itself to the inequalities of the channel in which it moves exactly as a semifluid or plastic substance would do. So thoroughly does a glacier behave in the manner of a viscous or plastic body that Professor Forbes was induced to believe that viscosity was a property of the ice, and that in virtue of this property it was enabled to move with a differential motion and accommodate itself to all the inequalities of its channel without losing its continuity just as a mass of mud or putty would do. But experience proves that ice is a hard and brittle substance far more resembling glass than putty. In fact it is one of the most brittle and unyielding substances in nature. So unyielding is a glacier that it will snap in two before it will stretch to any perceptible extent. This is proved by the fact that crevasses resulting from a strain on the glacier consist at first of a simple crack scarcely wide enough to admit the blade of a penknife.

All the effects which were considered to be due to the viscosity of the ice have been fully explained and accounted for on the principle of fracture and regelation discovered by Faraday. The principle of regelation explains why the ice moving with a differential motion and accommodating itself to the inequalities of its channel is yet enabled to retain its continuity, but it does not account for the *cause* of glacier motion. In fact it rather involves the question in deeper mystery than before. For it is far more difficult to conceive how the particles of a hard and brittle solid like that of ice can move with a differential motion, than it is to conceive how this may take place in the case of a soft and yielding substance. The particles of ice have all to be displaced one over another and alongside each other, and as those particles are rigidly fixed together this connection must be broken before the one can slide over the other. *Shearing-force,* as Canon Moseley shows, comes into play. Were ice a plastic substance there would not be much difficulty in understanding how the particles should move the one over the other, but it is totally different when we conceive ice to be a solid and unyielding substance. The difficulty in connection with glacier-motion is not to account for the continuity of the ice, for the principle of regelation fully explains this, but to show how it is that one particle succeeds in sliding over the over. The principle of regelation, instead of assisting to remove this difficulty, increases it tenfold. Regelation does not explain the cause of glacier-motion, but the reverse. It rather tends to show that a glacier should not move. What, then, is the cause of glacier-motion? According to the regelation theory, gravitation is the impelling cause.

But is gravitation sufficient to *shear* the ice in the manner in which it is actually done in a glacier?

I presume that few who have given much thought to the subject of glacier-motion have not had some slight misgivings in regard to the commonly received theory. There are some facts which I never could harmonize with this theory. For example, boulder clay is a far looser substance than ice; its shearing-force must be very much less than that of ice; yet immense masses of boulder clay will lie immovable for ages on the slope of a hill so steep that one can hardly venture to climb it, while a glacier will come crawling down a valley which by the eye we could hardly detect to be actually off the level. Again, a glacier moves faster during the day than during the night, and about twice as fast during summer as during winter. Professor Forbes, for example, found that the Glacier des Bois near its lower extremity moved sometimes in December only 11·5 inches daily, while during the month of July its rate of motion sometimes reached 52·1 inches per day. Why such a difference in the rate of motion between day and night, summer and winter? The glacier is not heavier during the day than it is during the night, or during the summer than it is during the winter; neither is the shearing-force of the great mass of the ice of a glacier sensibly less during day than night, or during summer than winter; for the temperature of the great mass of the ice does not sensibly vary with the seasons. If this be the case, then gravitation ought to be as able to shear the ice during the night as during the day, or during the winter as during the summer. At any rate, if there should be any difference it ought to be but trifling. It is true that, owing to the melting of the ice, the crevices of the glacier are more gorged with water during summer than winter; and this, as Professor Forbes maintains,* may tend to make the glacier move faster during the former than the latter season. But the advocates of the regelation theory cannot conclude, with Professor Forbes, that the water favours the motion of the glacier

* "Occasional Papers," pp. 166, 223.

by making the ice more soft and plastic. The melting of the ice, according to the regelation theory, cannot very materially aid the motion of the glacier.

The theory which has led to the general belief that the ice of a glacier is sheared by the force of gravity appears to be the following. It is supposed that the only forces to which the motion of a glacier can be referred are *gravitation* and *heat;* but as the great mass of a glacier remains constantly at the same uniform temperature it is concluded to be impossible that the motion of the glacier can be due to this cause, and therefore of course it must be attributed to gravitation, there being no other cause.

That gravitation is insufficient to shear the ice of a glacier has been clearly demonstrated by Canon Moseley.* He determined by experiment the amount of force required to shear one square inch of ice, and found it to be about 75 lb. By a process of calculation which will be found detailed in the Memoir referred to, he demonstrated that to descend by its own weight at the rate at which Professor Tyndall observed the ice of the Mer de Glace to be descending at the Tacul, the unit of shearing force of the ice could not have been more than 1·31931 lb. Consequently it will require a force more than 34 times the weight of the glacier to shear the ice and cause it to descend in the manner in which it is found to descend.

It is now six years since Canon Moseley's results were laid before the public, and no one, as far as I am aware, has yet attempted to point out any serious defect in his mathematical treatment of the question. Seeing the great amount of interest manifested in the question of glacier-motion, I think we are warranted to conclude that had the mathematical part of the memoir been inconclusive its defects would have been pointed out ere this time. The question, then, hinges on whether the experimental data on which his calculations are based be correct. Or, in other words, is the unit of shear of ice as much as 75 lbs.? This part of Mr. Moseley's researches has not passed

* Memoir read before the Royal Society, January 7, 1869.

unquestioned. Mr. Ball and Mr. Mathews, both of whom have had much experience among glaciers, and have bestowed considerable attention on the subject of glacier motion, have objected to the accuracy of Mr. Moseley's unit of shear. I have carefully read the interesting memoirs of Mr. Mathews and Mr. Ball in reply to Canon Moseley, but I am unable to perceive that anything which they have advanced materially affects his general conclusions as regards the commonly received theory. Mr. Mathews objects to Canon Moseley's experiments on the grounds that extraneous forces are brought to bear upon the substance submitted to operation, and that conditions are thus introduced which do not obtain in the case of an actual glacier. "It would throw," he says, "great light upon our inquiry if we were to change this method of procedure and simply to observe the deportment of masses of ice under the influence of no external forces but the gravitation of their own particles."* A plank of ice six inches wide and $2\frac{3}{8}$ inches in thickness was supported at each end by bearers six feet apart From the moment the plank was placed in position it began to sink, and continued to do so until it touched the surface over which it was supported. Mr. Mathews remarks that with this property of ice, viz., its power to change its form under strains produced by its own gravitation, combined with the sliding movement demonstrated by Hopkins, we have an adequate cause for glacier-motion. Mr. Mathews concludes from this experiment that the unit of shear in ice, instead of being 75 lbs., is less than $1\frac{3}{4}$ lbs.

There is, however, no parallel between the bending of the ice-plank and the shearing of a glacier. Mr. Mathews' experiment appears to prove too much, as will be seen from the following reply of Canon Moseley :—

"Now I will," he says, "suggest to Mr. Mathews a parallel experiment and a parallel explanation. If a bar of wrought iron 1 inch square and 20 feet long were supported at its extremities, it would *bend* by its weight alone, and would there-

* "Alpine Journal," February, 1870.

fore shear. Now the weight of such a rod would be about 67 lbs. According to Mr. Mathews's explanation in the case of the ice-plank, the unit of shear in wrought-iron should therefore be 67 lbs. per square inch. It is actually 50,000 lbs."*

Whatever theory we may adopt as to the cause of the motion of glaciers, the deflection of the plank in the way described by Mr. Mathews *follows as a necessary consequence.* Although no weight was placed upon the plank, it does not necessarily follow that the deflection was caused by the weight of the ice alone; for, according to Canon Moseley's own theory of the motion of glaciers by heat, the plank ought to be deflected in the middle, just as it was in Mr. Mathews's experiment. A solid body, when exposed to variations of temperature, will expand and contract transversely as well as longitudinally. Ice, according to Canon Moseley's theory, expands and contracts by heat. Then if the plank expands transversely, the upper half of the plank must rise and the lower half descend. But the side which rises has to perform work against gravity, whereas the side which descends has work performed upon it by gravity; consequently more of the plank will descend than rise, and this will, of course, tend to lower or deflect the plank in the middle. Again, when the plank contracts, the lower half will rise and the upper half will descend; but as gravitation, in this case also, favours the descending part and opposes the rising part, more of the plank will descend than rise, and consequently the plank will be lowered in the middle by contraction as well as by expansion. Thus, as the plank changes its temperature, it must, according to Mr. Moseley's theory, descend or be deflected in the middle, step by step—and this not by gravitation alone, but chiefly by the motive power of heat. I do not, of course, mean to assert that the descent of the plank was caused by heat; but I assert that Mr. Mathews's experiment does not necessarily prove (and this is all that is required in the meantime) that gravitation alone was the cause of the deflection of the plank. Neither does this experiment prove that the ice was

* Phil. Mag., January, 1872.

deflected without shearing; for although the weight of the plank was not sufficient to shear the ice, as Mr. Mathews, I presume, admits, yet Mr. Moseley would reply that the weight of the ice, assisted by the motive power of heat, was perfectly sufficient.

I shall now briefly refer to Mr. Ball's principal objections to Canon Moseley's proof that a glacier cannot shear by its weight alone. One of his chief objections is that Mr. Moseley has assumed the ice to be homogeneous in structure, and that pressures and tensions acting within it, are not modified by the varying constitution of the mass.* Although there is, no doubt, some force in this objection (for we have probably good reason to believe that ice will shear, for example, more easily along certain planes than others), still I can hardly think that Canon Moseley's main conclusion can ever be materially affected by this objection. The main question is this, Can the ice of the glacier shear by its own weight in the way generally supposed? Now the shearing force of ice, take it in whatever direction we may, so enormously exceeds that required by Mr. Moseley in order to allow a glacier to descend by its weight only, that it is a matter of indifference whether ice be regarded as homogeneous in structure or not. Mr. Ball objects also to Mr. Moseley's imaginary glacier lying on an even slope and in a uniform rectangular channel. He thinks that an irregular channel and a variable slope would be more favourable to the descent of the ice. But surely if the work by the weight of the ice be not equal to the work by the resistance in a glacier of uniform breadth and slope, it must be much less so in the case of one of irregular shape and slope.

That a relative displacement of the particles of the ice is involved in the motion of a glacier, is admitted, of course, by Mr. Ball; but he states that the amount of this displacement is but small, and that it is effected with extreme slowness. This may be the case; but if the weight of the ice be not able to overcome the mutual cohesion of the particles, then the weight

* Phil. Mag., July, 1870; February, 1871.

of the ice cannot produce the required displacement, however small it may be. Mr. Ball then objects to Mr. Moseley's method of determining the unit of shear on this ground:—The shearing of the ice in a glacier is effected with extreme slowness; but the shearing in Canon Moseley's experiment was effected with rapidity; and although it required 75 lbs. to shear one square inch of surface in his experiment, it does not follow that 75 lbs. would be required to shear the ice if done in the slow manner in which it is effected in the glacier. "In short," says Mr. Ball, "to ascertain the resistance opposed to very slow changes in the relative positions of the particles, so slight as to be insensible at short distances, Mr. Moseley measures the resistance opposed to rapid disruption between contiguous portions of the same substance."

There is force in this objection; and here we arrive at a really weak point in Canon Moseley's reasoning. His experiments show that if we want to shear ice quickly a weight of nearly 120 lbs. is required; but if the thing is to be done more slowly, 75 lbs. will suffice.* In short, the number of pounds required to shear the ice depends, to a large extent, on the length of time that the weight is allowed to act; the longer it is allowed to act, the less will be the weight required to perform the work. "I am curious to know," says Mr. Mathews, when referring to this point, "what weight would have sheared the ice if a *day* had been allowed for its operation." I do not know what would have been the weight required to shear the ice in Mr. Moseley's experiments had a day been allowed; but I feel pretty confident that, should the ice remain unmelted, and sufficient time be allowed, shearing would be produced without the application of any weight whatever. There are no weights placed upon a glacier to make it move, and yet the ice of the glacier shears. If the shearing is effected by weight, the only weight applied is the weight of the ice; and if the

* Philosophical Magazine for January, 1870, p. 8; Proceedings of the Royal Society for January, 1869.

weight of the ice makes the ice shear in the glacier, why may it not do the same thing in the experiment? Whatever may be the cause which displaces the particles of the ice in a glacier, they, as a matter of fact, are displaced without any weight being applied beyond that of the ice itself; and if so, why may not the particles of the ice in the experiment be also displaced without the application of weights? Allow the ice of the glacier to take its own time and its own way, and the particles will move over each other without the aid of external weights, whatever may be the cause of this; well, then, allow the ice in the experiment to take its own time and its own way, and it will probably do the same thing. There is something here unsatisfactory. If, by the unit of shear, be meant the pressure in pounds that must be applied to the ice to break the connection of one square inch of two surfaces frozen together and cause the one to slip over the other, then the amount of pressure required to do this will depend upon the time you allow for the thing being done. If the thing is to be done rapidly, as in some of Mr. Moseley's experiments, it will take, as he has shown, a pressure of about 120 lbs.; but if the thing has to be done more slowly, as in some other of his experiments, 75 lbs. will suffice. And if sufficient time be allowed, as in the case of glaciers, the thing may be done without any weight whatever being applied to the ice, and, of course, Mr. Moseley's argument, that a glacier cannot descend by its weight alone, falls to the ground. But if, by the unit of shear, be meant not the *weight* or *pressure* necessary to shear the ice, but the amount of *work* required to shear a square inch of surface *in a given time or at a given rate,* then he might be able to show that in the case of a glacier (say the Mer de Glace) the work of all the resistances which are opposed to its descent at the *rate* at which it is descending is greater than the work of its weight, and that consequently there must be some cause, in addition to the weight, urging the glacier forward. But then he would have no right to affirm that the glacier would not descend by its weight only; all that he could affirm would simply be that it could not descend by its weight alone at the *rate* at which it is descending.

Mr. Moseley's unit of shear, however, is not the amount of *work* performed in shearing a square inch of ice in a given time, but the amount of *weight* or *pressure* requiring to be applied to the ice to shear a square inch. But this amount of pressure depends upon the length of time that the pressure is applied. Here lies the difficulty in determining what amount of pressure is to be taken as the real unit. And here also lies the radical defect in Canon Moseley's result. Time as well as pressure enters as an element into the process. The key to the explanation of this curious circumstance will, I think, be found in the fact that the rate at which a glacier descends depends in some way or other upon the amount of heat that the ice is receiving. This fact shows that heat has something to do in the shearing of the ice of the glacier. But in the communication of heat to the ice *time* necessarily enters as an element. There are two different ways in which heat may be conceived to aid in shearing the ice: (1.) we may conceive that heat acts as a force along with gravitation in producing displacement of the particles of the ice; or (2.) we may conceive that heat does not act as a force in pushing the particles over each other, but that it assists the shearing processes by diminishing the cohesion of the particles of the ice, and thus allowing gravitation to produce displacement. The former is the function attributed to heat in Canon Moseley's theory of glacier-motion; the latter is the function attributed to heat in the theory of glacier-motion which I ventured to advance some time ago.* It results, therefore, from Canon Moseley's own theory, that the longer the time that is allowed for the pressure to shear the ice, the less will be the pressure required; for, according to his theory, a very large proportion of the displacement is produced by the motive power of heat entering the ice; and, as it follows of course, other things being equal, the longer the time during which the heat is allowed to act, the greater will be the propor-

* Philosophical Magazine for March, 1869.

tionate amount of displacement produced by the heat; consequently the less will require to be done by the weight applied. In the case of the glacier, Mr. Moseley concludes that at least thirty or forty times as much work is done by the motive power of heat in the way of shearing the ice as is done by mere pressure or weight. Then, if sufficient time be allowed, why may not far more be done by heat in shearing the ice in his experiment than by the weight applied? In this case how is he to know how much of the shearing is effected by the heat and how much by the weight? If the greater part of the shearing of the ice in the case of a glacier is produced, not by pressure, but by the heat which necessarily enters the ice, it would be inconceivable that in his experiments the heat entering the ice should not produce, at least to some extent, a similar effect. And if a portion of the displacement of the particles is produced by heat, then the weight which is applied cannot be regarded as the measure of the force employed in the displacement, any more than it could be inferred that the weight of the glacier is the measure of the force employed in the shearing of it. If the weight is not the entire force employed in shearing, but only a part of the force, then the weight cannot, as in Mr. Moseley's experiment, be taken as the measure of the force.

How, then, are we to determine what is the amount of force required to shear ice? in other words, how is the unit of shear to be determined? If we are to measure the unit of shear by the weight required to produce displacement of the particles of the ice, we must make sure that the displacement is wholly effected by the weight. We must be certain that heat does not enter as an element in the process. But if time be allowed to elapse during the experiment, we can never be certain that heat has not been at work. It is impossible to prevent heat entering the ice. We may keep the ice at a constant temperature, but this would not prevent heat from entering the ice and producing molecular work. True that, according to Moseley's theory of glacier-motion, if the temperature of the ice be not permitted to *vary*, then no displacement of the particles can take place from the influence of heat ; but according to the molecular theory of glacier-motion, which will shortly be considered, heat will aid the displacement of the particles whether the temperature be kept constant or not. In short, it is absolutely impossible in our experiments to be certain that heat is not in some way or other concerned in the displacement of the particles of the ice. But we can shorten the time, and thus make sure that the amount of heat entering the ice during the experiments is too small to affect materially the result. We cannot in this case say that all the displacement has been effected by the weight applied to the ice, but we can say that so little has been done by heat that, practically, we may regard it as all done by the weight.

This consideration, I trust, shows that the unit of shear adopted by Canon Moseley in his calculations is not too large. For if in half an hour, after all the work that may have been done by heat, a pressure of 75 lbs. is still required to displace the particles of one square inch, it is perfectly evident that if no work had been done by heat during that time, the force required to produce the displacement could not have been less than 75 lbs. It might have been more than that; but it could not have been less. Be this, however, as it may, in determining the unit of shear we cannot be permitted to prolong the experiment for any considerable length of time, because the weight under which the ice might then shear could not be taken as the measure of the force which is required to shear ice. By prolonging the experiment we might possibly get a unit smaller than that required by Canon Moseley for a glacier to descend by its own weight. But it would be just as much begging the whole question at issue to assume that, because the ice sheared under such a weight, a glacier might descend by its weight alone, as it would be to assume that, because a glacier shears without a weight being placed upon it, the glacier descends by its weight alone.

But why not determine the unit of shear of ice in the same way as we would the unit of shear of any other solid substance, such as iron, stone, or wood? If the shearing force of ice be determined in this manner, it will be found to be by far too great to allow of the ice shearing by its weight alone. We shall be obliged to admit either that the ice of the glacier does not shear (in the ordinary sense of the term), or if it does shear, that there must, as Canon Moseley concludes, be some other force in addition to the weight of the ice urging the glacier forward.

The fact that the rate of descent of a glacier depends upon the amount of heat which it receives, proves that heat must be regarded either as a cause or as a necessary condition of its motion; what, then, is the necessary relationship between heat and the motion of the glacier? If heat is to be regarded as a cause, in what way does the heat produce motion? I shall now briefly refer to one or two theories which have been advanced on the subject. Let us consider first that of Canon Moseley.

Canon Moseley's Theory.—He found, from observations and experiments, that sheets of lead, placed upon an inclined plane, when subjected to variations of temperature, tend to descend even when the slope is far less than that which would enable it to slide down under the influence of gravitation. The cause of the descent he shows to be this. When the temperature of the sheet is raised, it expands, and, in expanding, its upper portion moves up the slope, and its lower portion down the slope; but as gravitation opposes the upward and favours the downward motion, more of the sheet moves down than up, and consequently the centre of gravity of the sheet is slightly lowered. Again, when the sheet is cooled, it contracts, and in contracting the upper portion moves downwards and the lower portion upwards, and here again, for the same reason, more of the sheet moves downwards than upwards. Consequently, at every change of temperature there is a slight displacement of the sheet downwards. "Now a theory of the descent of glaciers," says Canon Moseley, "which I have ventured to propose myself, is that they descend, as the lead in this experiment does, by reason of the passage into them and the withdrawal of the sun's rays, and that the dilatation and contraction of the ice so produced is the proximate cause of their descent, as it is of that of the lead."*

The fundamental condition in Mr. Moseley's theory of the descent of solid bodies on an incline, is, not that heat should maintain these bodies at a high temperature, but that the temperature should vary. The rate of descent is proportionate, not simply to the amount of heat received, but to the extent and frequency of the variations of temperature. As a proof that glaciers are subjected to great variations of temperature, he adduces the following:—"All alpine travellers," he says, "from De Saussure to Forbes and Tyndall, have borne testimony to the intensity of the solar radiation on the surfaces of glaciers. 'I scarcely ever,' says Forbes, 'remember to have found the sun more piercing than at the Jardin.' This heat passes abruptly into a state of intense cold when any part of the glacier falls into shadow by an alteration of the position of the sun, or even by the passing over it of a cloud."†

Mr. Moseley is here narrating simply what the traveller feels, and not what the glacier experiences. The traveller is subjected to great variations of temperature; but there is no proof from this that the glacier experiences any changes of temperature. It is rather because the temperature of the glacier is not affected by the sun's heat that the traveller is so much chilled when the sun's rays are cut off. The sun shines down with piercing rays and the traveller is scorched; the glacier melts on the surface, but it still remains "cold as ice." The sun passes behind a cloud or disappears behind a neighbouring hill; the scorching rays are then withdrawn, and the traveller is now subjected to radiation on every side from surfaces at the freezing-point.

It is also a necessary condition in Mr. Moseley's theory that the heat should pass easily into and out of the glacier; for

* Proceedings of Bristol Naturalists' Society, p. 37 (1869).

† Ibid., vol. iv., p. 37 (new series).

unless this were the case sudden changes of temperature could produce little or no effect on the great mass of the glacier. How, then, is it possible that during the heat of summer the temperature of the glacier could vary much? During that season, in the lower valleys at least, everything, with the exception of the glacier, is above the freezing-point; consequently when the glacier goes into the shade there is nothing to lower the ice below the freezing-point; and as the sun's rays do not raise the temperature of the ice above the freezing-point, the temperature of the glacier must therefore remain unaltered during that season. It therefore follows that, instead of a glacier moving more rapidly during the middle of summer than during the middle of winter, it should, according to Moseley's theory, have no motion whatever during summer.

The following, written fifteen years ago by Professor Forbes on this very point, is most conclusive:—"But how stands the fact? Mr. Moseley quotes from De Saussure the following *daily ranges* of the temperature of the air in the month of July at the Col du Géant and at Chamouni, between which points the glacier lies:

At the Col du Géant	4°·257 Réaumur.
At Chamouni	10·092 ,,

And he assumes 'the same mean daily variation of temperature to obtain throughout the length' [and depth?] 'of the Glacier du Géant which De Saussure observed in July at the Col du Géant.' But between what limits does the temperature of the air oscillate? We find, by referring to the third volume of De Saussure's 'Travels,' that the mean temperature of the coldest hour (4 A.M.) during his stay at the Col du Géant was 33°·03 Fahrenheit, and of the warmest (2 P.M.) 42°·61 F. So that even upon that exposed ridge, between 2,000 and 3,000 feet above where the glacier can be properly said to commence, the air does not, on an average of the month of July, reach the freezing-point at any hour of the night. Consequently the *range of tempera-*

ture attributed to the glacier is between limits absolutely incapable of effecting the expansion of the ice in the smallest degree." *

Again, during winter, as Mr. Ball remarks, the glacier is completely covered with snow and thus protected both from the influence of cold and of heat, so that there can be nothing either to raise the temperature of the ice above the freezing-point or to bring it below that point; and consequently the glacier ought to remain immovable during that season also.

"There can be no doubt, therefore," Mr. Moseley states, "that the rays of the sun, which in those alpine regions are of such remarkable intensity, find their way into the depths of the glacier. They are a *power*, and there is no such thing as the loss of power. The mechanical work which is their equivalent, and into which they are converted when received into the substance of a solid body, accumulates and stores itself up in the ice under the form of what we call elastic force or tendency to dilate, until it becomes sufficient to produce actual dilatation of the ice in the direction in which the resistance is weakest, and by its withdrawal to produce contraction. From this expansion and contraction follows of necessity the descent of the glacier."† When the temperature of the ice is below the freezing-point, the rays which are absorbed will, no doubt, produce dilatation; but during summer, when the ice is not below the freezing-point, no dilatation can possibly take place. All physicists, so far as I am aware, agree that the rays that are then absorbed go to melt the ice, and not to expand it. But to this Mr. Moseley replied as follows:—"To this there is the obvious answer that radiant heat does find its way into ice as a matter of common observation, and that it does not melt it except at its surface. Blocks of ice may be seen in the windows of ice-shops with the sun shining full upon them, and melting nowhere but on their surfaces. And the experiment of the ice-lens shows that heat may stream through ice in abundance (of

* Phil. Mag., S. 4, vol. x., p. 303.

† Proceedings of the Bristol Naturalists' Society, vol. iv., p. 39 (new series).

which a portion is necessarily stopped in the passage) without melting it, except on its surface." But what evidence is there to conclude that if there is no melting of the ice in the interior of the lens there is a portion of the rays "necessarily stopped" in the interior? It will not do to assume a point so much opposed to all that we know of the physical properties of ice as this really is. It is absolutely essential to Mr. Moseley's theory of the motion of glaciers, during summer at least, that ice should continue to expand after it reaches the melting-point; and it has therefore to be shown that such is the case; or it need not be wondered at that we cannot accept his theory, because it demands the adoption of a conclusion contrary to all our previous conceptions. But, as a matter of fact, it is not strictly true that when rays pass through a piece of ice there is no melting of the ice in the interior. Experiments made by Professor Tyndall show the contrary.*

There is, however, one fortunate circumstance connected with Canon Moseley's theory. It is this: its truth can be easily tested by direct experiment. The ice, according to this theory, descends not simply in virtue of heat, but in virtue of *change of temperature.* Try, then, Hopkins's famous experiment, but keep the ice at a *constant temperature;* then, according to Moseley's theory, the ice will not descend. Let it be observed, however, that although the ice under this condition should descend (as there is little doubt but it would), it would show that Mr. Moseley's theory of the descent of glaciers is incorrect, still it would not in the least degree affect the conclusions which he lately arrived at in regard to the generally received theory of glacier-motion. It would not prove that the ice sheared, in the way generally supposed, by its weight only. It might be the heat, after all, entering the ice, which accounted for its descent, although gravitation (the weight of the ice) might be the impelling cause.

According to this theory, the glacier, like the sheet of lead, must expand and contract as one entire mass, and it is difficult

* See Philosophical Transactions, December, 1857.

to conceive how this could account for the differential motion of the particles of the ice.

Professor James Thomson's Theory.—It was discovered by this physicist that the freezing-point of water is lowered by pressure. The extent of the lowering is equal to ·0075° centigrade for every atmosphere of pressure. As glacier ice is generally about the melting-point, it follows that when enormous pressure is brought to bear upon any given point of a glacier a melting of the ice at that particular spot will take place in consequence of the lowering of the melting-point. The melting of the ice will, of course, tend to favour the descent of the glacier, but I can hardly think the liquefaction produced by pressure can account for the motion of glaciers. It will help to explain the giving way of the ice at particular points subjected to great pressure, but I am unable to comprehend how it can account for the general descent of the glacier. Conceive a rectangular glacier of uniform breadth and thickness, and lying upon an even slope. In such a glacier the pressure at each particular point would remain constant, for there would be no reason why it should be greater at one time than at another. Suppose the glacier to be 500 feet in thickness; the ice at the lower surface of the glacier, owing to pressure, would have its melting-point permanently lowered one-tenth of a degree centigrade below that of the upper surface; but the ice at the lower surface would not, on this account, be in the fluid state. It would simply be ice at a slightly lower temperature. True, when pressure is exerted the ice melts in consequence of the lowering of the melting-point, but in the case under consideration there would, properly speaking, be no exertion of pressure, but a constant statical pressure resulting from the weight of the ice. But this statical condition of pressure would not produce fluidity any more than a statical condition of pressure would produce heat, and consequently motion could not take place as a result of fluidity. In short, motion itself is required to produce the fluidity.

I need not here wait to consider the sliding theories of

De Saussure and Hopkins, as they are now almost universally admitted to be inadequate to explain the phenomena of glacier-motion, seeing that they do not account for the displacement of the particles of the ice over one another.

According to the dilatation theory of M. Charpentier, a glacier is impelled by the force exerted by water freezing in the fissures of the ice. A glacier he considers is full of fissures into which water is being constantly infiltrated, and when the temperature of the air sinks below the freezing-point it converts the water into ice. The water, in passing into ice, expands, and in expanding tends to impel the glacier in the direction of least resistance. This theory, although it does not explain glacier-motion, as has been clearly shown by Professor J. D. Forbes, nevertheless contains one important element which, as we shall see, must enter into the true explanation. The element to which I refer is the expansive force exerted on the glacier by water freezing.

CHAPTER XXXI.

THE PHYSICAL CAUSE OF THE MOTION OF GLACIERS.—THE MOLECULAR THEORY.

Present State of the Question.—Heat necessary to the Motion of a Glacier.—Ice does not shear in the Solid State.—Motion of a Glacier *molecular*.—How Heat is transmitted through Ice.—Momentary Loss of Shearing Force.—The *Rationale* of Regelation.—The Origin of "Crevasses."—Effects of Tension.—Modification of Theory.—Fluid Molecules crystallize in Interstices.—Expansive Force of crystallizing Molecules a Cause of Motion.—Internal molecular Pressure the chief Moving Power.—How Ice can excavate a Rock Basin.—How Ice can ascend a Slope.—How deep River Valleys are striated across.—A remarkable Example in the Valley of the Tay.—How Boulders can be carried from a lower to a higher Level.

THE condition which the perplexing question of the cause of the descent of glaciers has now reached seems to be something like the following. The ice of a glacier is not in a soft and plastic state, but is solid, hard, brittle, and unyielding. It nevertheless behaves in some respects in a manner very like what a soft and plastic substance would do if placed in similar circumstances, inasmuch as it accommodates itself to all the inequalities of the channel in which it moves. The ice of the glacier, though hard and solid, moves with a differential motion; the particles of the ice are displaced over each other, or, in other words, the ice shears as it descends. It had been concluded that the mere weight of the glacier is sufficient to shear the ice. Canon Moseley has investigated this point, and shown that it is not. He has found that for a glacier to shear in the way that it is supposed to do, it would require a force some thirty or forty times as great as the weight of the glacier. Consequently, for the glacier to descend, a force in addition to that of gravitation is required. What, then, is this force? It is found that the rate at which the glacier descends depends upon the amount of heat which it is receiving. This shows that the motion of the glacier is in some way or other dependent upon heat. Is heat, then, the force we are in search of? The answer to this, of course, is, since heat is a force necessarily required, we have no right to assume any other till we see whether or not heat will suffice. In what way, then, does heat aid gravitation in the descent of the glacier? In what way does heat assist gravitation in the shearing of the ice? There are two ways whereby we may conceive the thing to be done: the heat may assist gravitation to shear, by pressing the ice forward, or it may assist gravitation by diminishing the cohesion of the particles, and thus allow gravitation to produce motion which it otherwise could not produce. Every attempt which has yet been made to explain how heat can act as a force in pushing the ice forward, has failed. The fact that heat cannot expand the ice of the glacier may be regarded as a sufficient proof that it does not act as a force impelling the glacier forward; and we are thus obliged to turn our attention to the other conception, viz., that heat assists gravitation to shear the ice, not by direct pressure, but by diminishing the cohesive force of the particles, so as to enable gravitation to push the one past the other. But how is this done? Does heat diminish the cohesion by acting as an expansive force in separating the particles? Heat cannot do this, because it cannot expand the ice of a glacier; and besides, were it to do this, it would destroy the solid and firm character of the ice, and the ice of the glacier would not then, as a mass, possess the great amount of shearing-force which observation and experiment show that it does. In short it is because the particles are so firmly fixed together at the time the glacier is descending, that we are obliged to call in the aid of some other force in addition to the weight of the glacier to shear the ice. Heat does not cause displacement of the particles by making the ice soft and plastic; for we know that the ice of the glacier is not soft and plastic, but hard and brittle. The shearing-force of the ice of the moving glacier is found to be by at least from thirty to forty times too great to permit of the ice being sheared by the mere force of gravitation; how, then, is it that gravitation, without the direct assistance of any other force, can manage to shear the ice? Or to put the question under another form: heat does not reduce the shearing-force of the ice of a glacier to something like 1·3193 lb. per square inch of surface, the unit required by Mr. Moseley to enable a glacier to shear by its weight; the shearing-force of the ice, notwithstanding all the heat received, still remains at about 75 lbs.; how, then, can the glacier shear without any other force than its own weight pushing it forward? *This is the fundamental question; and the true answer to it must reveal the mystery of glacier-motion.* We are compelled in the present state of the problem to admit that glaciers do descend with a differential motion without any other force than their own weight pushing them forward; and yet the shearing-force of the ice is actually found to be thirty or forty times the maximum that would permit of the glacier shearing by its weight only. *The explanation of this apparent paradox will remove all our difficulties in reference to the cause of the descent of glaciers.*

There seems to be but one explanation (and it is a very obvious one), viz. that the motion of the glacier is *molecular.* The ice descends molecule by molecule. The ice of a glacier is in the hard crystalline state, but it does not descend in this state. Gravitation is a constantly acting force; if a particle of the ice lose its shearing-force, though but for the moment, it will descend by its weight alone. But a particle of the ice will lose its shearing-force for a moment if the particle loses its crystalline state for the moment. The passage of heat through ice, whether by conduction or by radiation, in all probability is a molecular process; that is, the form of energy termed heat is transmitted from molecule to molecule of the ice. A particle takes the energy from its neighbour A on the one side and hands it over to its neighbour

B on the opposite side. But the particle must be in a different state at the moment it is in possession of the energy from what it was before it received it from A, and from what it will be after it has handed it over to B. Before it became possessed of the energy, it was in the crystalline state—it was ice; and after it loses possession of the energy it will be ice; but at the moment that it is in possession of the passing energy is it in the crystalline or icy state? If we assume that it is not, but that in becoming possessed of the energy, it loses its crystalline form and for the moment becomes water, all our difficulties regarding the cause of the motion of glaciers are removed. We know that the ice of a glacier in the mass cannot become possessed of energy in the form of heat without becoming fluid; *if it can be shown that the same thing holds true of the ice particle, we have the key to the mystery of glacier-motion.* A moment's reflection will suffice to convince any one that if the glacier ice in the mass cannot receive energy in the form of heat without melting, the same must hold true of the ice particles, for it is inconceivable that the ice in the mass could melt and yet the ice particles themselves remain in the solid state. It is the solidity of the particles which constitutes the solidity of the mass. If the particles lose their solid form the mass loses its solid form, for the mass has no other solidity than that which is possessed by the particles.

The correctness of the conclusion, that the weight of the ice is not a sufficient cause, depends upon the truth of a certain element taken for granted in the reasoning, viz. that the *shearing-force* of the molecules of the ice remains *constant.* If this force remains constant, then Canon Moseley's conclusion is undoubtedly correct, but not otherwise; for if a molecule should lose its shearing-force, though it were but for a moment, if no obstacle stood in front of the molecule, it would descend in virtue of its weight.

The fact that the shearing-force of a mass of ice is found to be constant does not prove that the same is the case in regard to the individual molecules. If we take a mass of molecules in the aggregate, the shearing-force of the mass taken thus collectively may remain absolutely constant, while at the same time each individual molecule may be suffering repeated momentary losses of shearing-force. This is so obvious as to require no further elucidation. The whole matter, therefore, resolves itself into this one question, as to whether or not the shearing-force of a crystalline molecule of ice remains constant. In the case of ordinary solid bodies we have no reason to conclude that the shearing-force of the molecules ever disappears, but in regard to ice it is very different.

If we analyze the process by which heat is conducted through ice, we shall find that we have reason to believe *that while a molecule of ice is in the act of transmitting the energy received* (*say from a fire*), *it loses for the moment its shearing-force if the temperature of the ice be not under* 32° F. If we apply heat to the end of a bar of iron, the molecules at the surface of the end have their temperatures raised. Molecule A at the surface, whose temperature has been raised, instantly commences to transfer to B a portion of the energy received. The tendency of this process is to lower the temperature of A and raise that of B. B then, with its temperature raised, begins to transfer the energy to C. The result here is the same; B tends to fall in temperature, and C to rise. This process goes on from molecule to molecule until the opposite end of the bar is reached. Here in this case the energy or heat applied to the end of the bar is transmitted from molecule to molecule under the form of *heat or temperature.* The energy applied to the bar does *not change its character; it passes right along from molecule to molecule under the form of heat or temperature.* But the nature of the process must be wholly different if the transferrence takes place through a bar of ice at the temperature of 32°. Suppose we apply the heat of the fire to the end of the bar of ice at 32°, the molecules of the ice cannot possibly have their temperatures raised in the least degree. How, then, can molecule A take on, *under the form of heat,* the energy received from the fire without being heated or having its *temperature* raised? The thing is impossible. The energy of the fire must appear in A under a different form from that of heat. The same process of reasoning is equally applicable to B. The molecule B cannot accept of the energy from A under the form of heat; it must receive it under some other form. The same must hold equally true of all the other molecules till we reach the opposite end of the bar of ice. And yet, strange to say, the last molecule transmits in the form of heat its energy to the objects beyond; for we find that the heat applied to one side of a piece of ice will affect the thermal pile on the opposite side.

The question is susceptible of a clear and definite answer. When heat is applied to a molecule of ice at 32°, the heat applied does not raise the temperature of the molecule, it is consumed in work against the cohesive forces binding the atoms or particles together into the crystalline form. The energy then must exist in the dissolved crystalline molecule, under the statical form of an affinity—crystalline affinity, or whatever else we may call it. That is to say, the energy then exists in the particles as a power or tendency to rush together again into the crystalline form, and the moment they are allowed to do so they give out the energy that was expended upon them in their separation. This energy, when it is thus given out again, assumes the dynamical form of heat; in other words, the molecule gives out *heat* in the act of freezing. The heat thus given out may be employed to melt the next adjoining molecule. The ice-molecules take on energy from a heated body by melting. That peculiar form of motion or energy called heat disappears in forcing the particles of the crystalline molecule separate, and for the time being exists in the form of a tendency in the separated particles to come together again into the crystalline form.

But it must be observed that although the crystalline molecule, when it is acting as a conductor, takes on energy under this form from the heated body, it only exists in the molecule under such a form during the moment of transmission; that is to say, the molecule is melted, but only for the moment. When

B accepts of the energy from A, the molecule A instantly assumes the crystalline form. B is now melted; and when C accepts of the energy from B, then B also in turn assumes the solid state. This process goes on from molecule to molecule till the energy is transmitted through to the opposite side and the ice is left in its original solid state. This, as will be shown in the Appendix, is the *rationale* of Faraday's property of regelation.

This is no mere theory or hypothesis; it is a necessary consequence from known facts. We know that ice at 32° cannot take on energy from a heated body without melting; and we know also equally well that a slab of ice at 32°, notwithstanding this, still, as a mass, retains its solid state while the heat is being transmitted through it. This proves that every molecule resumes its crystalline form the moment after the energy is transferred to the adjoining molecule.

This point being established, every difficulty regarding the descent of the glacier entirely disappears; for a molecule the moment that it assumes the fluid state is completely freed from shearing-force, and can descend by virtue of its own weight without any impediment. All that the molecule requires is simply room or space to advance in. If the molecule were in absolute contact with the adjoining molecule below, it would not descend unless it could push that molecule before it, which it probably would not be able to do. But the molecule actually has room in which to advance; for in passing from the solid to the liquid state its volume is diminished by about $\frac{1}{10}$, and it consequently can descend. True, when it again assumes the solid form it will regain its former volume; but the question is, will it go back to its old position? If we examine the matter thoroughly we shall find that it cannot. If there were only this one molecule affected by the heat, this molecule would certainly not descend; but all the molecules are similarly affected, although not all at the same moment of time.

Let us observe what takes place, say, at the lower end of the glacier. The molecule A at the lower end, say, of the surface, receives heat from the sun's rays; it melts, and in melting not only loses its shearing-force and descends by its own weight, but it contracts also. B immediately above it is now, so far as A is concerned, at liberty to descend, and will do so the moment that it assumes the liquid state. A by this time has become solid, and again fixed by shearing-force; but it is not fixed in its old position, but a little below where it was before. If B has not already passed into the fluid state in consequence of heat derived from the sun, the additional supply which it will receive from the solidifying of A will melt it. The moment that B becomes fluid it will descend till it reaches A. B then is solidified a little below its former position. The same process of reasoning is in a similar manner applicable to every molecule of the glacier. Each molecule of the glacier consequently descends step by step as it melts and solidifies, and hence the glacier, considered as a mass, is in a state of constant motion downwards. The fact observed by Professor Tyndall that there are certain planes in the ice along which melting takes place more readily than others will perhaps favour the descent of the glacier.

We have in this theory a satisfactory explanation of the origin of "crevasses" in glaciers. Take, for example, the transverse crevasses formed at the point where an increase in the inclination of the glacier takes place. Suppose a change of inclination from, say, 4° to 8° in the bed of the glacier. The molecules on the slope of 8° will descend more rapidly than those above on the slope of 4°. A state of tension will therefore be induced at the point where the change of inclination occurs. The ice on the slope of 8° will tend to pull after it the mass of the glacier moving more slowly on the slope above. The pull being continued, the glacier will snap asunder the moment that the cohesion of the ice is overcome. The greater the change of inclination is, the more readily will the rupture of the ice take place. Every species of crevasse can be explained upon the same principle.*

* There is one circumstance tending slightly to prevent the rupture of the glacier, when under tension, which I do not remember to have seen noticed;

This theory explains also why a glacier moves at a greater rate during summer than during winter; for as the supply of heat to the glacier is greater during the former season than during the latter, the molecules will pass oftener into the liquid state.

As regards the denuding power of glaciers, I may observe that, though a glacier descends molecule by molecule, it will grind the rocky bed over which it moves as effectually as it would do did it slide down in a rigid mass in the way generally supposed; for the grinding-effect is produced not by the ice of the glacier, but by the stones, sand, and other materials forced along under it. But if all the resistances opposing the descent of a glacier, internal and external, are overcome by the mere weight of the ice alone, it can be proved that in the case of one descending with a given velocity the amount of work performed in forcing the grinding materials lying under the ice forward must be as great, supposing the motion of the ice to be molecular in the way I have explained, as it would be supposing the ice descended in the manner generally supposed.

Of course, a glacier could not descend by means of its weight as rapidly in the latter case as in the former; for, in fact, as Canon Moseley has shown, it would not in the latter case descend at all; but assuming for the sake of argument the rate of descent in both cases to be the same, the conclusion I have stated would follow. Consequently whatever denuding effects may have been attributed to the glacier, according to the ordinary theory, must be equally attributable to it according to the present explanation.

This theory, however, explains, what has always hitherto excited astonishment, viz., why a glacier can descend a slope almost horizontal, or why the ice can move off the face of a continent perfectly level.

that is, the cooling effect which is produced in solids, such as ice, when subjected to tension. Tension would tend to lower the temperature of the ice-molecules, and this lowering of temperature would have the tendency of freezing them more firmly together. The cause of this cooling effect will be explained in the Appendix.

This is the form in which my explanation was first stated about half-a-dozen years ago.* There is, however another element which must be taken into account. It is one which will help to cast additional light on some obscure points connected with glacial phenomena.

Ice is evidently not absolutely solid throughout. It is composed of crystalline particles, which, though in contact with one another, are, however, not packed together so as to occupy the least possible space, and, even though they were, the particles would not fit so closely together as to exclude interstices. The crystalline particles are, however, united to one another at special points determined by their polarity, and on this account they require more space; and this in all probability is the reason, as Professor Tyndall remarks, why ice, volume for volume, is less dense than water.

"They (the molecules) like the magnets," says Professor Tyndall, "are acted upon by two distinct forces; for a time, while the liquid is being cooled, they approach each other, in obedience to their general attraction for each other. But at a certain point new forces, some attractive some repulsive, *emanating from special points* of the molecules, come into play. The attracted points close up, the repelled points retreat. Thus the molecules turn and rearrange themselves, demanding as they do so more space, and overcoming all ordinary resistance by the energy of their demand. This, in general terms, is an explanation of the expansion of water in solidifying." †

It will be obvious, then, that when a crystalline molecule melts, it will not merely descend in the manner already described, but capillary attraction will cause it to flow into the interstices between the adjoining molecules. The moment that it parts with the heat received, it will of course resolidify, as has been shown, but it will not solidify so as to fit the cavity which it occupied when in the fluid state. For the liquid molecule in solidifying assumes the crystalline form, and of

* Phil. Mag., March, 1869; September, 1870.

† "Forms of Water," p. 127.

course there will be a definite proportion between the length, breadth, and thickness of the crystal; consequently it will always happen that the interstice in which it solidifies will be too narrow to contain it. The result will be that the fluid molecule in passing into the crystalline form will press the two adjoining molecules aside in order to make sufficient room for itself between them, and this it will do, no matter what amount of space it may possess in all other directions. The crystal will not form to suit the cavity, the cavity must be made to contain the crystal. And what holds true of one molecule, holds true of every molecule which melts and resolidifies. This process is therefore going on incessantly in every part of the glacier, and in proportion to the amount of heat which the glacier is receiving. This internal molecular pressure, resulting from the solidifying of the fluid molecules in the interstices of the ice, acts on the mass of the ice as an expansive force, tending to cause the glacier to widen out laterally in all directions.

Conceive a mass of ice lying on a flat horizontal surface, and receiving heat on its upper surface, say from the sun; as the heat passes downwards through the mass, the molecules, acting as conductors, melt and resolidify. Each fluid molecule solidifies in an interstice, which has to be widened in order to contain it. The pressure thus exerted by the continual resolidifying of the molecules will cause the mass to widen out laterally, and of course as the mass widens out it will grow thinner and thinner if it does not receive fresh acquisition on its surface. In the case of a glacier lying in a valley, motion, however, will only take place in one direction. The sides of the valley prevent the glacier from widening; and as gravitation opposes the motion of the ice up, and favours its motion down the valley, the path of least resistance to molecular pressure will always be down the slope, and consequently in this direction molecular displacement will take place. Molecular pressure will therefore produce motion in the same direction as that of gravity. In other words, it will tend to cause the glacier to descend the valley.

The lateral expansion of the ice from internal molecular pressure explains in a clear and satisfactory manner how rock-basins may be excavated by means of land-ice. It also removes the difficulties which have been felt in accounting for the ascent of ice up a steep slope. The main difficulty besetting the theory of the excavation of rock-basins by ice is to explain how the ice after entering the basin manages to get out again—how the ice at the bottom is made to ascend the sloping sides of the basin. Pressure acting from behind, it has been argued by some; but if the basin be deep and its sides steep, this will simply cause the ice lying above the level of the basin to move forward over the surface of the mass filling it. This conclusion is, however, incorrect. The ice filling the basin and the glacier overlying it are united in one solid mass, so that the latter cannot move over the former without shearing; and although the resistance to motion offered by the sloping sides of the basin may be much greater than the resistance to shear, still the ice will be slowly dragged out of the basin. However, in order to obviate this objection to which I refer, the advocates of the glacial origin of lake-basins point out that the length of those basins in proportion to their depth is so great that the slope up which the ice has to pass is in reality but small. This no doubt is true of lake-basins in general, but it does not hold universally true. But the theory does not demand that an ice-formed lake-basin cannot have steep sides. We have incontestable evidence that ice will pass up a steep slope; and, if ice can pass up a steep slope, it can excavate a basin with a steep slope. That ice will pass up a steep slope is proved by the fact that comparatively deep and narrow river-valleys are often found striated across, while hills which stood directly in the path of the ice of the glacial epoch are sometimes found striated *upwards* from their base to their summit. Some striking examples of striæ running up-hill are given by Professor Geikie in his "Glacial Drift of Scotland." I have myself seen a slope striated upwards so steep that one could not climb it.

A very good example of a river-valley striated across came

under my observation during the past summer. The Tay, between Cargill and Stanley (in the centre of the broad plain of Strathmore), has excavated, through the Old Red Sandstone, a channel between 200 and 300 feet in depth. The channel here runs at right angles to the path taken during the glacial epoch by the great mass of ice coming from the North-west Highlands. At a short distance below Cargill, the trap rising out of the bed of the river is beautifully ice-grooved and striated, at right angles to the stream. A trap-dyke, several miles in length, crosses the river about a mile above Stanley, forming a rapid, known as the Linn of Campsie. This dyke is *moutonnée* and striated from near the Linn up the sloping bank to the level of the surrounding country, showing that the ice must have ascended a gradient of one in seven to a height of 300 feet.

From what has been already stated in reference to the resolidifying of the molecules in the interstices of the ice, the application of the molecular theory to the explanation of the effects under consideration will no doubt be apparent. Take the case of the passage of the ice-sheet across a river-valley. As the upper surface of the ice-sheet is constantly receiving heat from the sun and the air in contact with it, there is consequently a transferrence of heat from above downwards to the bottom of the sheet. This transferrence of heat from molecule to molecule is accompanied by the melting and resolidifying of the successive molecules in the manner already detailed. As the fluid molecules tend to flow into adjoining interstices before solidifying and assuming the crystalline form, the interstices of the ice at the bottom of the valley are constantly being filled by fluid molecules from above. These molecules no sooner enter the interstices than they pass into the crystalline form, and become, of course, separated from their neighbours by fresh interstices, which new interstices become filled by fluid molecules, which, in turn, crystallize, forming fresh interstices, and so on. The ice at the bottom of the valley, so long as this process continues, is constantly receiving fresh additions from above. The ice must therefore expand laterally to make room for these additions, which it must do unless the resistance to lateral expansion be greater than the force exerted by the molecules in crystallizing. But a resistance sufficient to do this must be enormous. The ice at the bottom of the valley cannot expand laterally without passing up the sloping sides. In expanding it will take the path of least resistance, but the path of least resistance will always be on the side of the valley towards which the general mass of the ice above is flowing.

It has been shown (Chapter XXVII.) that the ice passing over Strathmore must have been over 2,000 feet in thickness. An ice-sheet 2,000 feet in thickness exerts on its bed a pressure of upwards of 51 tons per square foot. When we reflect that ice under so enormous a pressure, with grinding materials lying underneath, was forced by irresistible molecular energy up an incline of one in seven, it is not at all surprising that the hard trap should be ground down and striated.

We can also understand how the softer portions of the rocky surface over which the ice moved should have been excavated into hollow basins. We have also an explanation of the transport of boulders from a lower to a higher level, for if ice can move from a lower to a higher level, it of course can carry boulders along with it.

The bearing which the foregoing considerations of the manner in which heat is transmitted through ice have on the question of the cause of regelation will be considered in the Appendix.

APPENDIX.

I.

OPINIONS EXPRESSED PREVIOUS TO 1864 REGARDING THE INFLUENCE OF THE ECCENTRICITY OF THE EARTH'S ORBIT ON CLIMATE.*

M. DE MAIRAN.

M. de Mairan, in an article in the *Memoirs of the Royal Academy of France* † "On the General Cause of Heat in Summer and Cold in Winter, in so far as depends on the internal and permanent Heat of the Earth," makes the following remarks on the influence of the difference of distance of the sun in apogee and perigee :—

"Cet élément est constant pour les deux solstices ; tandis que les autres (height of the sun and obliquity of his rays) y varient à raison des latitudes locales ; et il y a encore cela de particulier, qu'il tend à diminuer la valeur de notre été, et à augmenter celle de notre hiver dans l'hémisphère boréal où nous sommes, et tout au contraire dans l'austral. Remarquons cependant que de ces mêmes distances, qui constituent ce troisième élément, naît en partie un autre principe de chaleur tout opposé, et qui semble devoir tempérer les effets du précédent ; sçavoir, la lenteur et la vitesse réciproques du mouvement annuel apparent, en vertu duquel et du réel qui s'y mêle, le soleil emploie 8 jours de plus à parcourir les signes septentrionaux. C'est-à-dire, que le soleil passe 186½ jours dans notre hémisphère, et seulement 178½ dans l'hémisphère opposé. Ce qui, en général, ne peut manquer de répandre un peu plus de chaleur sur l'été du premier, et un peu moins sur son hiver."

* See text, p. 10.

† Mathematical and Physical Series, vol. xxxvi. (1765).

MR. RICHARD KIRWAN.

"Œpinus,* reasoning on astronomical principles, attributes the inferior temperature of the southern hemisphere to the shorter abode of the sun in the southern tropic, shorter by seven days, which produces a difference of fourteen days in favour of the northern hemisphere, during which more heat is accumulated, and hence he infers that the temperature of the northern hemisphere is to that of the southern, as 189·5 to 175·5, or as 14 to 13."—*Trans. of the Royal Irish Academy*, vol. viii., p. 417. 1802.

SIR CHARLES LYELL.

"Before the amount of difference between the temperature of the two hemispheres was ascertained, it was referred by astronomers to the acceleration of the earth's motion in its perihelion; in consequence of which the spring and summer of the southern hemisphere are shorter by nearly eight days than those seasons north of the equator. A sensible effect is probably produced by this source of disturbance, but it is quite inadequate to explain the whole phenomena. It is, however, of importance to the geologist to bear in mind that in consequence of the precession of the equinoxes, the two hemispheres receive alternately, each for a period of upwards of 10,000 years, a greater share of solâr light and heat. This cause may sometimes tend to counterbalance inequalities resulting from other circumstances of a far more influential nature; but, on the other hand, it must sometimes tend to increase the extreme of deviation, which certain combinations of causes produce at distant epochs."—*Principles*, First Edition, 1830, p. 110, vol. i.

SIR JOHN F. HERSCHEL, Bart.

The following, in so far as it relates to the effects of eccentricity, is a copy of Sir John Herschel's memoir, "On the Astronomical Causes which may influence Geological Phenomena," read before the Geological Society, Dec. 15th, 1830.—*Trans. Geol. Soc.*, vol. iii., p. 293, Second Series:—

" . . . . Let us next consider the changes arising in the orbit of the earth itself about the sun, from the disturbing action of the planets. In so doing it will be obviously unnecessary to consider the effect produced on the solar tides, to which the above reasoning applies much more forcibly than in the case of the lunar. It is,

* "Memoirs of St. Petersburg Academy," 1761.

therefore, only the variations in the supply of light and heat received from the sun that we have now to consider.

"Geometers having demonstrated the absolute invariability of the *mean* distance of the earth from the sun, it would seem to follow that the mean annual supply of light and heat derived from that luminary would be alike invariable; but a closer consideration of the subject will show that this would not be a legitimate conclusion, but that, on the contrary, the *mean* amount of solar radiation is dependent on the eccentricity of the orbit, and therefore liable to variation. Without going at present into any geometrical investigations, it will be sufficient for the purpose here to state it as a theorem, of which any one may easily satisfy himself by no very abstruse geometrical reasoning, that '*the eccentricity of the orbit varying, the* total *quantity of heat received by the earth from the sun in one revolution is inversely proportional to the* minor *axis of the orbit.*' Now since the major axis is, as above observed, invariable, and therefore, of course, the absolute length of the year, it will follow that the *mean annual* average of heat will also be in the same inverse ratio of the *minor* axis; and thus we see that the very circumstance which on a cursory view we should have regarded as demonstrative of the constancy of our supply of solar heat, forms an essential link in the chain of strict reasoning by which its variability is proved.

"The eccentricity of the earth's orbits is actually diminishing, and has been so for ages, beyond the records of history. In consequence, the ellipse is in a state of approach to a circle, and its minor axis being, therefore, on the increase, the annual average of solar radiation is actually on the *decrease.*

"So far this is in accordance with the testimony of geological evidence, which indicates a general refrigeration of climate; but when we come to consider the amount of diminution which the eccentricity must be supposed to have undergone to render an account of the variation which has taken place, we have to consider that, in the first place, a great diminution of the eccentricity is required to produce any sensible increase of the minor axis. This is a purely geometrical conclusion, and is best shown by the following table:—

Eccentricity.	Minor Axis.	Reciprocal or Ratio of Heat received.
0·00	1·000	1·000
0·05	0·999	1·002
0·10	0·995	1·005
0·15	0·989	1·011
0·20	0·980	1·021
0·25	0·968	1·032
0·30	0·954	1·048

By this it appears that a variation of the eccentricity of the orbit from the circular form, to that of an ellipse, having an eccentricity of one-fourth of the major axis, would produce only a variation of 3 per cent. on the *mean* annual amount of solar radiation, and this variation takes in the whole range of the planetary eccentricities, from that of Pallas and Juno downwards.

"I am not aware that the limit of increase of the eccentricity of the earth's orbit has ever been determined. That it has a limit has been satisfactorily proved; but the celebrated theorem of Laplace, which is usually cited as demonstrating that none of the planetary orbits can ever deviate materially from the circular form, leads to no such conclusion, except in the case of the great preponderant planets Jupiter and Saturn, while for anything that theorem proves to the contrary, the orbit of the earth may become elliptic to any amount.

"In the absence of calculations which though practicable have, I believe, never been made,* and would be no slight undertaking, we may assume that eccentricities which exist in the orbits of planets, both interior and exterior to that of the earth, may *possibly* have been attained, and may be attained again by that of the earth itself. It is clear that such eccentricities *existing* they cannot be incompatible with the stability of the system generally, and that, therefore, the question of the possibility of such an amount in the particular case of the earth's orbit will depend on the particular data belonging to that case, and can only be determined by executing the calculations alluded to, having regard to the simultaneous effects of at least the four most influential planets, Venus, Mars, Jupiter, and Saturn, *not only on the orbit of the earth, but on those of each other*. The principles of this calculation are detailed in the article of Laplace's work cited. But before entering on a work of so much labour, it is quite necessary to inquire what prospect of advantage there is to induce any one to undertake it.

"Now it certainly at first sight seems clear that a variation of

* The calculations here referred to were made by Lagrange nearly half a century previous to the appearance of this paper, and published in the "Mémoires de l'Académie de Berlin," for 1782, p. 273. Lagrange's results differ but slightly from those afterwards obtained by Leverrier, as will be seen from the following table; but as he had assigned erroneous values to the masses of the smaller planets, particularly that of Venus, the mass of which he estimated at one-half more than its true value, full confidence could not be placed in his results.

Superior limits of eccentricity as determined by Lagrange, Leverrier, and Mr. Stockwell:—

	By Lagrange.	By Leverrier.	By Mr. Stockwell.
Mercury	0·22208	0·225646	0·2317185
Venus	0·08271	0·086716	0·0706329
Earth	0·07641	0·077747	0·0693888
Mars	0·14726	0·142243	0·139655
Jupiter	0·06036	0·061548	0·0608274
Saturn	0·08408	0·084919	0·0843289
Uranus	..	0·064666	0·0779652
Neptune	..	..	0·0145066

[J. C.]

3 per cent. only in the mean annual amount of solar radiation, and that arising from an extreme supposition, does *not* hold out such a prospect. Yet it might be argued that the effects of the sun's heat is to maintain the temperature of the earth's surface at its actual mean height, not above the zero of Fahrenheit's or any other thermometer, but above the temperature of the celestial spaces, out of the reach of the sun's influence, and what that temperature is may be a matter of much discussion. M. Fourier has considered it as demonstrated that it is not greatly inferior to that of the polar regions of our own globe, but the grounds of this decision appear to me open to considerable objection.* If those regions be really void of matter, their temperature can only arise, according to M. Fourier's own view of the subject, from the radiation of the stars. It ought, therefore, to be as much inferior to that due to solar radiation, as the light of a starlight night is to that of the brightest noon day, in other words it should be very nearly a total privation of heat—almost the *absolute zero* respecting which so much difference of opinion exists, some placing it at 1,000°, some at 5,000° of Fahrenheit below the freezing-point, and some still lower, in which case a single unit per cent. in the mean annual amount of radiation would suffice to produce a change of climate fully commensurate to the demands of geologists.†

"Without attempting, however, to enter further into the perplexing difficulties in which this point is involved, which are far greater than appear on a cursory view, let us next consider, not the *mean*, but the *extreme* effects which a variation in the eccentricity of the earth's orbit may be expected to produce in the summer and winter climates in particular regions of its surface, and under the influence of circumstances favouring a difference of effect. And here, if I mistake not, it will appear that an amount of variation, which we need not hesitate to admit (at least, provisionally) as a possible one, may be productive of considerable diversity of climate, and may operate during great periods of time either to mitigate or to exaggerate the difference of winter and summer temperatures, so as to produce alternately, in the same latitude of either hemisphere, a perpetual spring, or the extreme vicissitudes of a burning summer and a rigorous winter.

* "Mém. de l'Acad. royale des Sciences." 1827. Tom. vii., p. 598.

† Absolute zero is now considered to be only 493° Fah. below the freezing-point, and Herschel himself has lately determined 271° below the freezing-point to be the temperature of space. Consequently, a decrease, or an increase of one per cent. in the mean annual amount of radiation would not produce anything like the effect which is here supposed. But the mean annual amount of heat received cannot vary much more than one-tenth part of one per cent. In short, the effect of eccentricity on the mean annual supply of heat received from the sun, in so far as geological climate is concerned, may be practically disregarded.—[J. C.]

"To show this, let us at once take the extreme case of an orbit as eccentric as that of Juno or Pallas, in which the greatest and least distances of the sun are to each other as 5 to 3, and consequently the radiations at those distances as 25 to 9, or very nearly as 3 to 1. To conceive what would be the *extreme* effects of this great variation of the heat received at different periods of the year, let us first imagine in our latitude the place of the perigee of the sun to coincide with the summer solstice. In that case, the difference between the summer and winter temperature would be exaggerated in the same degree as if three suns were placed side by side in the heavens in the former season and only one in the latter, which would produce a climate perfectly intolerable. On the other hand, were the perigee situated in the winter solstice our three suns would combine to warm us in the winter, and would afford such an excess of winter radiation as would probably more than counteract the effect of short days and oblique sunshine, and throw the summer season into the winter months.

"The actual diminution of the eccentricity is so slow, that the transition from a state of the orbit such as we have assumed to the present nearly circular figure would occupy upwards of 600,000 years, supposing it uniformly changeable—this, of course, would not be the case; when near the maximum, however, it would vary slower still, so that at that point it is evident a period of 10,000 years would elapse without any perceptible change in the state of the data of the case we are considering.

"Now this adopting the very ingenious idea of Mr. Lyell* would suffice, by reason of the combined effect of the precession of the equinoxes and the motion of the apsides of the orbit itself, to transfer the perigee from the summer to the winter solstice, and thus to produce a transition from the one to the other species of climate in a period sufficiently great to give room for a material change in the botanical character of country.

"The supposition above made is an extreme, but it is not demonstrated to be an impossible one, and should even an approach to such a state of things be possible, the same conse-

* "Principles of Geology," p. 110. "Mr. Lyell, however, in stating the actual excess of eight days in the duration of the sun's presence in the northern hemisphere over that in the southern as productive of an excess of light and heat annually received by the one over the other hemisphere, appears to have misconceived the effect of elliptic motion in the passage here cited, since it is demonstrable that whatever be the ellipticity of the earth's orbit the two hemispheres must receive equal absolute quantities of light and heat per annum, the proximity of the sun in perigee exactly compensating the effect of its swifter motion. This follows from a very simple theorem, which may be thus stated: 'The amount of heat received by the earth from the sun while describing any part of its orbit is proportional to the angle described round the sun's centre,' so that if the orbit be divided into two portions by a line drawn *in any direction* through the sun's centre, the heats received in describing the two unequal segments of the ellipse so produced will be equal."

quences, in a mitigated degree, would follow. But if, on executing the calculations, it should appear that the limits of the eccentricity of the earth's orbit are really narrow, and if, on a full discussion of the very difficult and delicate point of the actual effect of solar radiation, it should appear that the mean, as well as the extreme, temperature of our climates would *not* be materially affected,—it will be at least satisfactory to *know* that the causes of the phenomena in question are to be sought elsewhere than in the relations of our planet to the system to which it belongs, since there does not appear to exist any other conceivable connections between these relations and the facts of geology than those we have enumerated, the obliquity of the ecliptic being, as we know, confined within too narrow limits for its variation to have any sensible influence."—*J. F. W. Herschel.*

The influence which this paper might have had on the question as to whether eccentricity may be regarded as a cause of changes in geological climate appears to have been completely neutralized by the following, which appeared shortly afterwards both in his "Treatise" and "Outlines of Astronomy," showing evidently that he had changed his mind on the subject.

"It appears, therefore, from what has been shown, the supplies of heat received from the sun will be equal in the two segments, in whatever direction the line PTQ be drawn. They will, indeed, be described in unequal times: that in which the perihelion A lies in a shorter, and the other in a longer, in proportion to their unequal area; but the greater proximity of the sun in the smaller segment compensates exactly for its more rapid description, and thus an equilibrium of heat is, as it were, maintained.

"Were it not for this the eccentricity of the orbit would materially influence the transition of seasons. The fluctuation of distance amounts to nearly 1-30th of the mean quantity, and, consequently, the fluctuation of the sun's direct heating power to double this, or 1-15th of the whole. . . . . Were it not for the compensation we have just described, the effect would be to exaggerate the difference of summer and winter in the southern hemisphere, and to moderate it in the northern; thus producing a more violent alternation of climate in the one hemisphere and an approach to perpetual spring in the other. *As it is, however, no such inequality subsists*, but an equal and impartial distribution of heat and light is accorded to both."—"*Treatise of Astronomy,*" *Cabinet Cyclopædia*, § 315; *Outlines of Astronomy*, § 368.

"The fact of a great change in the general climate of large tracts of the globe, if not of the whole earth, and of a diminution of general temperature, having been recognised by geologists, from their examination of the remains of animals and vegetables of former ages enclosed in the strata, various causes for such diminution of temperature have been assigned. . . . . It is evident that the *mean* temperature of the whole surface of the globe, in so far as it is maintained by the action of the sun at a higher degree than it would have were the sun extinguished, must depend on the mean quantity of the sun's rays which it receives, or, which comes to the same thing, on the *total* quantity received in a given invariable time; and the length of the year being unchangeable in all the fluctuations of the planetary system, it follows that the total *annual* amount of solar radiation will determine, *cæteris paribus*, the general climate of the earth. Now, it is not difficult to show that this amount is inversely proportional to the minor axis of the ellipse described by the earth about the sun, regarded as slowly variable; and that, therefore, the major axis remaining, as we know it to be, constant, and the orbit being actually in a state of approach to a circle, and consequently the minor axis being on the *increase*, the mean annual amount of solar radiation received by the whole earth must be actually on the *decrease*. We have here, therefore, an evident real cause of sufficient universality, and acting *in the right direction*, to account for the phenomenon. Its adequacy is another consideration."*—*Discourse on the Study of Natural Philosophy*, pp. 145—147 (1830).

SIR CHARLES LYELL, Bart.

"*Astronomical Causes of Fluctuations in Climate.*—Sir John Herschel has lately inquired, whether there are any astronomical causes which may offer a possible explanation of the difference between the actual climate of the earth's surface, and those which formerly appear to have prevailed. He has entered upon this subject, he says, 'impressed with the magnificence of that view of geological revolutions, which regards them rather as regular and necessary effects of great and general causes, than as resulting from a series of convulsions and catastrophes, regulated by no laws, and reducible to no fixed principles.' Geometers, he adds, have demonstrated the absolute invariability of the mean distance of the earth from the sun; whence it would seem to follow that the mean annual supply of light and heat derived from that luminary would be alike invariable; but a closer consideration of the subject will show that this would not be a legitimate conclusion, but that, on the contrary, the *mean* amount of solar radiation is dependent on the eccentricity of the earth's orbit, and, therefore, liable to variation.

"Now, the eccentricity of the orbit, he continues, is actually

* When the eccentricity of the earth's orbit is at its superior limit, the absolute quantity of heat received by the globe during one year will be increased by only 1-300th part; an amount which could produce no sensible influence on climate.—[J. C.]

diminishing, and has been so for ages beyond the records of history. In consequence, the ellipse is in a state of approach to a circle, and the annual average of solar heat radiated to the earth is actually on the *decrease*. So far, this is in accordance with geological evidence, which indicates a general refrigeration of climate; but the question remains, whether the amount of diminution which the eccentricity may have ever undergone can be supposed sufficient to account for any sensible refrigeration.* The calculations necessary to determine this point, though practicable, have never yet been made, and would be extremely laborious; for they must embrace all the perturbations which the most influential planets, Venus, Mars, Jupiter, and Saturn, would cause in the earth's orbit and in each other's movements round the sun.

"The problem is also very complicated, inasmuch as it depends not merely on the ellipticity of the earth's orbit, but on the assumed temperature of the celestial spaces beyond the earth's atmosphere; a matter still open to discussion, and on which M. Fourier and Sir J. Herschel have arrived at very different opinions. But if, says Herschel, we suppose an extreme case, as if the earth's orbit should ever become as eccentric as that of the planet Juno or Pallas, a great change of climate might be conceived to result, the winter and summer temperatures being sometimes mitigated and at others exaggerated, in the same latitudes.

"It is much to be desired that the calculations alluded to were executed, as even if they should demonstrate, as M. Arago thinks highly probable, that the mean of solar radiation can never be materially affected by irregularities in the earth's motion, it would still be satisfactory to ascertain the point."—*Principles of Geology*, Ninth Edition, 1853, p. 127.

M. ARAGO.

"*Can the variations which certain astronomical elements undergo sensibly modify terrestrial climates?*

"The sun is not always equally distant from the earth. At this time its least distance is observed in the first days of January, and the greatest, six months after, or in the first days of July. But, on the other hand, a time will come when the *minimum* will occur in July, and the *maximum* in January. Here, then, this

* Sir Charles has recently, to a certain extent, adopted the views advocated in the present volume, viz., that the cold of the glacial epoch was brought about not by a *decrease*, but by an *increase* of eccentricity. (See vol. i. of "Principles," tenth and eleventh editions.) The decrease in the mean annual quantity of heat received from the sun, resulting from the decrease in the eccentricity of the earth's orbit—the astronomical cause to which he here refers—could have produced no sensible effect on climate.—[J. C.]

interesting question presents itself,—Should a summer such as those we now have, in which the *maximum* corresponds to the solar distance, differ sensibly, from a summer with which the *minimum* of this distance should coincide?

"At first sight every one probably would answer in the affirmative; for, between the *maximum* and the *minimum* of the sun's distance from the earth there is a remarkable difference, a difference in round numbers of a thirtieth of the whole. Let, however, the consideration of the velocities be introduced into the problem, elements which cannot fairly be neglected, and the result will be on the side opposite to that we originally imagined.

"The part of the orbit where the sun is found nearest the earth, is, at the same time, the point where the luminary moves most rapidly along. The demi-orbit, or, in other words, the 180° comprehended betwixt the two equinoxes of spring-time and autumn, will then be traversed in the least possible time, when, in moving from the one of the extremities of this arc to the other, the sun shall pass, near the middle of this course of six months, at the point of the smallest distance. To resume—the hypothesis we have just adopted would give, on account of the lesser distance, a spring-time and summer hotter than they are in our days; but on account of the greater rapidity, the sum of the two seasons would be shorter by about seven days. Thus, then, all things considered, the compensation is mathematically exact. After this it is superfluous to add, that the point of the sun's orbit corresponding to the earth's least distance changes very gradually; and that since the most distant periods, the luminary has always passed by this point, either at the end of autumn or beginning of winter.

"We have thus seen that the changes which take place in the *position* of the solar orbit, *have no power in modifying the climate of our globe.* We may now inquire, if it be the same concerning the variations which this orbit experiences in its *form*. . . . .

"Herschel, who has recently been occupying himself with this problem, in the hope of discovering the explanation of several geological phenomena, allows that the succession of ages might bring the eccentricity of the terrestrial orbit to the proportion of that of the planet Pallas, that is to say, to be the $\frac{25}{100}$ of a semi-greater axis. It is exceedingly improbable that in these periodical changes the eccentricity of our orbit should ever experience such enormous variations, and even then these twenty-five hundredth parts $\left(\frac{25}{100}\right)$, would not augment the *mean* annual solar radiation except by about one hundredth part $\left(\frac{1}{100}\right)$. To repeat, an eccentricity of $\frac{25}{100}$ *would not alter in any appreciated manner the mean thermometrical state of the globe.* . . . .

"The changes of the form, and of the position, of the terrestrial orbit are mathematically inoperative, or, at most, their influence is so minute that it is not indicated by the most delicate instruments. For the explanation of the changes of climates, then, there only remains to us either the local circumstances, or some alteration in the heating or illuminating power of the sun. But of these two causes, we may continue to reject the last. And thus, in fact, all the changes would come to be attributed to agricultural operations, to the clearing of plains and mountains from wood, the draining of morasses, &c.

"Thus, at one swoop, to confine, the whole earth, the variations of climates, past and future, within the limits of the naturally very narrow influence which the labour of man can effect, would be a meteorological result of the very last importance."—pp. 221—224, *Memoir on the "Thermometrical State of the Terrestrial Globe," in the Edinburgh New Philosophical Journal*, vol. xvi., 1834.

BARON HUMBOLDT.

"The question," he says, "has been raised as to whether the increasing value of this ellipticity is capable during thousands of years of modifying to any considerable extent the temperature of the earth, in reference to the daily and annual quantity and distribution of heat? Whether a partial solution of the great geological problem of the imbedding of tropical vegetable and animal remains in the now cold zones may not be found in these *astronomical* causes proceeding regularly in accordance with eternal laws? . . . . It might at the first glance be supposed that the occurrence of the perihelion at an opposite time of the year (instead of the winter, as, is now the case, in summer) must necessarily produce great climatic variations; but, on the above supposition, the sun will no longer remain seven days longer in the northern hemisphere; no longer, as is now the case, traverse that part of the ecliptic from the autumnal equinox to the vernal equinox, in a space of time which is one week shorter than that in which it traverses the other half of its orbit from the vernal to the autumnal equinox.

"The difference of temperature which is considered as the consequence to be apprehended from the turning of the major axis, *will on the whole disappear*, principally from the circumstance that the point of our planet's orbit in which it is nearest to the sun is at the same time always that over which it passes with the greatest velocity. . . . .

"As the altered position of the major axis is capable of exerting only a very *slight influence upon the temperature of the earth;* so likewise the *limit* of the probable changes in the elliptical form of the earth's orbit are, according to Arago and Poisson, so narrow that these changes could *only very slightly* modify the climates of the individual zones, and that in very long periods." * — *Cosmos*, vol. iv., pp. 458, 459. Bohn's Edition. 1852.

SIR HENRY T. DE LA BECHE.

"Mr. Herschel, viewing this subject with the eye of an astronomer, considers that a diminution of the surface-temperature might arise from a change in ellipticity of the earth's orbit, which, though slowly, gradually becomes more circular. No calculations having yet been made as to the probable amount of decreased temperature from this cause, it can at present be only considered as a possible explanation of those geological phenomena which point to considerable alterations in climates." — *Geological Manual*. Third Edition. 1833. p. 8.

PROFESSOR PHILLIPS.

"*Temperature of the Globe.—Influence of the Sun.*—No proposition is more certain than the fundamental dependence of the temperature of the surface of the globe on the solar influence.

"It is, therefore, very important for geologists to inquire whether this be variable or constant; whether the amount of solar heat communicated to the earth is and has always been the same in every annual period, or what latitude the laws of planetary movements permit in this respect.

"Sir John Herschel has examined this question in a satisfactory manner, in a paper read to the Geological Society of London. The total amount of solar radiation which determines the general climate of the earth, the year being of invariable length, is inversely proportional to the minor axis of the ellipse described by the earth about the sun, regarded as slowly variable; the major axis remaining constant and the orbit being actually in a state of approach to a circle, and, consequently, the minor axis being on the increase, it follows that the mean annual amount of solar radiation received by the whole earth must be actually on the decrease. The limits of the variation in the eccentricity of the earth's orbit are not known. It is, therefore, impossible to say accurately what may have been in former periods of time, the amount of solar radiation; it is, however, certain that if the ellipticity has ever been so great as that of the orbit of Mercury or Pallas, the temperature of the earth must have been sensibly higher than it is at present. But the difference of a few degrees of temperature thus occasioned, is of too small an order to be

* It is singular that both Arago and Humboldt should appear to have been unaware of the researches of Lagrange on this subject.

employed in explaining the growth of tropical plants and corals in the polar or temperate zones, and other great phenomena of Geology."—*From A Treatise on Geology*, p. 11, *forming the article under that head in the seventh edition of the Encyclopædia Britannica.* 1837.

MR. ROBERT BAKEWELL.

"A change in the form of the earth's orbit, if considerable, might change the temperature of the earth, by bringing it nearer to the sun in one part of its course. The orbit of the earth is an ellipsis approaching nearly to a circle; the distance from the centre of the orbit to either focus of the ellipsis is called by astronomers 'the eccentricity of the orbit.' This eccentricity has been for ages slowly decreasing, or, in other words, the orbit of the earth has been approaching nearer to the form of a perfect circle; after a long period it will again increase, and the possible extent of the variation has not been yet ascertained. From what is known respecting the orbits of Jupiter and Saturn, it appears highly probable that the eccentricity of the earth's orbit is confined within limits that preclude the belief of any great change in the mean annual temperature of the globe ever having been occasioned by this cause."—*Introduction to Geology*, p. 600. 1838. Fifth Edition.

MRS. SOMERVILLE.

"Sir John Herschel has shown that the elliptical form of the earth's orbit has but a trifling share in producing the variation of temperature corresponding to the difference of the seasons."—*Physical Geography*, vol. ii., p. 20. Third Edition.

MR. L. W. MEECH, A.M.

"Let us, then, look back to that primeval epoch when the earth was in aphelion at midsummer, and the eccentricity at its maximum value—assigned by Leverrier near to 0777. Without entering into elaborate computation, it is easy to see that the extreme values of diurnal intensity, in Section IV., would be altered as by the multiplier $\left(\frac{1 \pm e}{1 \pm e'}\right)^2$, that is 1—0·11 in summer, and 1 + 0·11 in winter. This would diminish the midsummer intensity by about 9°, and increase the midwinter intensity by 3° or 4°; the temperature of spring and autumn being nearly unchanged. But this does not appear to be of itself adequate to the geological effects in question.

"It is not our purpose, here, to enter into the inquiry whether the atmosphere was once more dense than now, whether the earth's axis had once a different inclination to the orbit, or the sun a greater emissive power of heat and light. Neither shall we attempt to speculate upon the primitive heat of the earth, nor of planetary space, nor of the supposed connection of terrestrial heat and magnetism; nor inquire how far the existence of coal-fields in this latitude, of fossils, and other geological remains, have depended upon existing causes. The preceding discussion seems to prove simply that, under the present system of physical astronomy, the sun's intensity could never have been materially different from what is manifested upon the earth at the present day. *The causes of notable geological changes must be other than the relative position of the sun and earth, under their present laws of motion.*"—"*On the Relative Intensity of the Heat and Light of the Sun.*" *Smithsonian Contributions to Knowledge*, vol. ix.

M. JEAN REYNAUD.

"La révolution qui pourrait y causer les plus grands changements thermométriques, celle qui porte l'orbite à s'élargir et à se rétrécir alternativement et, par suite, la planète à passer, aux époques de périhélie, plus ou moins près du soleil, embrasse une periode de plus de cent mille années terrestres et demeure comprise dans de si étroites limites que les habitants doivent être à peine avertis que la chaleur décroît, par cette raison, depuis une haute antiquité et décroîtra encore pendant des siècles en variant en même temps dans sa répartition selon les diverses époques de l'année. . . . . Enfin, le tournoiement de l'axe du globe s'empreint également d'une manière particulière sur l'ètablissement des saisons qui, à tour de rôle, dans chacun des deux hémisphères, deviennent graduellement, durant une période d'environ vingt-cinq mille ans, de plus en plus uniformes, ou, à l'inverse, de plus en plus dissemblables. C'est actuellement dans l'hémisphère boréal que règne l'uniformité, et quoique les étés et les hivers y tendent, dès à présent, à se trancher de plus en plus, il ne paraît pas douteux que la modération des saisons n'y produise, pendant longtemps encore, des effets appréciables. En résumé, de tous ces changements il n'en est donc aucun ni qui suive un cours précipité, ni qui s'élève jamais à des valeurs considérables; ils se règlent tous sur un mode de développement presque insensible, et il s'ensuit que les années de la terre, malgré leur complexité virtuelle, se distinguent par le constance de leurs caractères non-seulement de ce qui peut avoir lieu, en vertu des mêmes principes, dans les autres systèmes planétaires de l'univers, mais même de ce qui s'observe dans plusieurs des mondes qui composent le nôtre."—*Philosophie Religieuse: Terre et Ciel.*

M. ADHÉMAR.

Adhémar does not consider the effects which ought to result from a change in the eccentricity of the earth's orbit; he only concerns himself with those which, in his opinion, arise from the present amount of such eccentricity. He admits, of course, that both hemispheres receive from the sun equal quantities of heat per annum; but, as the southern hemisphere has a winter longer by 168 hours than the corresponding season in the northern hemisphere, an accumulation of heat necessarily takes place in the latter, and an accumulation of cold in the former. Adhémar also measures the loss of heat sustained by the southern hemisphere in a year by the number of hours by which the southern exceeds the northern winter. "The south pole," he says, "loses in one year more heat than it receives, because the total duration of its nights surpasses that of the days by 168 hours; and the contrary takes place for the north pole. If, for example, we take for unity the mean quantity of heat which the sun sends off in one hour, the heat accumulated at the end of the year at the north pole will be expressed by 168, while the heat lost by the south pole will be equal to 168 times what the radiation lessens it by in one hour; so that at the end of the year the difference in the heat of the two hemispheres will be represented by 336 times what the earth receives from the sun or loses in an hour by radiation,"* and at the end of 100 years the difference will be 33,600 times, and at the end of 1,000 years 336,000 times, or equal to what the earth receives from the sun in 38½ years, and so on during the 10,000 years that the southern winter exceeds in length the northern. This, in his opinion, is all that is required to melt the ice off the arctic regions, and cover the antarctic regions with an enormous ice-cap. He further supposes that in about 10,000 years, when our northern winter will occur in aphelion and the southern in perihelion, the climatic conditions of the two hemispheres will be reversed; that is to say, the ice will melt at the south pole, and the northern hemisphere will become enveloped in one continuous mass of ice, leagues in thickness, extending down to temperate regions.

This theory, as shown in Chapter V., is based upon a misconception regarding the laws of radiant heat. The loss of heat sustained by the southern hemisphere from radiation, resulting from the greater length of the southern winter, is vastly overestimated by M. Adhémar, and could not possibly produce the effects which he supposes. But I need not enter into this subject here, as the reader will find the whole question discussed at length in the chapter above referred to. By far the most important part of Adhémar's theory, however, is his conception of the sub-

* "Révolutions de la Mer," p. 37. Second Edition.

mergence of the land by means of a polar ice-cap. He appears to have been the first to put forth the idea that a mass of ice placed on the globe, say, for example, at the south pole, will shift the earth's centre of gravity a little to the south of its former position, and thus, as a physical consequence, cause the sea to sink at the north pole and to rise at the south. According to Adhémar, as the one hemisphere cools and the other grows warmer, the ice at the pole of the former will increase in thickness and that at the pole of the latter diminish.

The sea, as a consequence, will sink on the warm hemisphere where the ice is decreasing and rise on the cold hemisphere where the ice is increasing. And, again, in 10,000 years, when the climatic conditions of the two hemispheres are reversed, the sea will sink on the hemisphere where it formerly rose, and rise on the hemisphere where it formerly sank, and so on in like manner through indefinite ages.

Adhémar, however, acknowledges to have derived the grand conception of a submergence of the land from the shifting of the earth's centre of gravity from the following wild speculation of one Bertrand, of Hamburgh:—

"Bertrand de Hambourg, dans un ouvrage imprimé en 1799 et qui a pour titre: *Renouvellement périodique des Continents*, avait déjà émis cette idée, que la masse des eaux pouvait être alternativement entraînée d'un hémisphère à l'autre par le déplacement du centre de gravité du globe. Or, pour expliquer ce déplacement, il supposait que la terre était creuse et qu'il y avait dans son intérieur un gros noyau d'aimant auquel les comètes par leur attraction communiquaient un mouvement de va-et-vient analogue à celui du pendule."—*Révolutions de la Mer*, p. 41.

The somewhat extravagant notions which Adhémar has advanced in connection with his theory of submergence have very much retarded its acceptance. Amongst other remarkable views he supposes the polar ice-cap to rest on the bottom of the ocean, and to rise out of the water to the enormous height of twenty leagues. Again, he holds that on the winter approaching perihelion and the hemisphere becoming warm the ice waxes soft and rotten from the accumulated heat, and the sea now beginning to eat into the base of the cap, this is so undermined as, at last, to be left standing upon a kind of gigantic pedestal. This disintegrating process goes on till the fatal moment at length arrives, when the whole mass tumbles down into the sea in huge fragments which become floating icebergs. The attraction of the opposite ice-cap, which has by this time nearly reached its maximum thickness, becomes now predominant. The earth's centre of gravity suddenly crosses the plain of the equator, dragging the ocean along with it, and carrying death and destruction to everything on the surface of the globe. And these catastrophes, he asserts, occur alternately on the two hemispheres every 10,000 years.—*Révolutions de la Mer*, pp. 316—328.

Adhémar's theory has been advocated by M. Le Hon, of Brussels, in a work entitled *Périodicité des Grands Déluges*. Bruxelles et Leipzig, 1858.

II.

ON THE NATURE OF HEAT-VIBRATIONS.*

From the *Philosophical Magazine* for May, 1864.

In a most interesting paper on "Radiant Heat," by Professor Tyndall, read before the Royal Society in March last, it is shown conclusively that the *period* of heat-vibrations is not affected by the state of aggregation of the molecules of the heated body; that is to say, whether the substance be in the gaseous, the liquid, or, perhaps, the solid condition, the tendency of its molecules to vibrate according to a given period remains unchanged. The force of cohesion binding the molecules together exercises no effect on the rapidity of vibration.

I had arrived at the same conclusion from theoretical considerations several years ago, and had also deduced some further conclusions regarding the nature of heat-vibrations, which seem to be in a measure confirmed by the experimental results of Professor Tyndall. One of these conclusions was, that the heat-vibration does not consist in a motion of an aggregate mass of molecules, but in a motion of the individual molecules themselves. Each molecule, or rather we should say each atom, acts as if there were no other in existence but itself. Whether the atom stands by itself as in the gaseous state, or is bound to other atoms as in the liquid or the solid state, it behaves in exactly the same manner. The deeper question then suggested itself, viz., what is the nature of that mysterious motion called heat assumed by the atom? Does it consist in excursions across centres of equilibrium external to the atom itself? It is the generally received opinion among physicists that it does. But I think that the experimental results arrived at by Professor Tyndall, as well as some others which will presently be noticed, are entirely hostile to such an opinion. The relation of an atom to its centre of equilibrium depends entirely on the state of aggregation. Now if heat-vibrations consist in

* See text, p. 37.

excursions to and fro across these centres, then the *period* ought to be affected by the state of aggregation. The higher the *tension* of the atom in regard to the centre, the more rapid ought its movement to be. This is the case in regard to the vibrations constituting sound. The harder a body becomes, or, in other words, the more firmly its molecules are bound together, the higher is the *pitch*. Two harp-cords struck with equal force will vibrate with equal force, however much they may differ in the rapidity of their vibrations. The *vis viva* of vibration depends upon the force of the stroke; but the rapidity depends, not on the stroke, but upon the tension of the cord.

That heat-vibrations do not consist in excursions of the molecules or atoms across centres of equilibrium, follows also as a necessary consequence from the fact that the real specific heat of a body remains unchanged under all conditions. All changes in the specific heat of a body are due to differences in the amount of heat consumed in molecular work against cohesion or other forces binding the molecules together. Or, in other words, to produce in a body no other effect than a given rise of temperature, requires the same amount of force, whatever may be the physical condition of the body. Whether the body be in the solid, the fluid, or the gaseous condition, the same rise of temperature always indicates the same quantity of force consumed in the simple production of the rise. Now, if heat-vibrations consist in excursions of the atom to and fro across a centre of equilibrium *external to itself*, as is generally supposed, then the *real* specific heat of a solid body, for example, *ought to decrease with the hardness of the body*, because an increase in the strength of the force binding the molecules together would in such a case tend to favour the rise in the rapidity of the vibrations.

These conclusions not only afford us an insight into the hidden nature of heat-vibrations, but they also appear to cast some light on the physical constitution of the atom itself. They seem to lead to the conclusion that the ultimate atom itself is *essentially elastic*.* For if heat-vibrations do not consist in excursions of the atom, then it must consist in alternate expansions and contractions of the atom itself. This again is opposed to the ordinary idea that the atom is essentially solid and impenetrable. But it favours the modern idea, that matter consists of forces of resistance acting from a centre.

Professor Tyndall in a memoir read before the Royal Society "On a new Series of Chemical Reactions produced by Light," has subsequently arrived at a similar conclusion in reference to the atomic nature of heat-vibrations. The following are his views on the subject:—

* See *Philosophical Magazine* for December, 1867, p. 457.

"A question of extreme importance in molecular physics here arises:—What is the real mechanism of this absorption, and where is its seat?

"I figure, as others do, a molecule as a group of atoms, held together by their mutual forces, but still capable of motion among themselves. The vapour of the nitrite of amyl is to be regarded as an assemblage of such molecules. The question now before us is this:—In the act of absorption, is it the *molecules* that are effective, or is it their constituent *atoms?* Is the *vis viva* of the intercepted waves transferred to the molecule as a whole, or to its constituent parts?

"The molecule, as a whole, can only vibrate in virtue of the forces exerted between it and its neighbour molecules. The intensity of these forces, and consequently the rate of vibration, would, in this case, be a function of the distance between the molecules. Now the identical absorption of the liquid and of the vaporous nitrite of amyl indicates an identical vibrating period on the part of liquid and vapour, and this, to my mind, amounts to an experimental demonstration that the absorption occurs in the main *within* the molecule. For it can hardly be supposed, if the absorption were the act of the molecule as a whole, that it could continue to affect waves of the same period after the substance had passed from the vaporous to the liquid state."—*Proc. of Roy. Soc.*, No. 105. 1868.

Professor W. A. Norton, in his memoir on "Molecular Physics,"* has also arrived at results somewhat similar in reference to the nature of heat-vibrations. "It will be seen," he says, "that these (Mr. Croll's) ideas are in accordance with the conception of the constitution of a molecule adopted at the beginning of the present memoir (p. 193), and with the theory of heat-vibrations or heat-pulses deduced therefrom (p. 196)." †

* Silliman's American Journal for July, 1864. *Philosophical Magazine* for September, 1864, pp. 193, 196.

† *Philosophical Magazine* for August, 1865, p. 95.

III.

ON THE REASON WHY THE DIFFERENCE OF READING BETWEEN A THERMOMETER EXPOSED TO DIRECT SUNSHINE AND ONE SHADED DIMINISHES AS WE ASCEND IN THE ATMOSPHERE.*

From the *Philosophical Magazine* for March, 1867.

The remarkable fact was observed by Mr. Glaisher, that the difference of reading between a black-bulb thermometer exposed to the direct rays of the sun and one shaded diminishes as we ascend in the atmosphere. On viewing the matter under the light of Professor Tyndall's important discovery regarding the influence of aqueous vapour on radiant heat, the fact stated by Mr. Glaisher appears to be in perfect harmony with theory. The following considerations will perhaps make this plain.

The shaded thermometer marks the temperature of the surrounding air; but the exposed thermometer marks not the temperature of the air, but that of the bulb heated by the direct rays of the sun. The temperature of the bulb depends upon two elements: (1) the rate at which it receives heat by direct *radiation* from the sun above, the earth beneath, and all surrounding objects, and by *contact* with the air; (2) the rate at which it loses heat by radiation and by contact with the air. As regards the heat gained and lost by contact with the surrounding air, both thermometers are under the same conditions, or nearly so. We therefore require only to consider the element of radiation.

We begin by comparing the two thermometers at the earth's surface, and we find that they differ by a very considerable number of degrees. We now ascend some miles into the air, and on again comparing the thermometers we find that the difference between them has greatly diminished. It has been often proved, by direct observation, that the intensity of the sun's rays increases as we rise in the atmosphere. How then does the exposed thermometer sink more rapidly than the shaded one as we ascend? The reason is obviously this. The temperature of the thermometers depends as much upon the rate at which they are losing their heat as upon the rate at which they are gaining it. The higher temperature of the exposed thermometer is the result of *direct radiation* from the sun. Now, although this thermometer receives by radiation more heat

* See text, p. 80.

from the sun at the upper position than at the lower, it does not necessarily follow on this account that its temperature ought to be higher. Suppose that at the upper position it should receive one-fourth more heat from the sun than at the lower, yet if the rate at which it loses its heat by radiation into space be, say, one-third greater at the upper position than at the lower, the temperature of the bulb would sink to a considerable extent, notwithstanding the extra amount of heat received. Let us now reflect on how matters stand in this respect in regard to the actual case under our consideration. When the exposed thermometer is at the higher position, it receives more heat from the sun than at the lower, but it receives less from the earth; for a considerable part of the radiation from the earth is cut off by the screen of aqueous vapour intervening between the thermometer and the earth. But, on the whole, it is probable that the total quantity of radiant heat reaching the thermometer is greater in the higher position than in the lower. Compare now the two positions in regard to the rate at which the thermometer loses its heat by radiation. When the thermometer is at the lower position, it has the warm surface of the ground against which to radiate its heat downwards. The high temperature of the ground thus tends to diminish the rate of radiation. Above, there is a screen of aqueous vapour throwing back upon the thermometer a very considerable part of the heat which the instrument is radiating upwards. This, of course, tends greatly to diminish the loss from radiation. But at the upper position this very screen, which prevented the thermometer from throwing off its heat into the cold space above, now affects the instrument in an opposite manner; for the thermometer has now to radiate its heat downwards, not upon the warm surface of the ground as before, but upon the cold upper surface of the aqueous screen intervening between the instrument and the earth. This of course tends to lower the mercury. We are now in a great measure above the aqueous screen, with nothing to protect the thermometer from the influence of cold stellar space. It is true that the air above is at a temperature little below that of the thermometer itself; but then the air is dry, and, owing to its diathermancy, it does not absorb the heat radiated from the thermometer, and consequently the instrument radiates its heat directly into the cold stellar space above, some hundreds of degrees below zero, almost the same as it would do were the air entirely removed. The enormous loss of heat which the thermometer now sustains causes it to fall in temperature to a great extent. The molecules of the comparatively dry air at this elevation, being very bad radiators, do not throw off their heat into space so rapidly as the bulb of the exposed thermometer; consequently their temperature does not (for this reason) tend to sink so rapidly as that of the bulb. Hence the shaded thermometer, which indicates the temperature of those molecules, is not affected to such an extent as the exposed one. Hence also the difference of reading between the two instruments must diminish as we rise in the atmosphere.

This difference between the temperature of the two thermometers evidently does not go on diminishing to an indefinite extent. Were we able to continue our ascent in the atmosphere, we should certainly find that a point would be reached beyond which the difference of reading would begin to increase, and would continue to do so till the outer limits of the atmosphere were reached. The difference between the temperatures of the two thermometers beyond the limits of the atmosphere would certainly be enormous. The thermometer exposed to the direct rays of the sun would no doubt be much colder than it had been when at the earth's surface; but the shaded thermometer would now indicate the temperature of space, which, according to Sir John Herschel and M. Pouillet, is more than 200° Fahrenheit below zero.

It follows also, from what has been stated, that even under direct sunshine the removal of the earth's atmosphere would tend to lower the temperature of the earth's surface to a great extent. This conclusion also follows as an immediate inference from the fact that the earth's atmosphere, as it exists at present charged with aqueous vapour, affects terrestrial radiation more than it does radiation from the sun; for the removal of the atmosphere would increase the rate at which the earth throws off its heat into space more than it would increase the rate at which it receives heat from the sun; therefore its temperature would necessarily fall until the rate of radiation *from* the earth's surface exactly equalled the rate of radiation *to* the surface. Let the atmosphere again envelope the earth, and terrestrial radiation would instantly be diminished; the temperature of the earth's surface would therefore necessarily begin to rise, and would continue to do so till the rate of radiation from the surface would equal the rate of radiation received by the surface. Equilibrium being thus restored, the temperature would remain stationary. It is perfectly obvious that if we envelope the earth with a substance such as our atmosphere, that offers more resistance to terrestrial radiation than to solar, the temperature of the earth's surface must necessarily rise until the heat which is being radiated off equals that which is being received from the sun. Remove the air and thus get quit of the resistance, and the temperature of the surface would fall, because in this case a lower temperature would maintain equilibrium.

It follows, therefore, that the moon, which has no atmosphere, must be much colder than our earth, even on the side exposed to the sun. Were our earth with its atmosphere as it exists at present removed to the orbit of Venus or Mars, for example, it certainly would not be habitable, owing to the great change of temperature that would result. But a change in the physical constitution of the atmospheric envelope is really all that would be necessary to retain the earth's surface at its present temperature in either position.

IV.

REMARKS ON MR. J. Y. BUCHANAN'S THEORY OF THE VERTICAL DISTRIBUTION OF TEMPERATURE OF THE OCEAN.*

Since the foregoing was in type, a paper on the "Vertical Distribution of Temperature of the Ocean," by Mr. J. Y. Buchanan, chemist on board the *Challenger*, has been read before the Royal Society.† In that paper Mr. Buchanan endeavours to account for the great depth of warm water in the middle of the North Atlantic compared with that at the equator, without referring it to horizontal circulation of any kind.

The following is the theory as stated by Mr. Buchanan:—

"Let us assume the winter temperature of the surface-water to be 60° F. and the summer temperature to be 70° F. If we start from midwinter, we find that, as summer approaches, the surface-water must get gradually warmer, and that the temperature of the layers below the surface must decrease at a very rapid rate, until the stratum of winter temperature, or 60° F., is reached; in the language of the isothermal charts, the isothermal line for degrees between 70° F. (if we suppose that we have arrived at midsummer) and 60° F. open out or increase their distance from each other as the depth increases. Let us now consider the conditions after the summer heat has begun to waver. During the whole period of heating, the water, from its increasing temperature, has been always becoming lighter, so that heat communication by convection with the water below has been entirely suspended during the whole period. The heating of the surface-water has, however, had another effect, besides increasing its volume; it has, by evaporation, rendered it denser than it was before, at the same temperature. Keeping in view this double effect of the summer heat upon the surface-water, let us consider the effect of the winter cold upon it. The superficial water having assumed the atmospheric temperature of, say 60° F., will sink through the warmer water below it, until it reaches the stratum of water having the same temperature as itself. Arrived here, however, although it has the same tem-

* See text, p. 222. † Proc. Roy. Soc., No. 157, 1875.

perature as the surrounding water, the two are no longer in equilibrium, for the water which has come from the surface, has a greater density than that below at the same temperature. It will therefore not be arrested at the stratum of the same temperature, as would have been the case with fresh water; but it will continue to sink, carrying of course its higher temperature with it, and distributing it among the lower layers of colder water. At the end of the winter, therefore, and just before the summer heating recommences, we shall have at the surface a more or less thick stratum of water having a nearly uniform temperature of 60° F., and below this the temperature decreasing at a considerable but less rapid rate than at the termination of the summer heating. If we distinguish between *surface-water*, the temperature of which rises with the atmospheric temperature (following thus, in direction at least, the variation of the seasons), and *subsurface*-water, or the stratum immediately below it, we have for the latter the, at first sight, paradoxical effect of summer cooling and winter heating. The effect of this agency is to diffuse the same heat to a greater depth in the ocean, the greater the yearly range of atmospheric temperature at the surface. This effect is well shown in the chart of isothermals, on a vertical section, between Madeira and a position in lat. 3° 8′ N., long. 14° 49′ W. The isothermal line for 45° F. rises from a depth of 740 fathoms at Madeira to 240 fathoms at the above-mentioned position. In equatorial regions there is hardly any variation in the surface-temperature of the sea; consequently we find cold water very close to the surface all along the line. On referring to the temperature section between the position lat. 3° 8′ N., long. 14° 49′ W., and St. Paul's Rocks, it will be seen that, with a surface-temperature of from 75° F. to 79° F., water at 55° F. is reached at distances of less than 100 fathoms from the surface. Midway between the Azores and Bermuda, with a surface-temperature of 70° F., it is only at a depth of 400 fathoms that we reach water of 55° F."

What Mr. Buchanan states will explain why the mean annual temperature of the water at the surface extends to a greater depth in the middle of the North Atlantic than at the equator. It also explains why the temperature from the surface downwards decreases more rapidly at the equator than in the middle of the North Atlantic; but, if I rightly understand the theory, it does not explain (and this is the point at issue) why at a given depth the temperature of the water in the North Atlantic should be higher than the temperature at a corresponding depth at the equator. Were there no horizontal circulation the greatest thickness of warm water would certainly be found at the equator and the least at the poles. The isothermals would in such a case gradually slope downwards from the poles to the equator. The slope might not be uniform, but still it would be a continuous downward slope.

V.

ON THE CAUSE OF THE COOLING EFFECT PRODUCED ON SOLIDS BY TENSION.*

From the *Philosophical Magazine* for May, 1864.

From a series of experiments made by Dr. Joule with his usual accuracy, he found that when bodies are subjected to tension, a cooling effect takes place. "The quantity of cold," he says, "produced by the application of tension was sensibly equal to the heat evolved by its removal; and further, that the thermal effects were proportional to the weight employed." † He found that when a weight was applied to compress a body, a certain amount of heat was evolved; but the same weight, if applied to stretch the body, produced a corresponding amount of cold.

This, although it does not appear to have been remarked, is a most singular result. If we employ a force to compress a body, and then ask what has become of the force applied, it is quite a satisfactory answer to be told that the force is converted into heat, and reappears in the molecules of the body as such; but if the same force be employed to stretch the body, it will be no answer to be told that the force is converted into cold. Cold cannot be the force under another form, for cold is a privation of force. If a body, for example, is compressed by a weight, the *vis viva* of the descending weight is transmitted to the molecules of the body and reappears under that form of force called heat; but if the same weight is applied so as to stretch or expand the body, not only does the force of the weight disappear without producing heat, but the molecules which receive the force lose part of that which they already possessed. Not only does the force of the weight disappear, but along with it a portion of the force previously existing in the molecules under the form of heat. We have therefore to inquire, not merely into what becomes of the force imparted by the weight, but also what becomes of the force in the form of heat which disappears from the molecules of the body itself. That the *vis viva* of the descending weight should disappear without increasing the heat of the molecules is not so surprising, because it may be transformed into some other form of force different from that of heat. For it is by no means evident *à priori* that heat should be the only form under which it may exist. But it is somewhat strange that it should cause the force previously existing in the molecules in the form of heat also to change into some other form.

When a weight, for example, is employed to stretch a solid body,

* See text, p. 522.

† Phil. Trans. for 1859, p. 91.

it is evident that the force exerted by the weight is consumed in work against the cohesion of the particles, for the entire force is exerted so as to pull them separate from each other. But the cooling effect which takes place shows that more force disappears than simply what is exerted by the weight; for the cooling effect is caused by the disappearance of force in the shape of heat from the body itself. The force exerted by the weight disappears in performing work against the cohesion of the particles of the body stretched. But what becomes of the energy in the form of heat which disappears from the body at the same time? It must be consumed in performing work of some kind or other. The force exerted by the weight cannot be the cause of the cooling effect. The transferrence of force from the weight to the body may be the cause of a heating effect—an increase of force in the body; but this transferrence of force to the body cannot be the cause of a decrease of force in the body. If a decrease of force actually follows the application of tension, the weight can only be the occasion, not the cause of the decrease.

In what manner, then, does the stretching of the body by the weight become the occasion of its losing energy in the shape of heat? Or, in other words, what is the cause of the cooling effects which result from tension? The probable explanation of the phenomenon seems to be this: if the molecules of a body are held together by any force, of whatever nature it may be, which prevents any further separation taking place, then the entire heat applied to such a body will appear as temperature; but if this binding force becomes lessened so as to allow further expansion, then a portion of the heat applied will be lost in producing expansion. All solids at any given temperature expand until the expansive force of their heat exactly balances the cohesive force of their molecules, after which no further expansion at the same temperature can possibly take place while the cohesive force of the molecules remains unchanged. But if, by some means or other, the cohesive force of the molecules become reduced, then instantly the body will expand under the heat which it possesses, and of course a portion of the heat will be consumed in expansion, and a cooling effect will result. Now tension, although it does not actually lessen the cohesive force of the molecules of the stretched body, yet produces, by counteracting this force, the same effect; for it allows the molecules an opportunity of performing work of expansion, and a cooling effect is the consequence. If the piston of a steam-engine, for example, be loaded to such an extent that the steam is unable to move it, the steam in the interior of the cylinder will not lose any of its heat; but if the piston be raised by some external force, the molecules of the steam will assist this force, and consequently will suffer loss of heat in proportion to the amount of work which they perform. The very same occurs when tension is applied to a solid. Previous to the application of tension, the heat existing in the molecules is unable to produce any expansion against the force of cohesion. But when the influence of cohesion is partly counteracted by the tension applied, the heat then becomes enabled to perform work of expansion, and a cooling effect is the result.

VI.

THE CAUSE OF REGELATION.*

There are two theories which have been advanced to explain Regelation, the one by Professor Faraday, and the other by Professor James Thomson.

According to Professor James Thomson, pressure is the cause of regelation. Pressure applied to ice tends to lower the melting-point, and thus to produce liquefaction, but the water which results is colder than the ice, and refreezes the moment it is relieved from pressure. When two pieces of ice are pressed together, a melting takes place at the points in contact, resulting from the lowering of the melting-point; the water formed, re-freezing, joins the two pieces together.

The objection which has been urged against this theory is that regelation will take place under circumstances where it is difficult to conceive how pressure can be regarded as the cause. Two pieces of ice, for example, suspended by silken threads in an atmosphere above the melting-point, if but simply allowed to touch each other, will freeze together. Professor J. Thomson, however, attributes the freezing to the pressure resulting from the capillary attraction of the two moist surfaces in contact. But when we reflect that it requires the pressure of a mile of ice—135 tons on the square foot—to lower the melting-point one degree, it must be obvious that the lowering effect resulting from capillary attraction in the case under consideration must be infinitesimal indeed.

The following clear and concise account of Faraday's theory, I quote from Professor Tyndall's "Forms of Water:"—

"Faraday concluded that *in the interior* of any body, whether solid or liquid, where every particle is grasped, so to speak, by the surrounding particles, and grasps them in turn, the bond of cohesion is so strong as to require a higher temperature to change the state of aggregation than is necessary *at the surface*. At the surface of a piece of ice, for example, the molecules are free on one side from

* See text, p. 527.

the control of other molecules; and they therefore yield to heat more readily than in the interior. The bubble of air or steam in overheated water also frees the molecules on one side; hence the ebullition consequent upon its introduction. Practically speaking, then, the point of liquefaction of the interior ice is higher than that of the superficial ice. . . . .

"When the surfaces of two pieces of ice, covered with a film of the water of liquefaction, are brought together, the covering film is transferred from the surface to the centre of the ice, where the point of liquefaction, as before shown, is higher than at the surface. The special solidifying power of ice upon water is now brought into play *on both sides of the film.* Under these circumstances, Faraday held that the film would congeal, and freeze the two surfaces together."—*The Forms of Water*, p. 173.

The following appears to be a more simple explanation of the phenomena than either of the preceding:—

The freezing-point of water, and the melting-point of ice, as Professor Tyndall remarks, touch each other as it were at this temperature. At a hair's-breadth lower water freezes; at a hair's-breadth higher ice melts. Now if we wish, for example, to freeze water, already just about the freezing-point, or to melt a piece of ice already just about the melting-point, we can do this either by a change of temperature or by a change of the melting-point. But it will be always much easier to effect this by the former than by the latter means. Take the case already referred to, of the two pieces of ice suspended in an atmosphere above the melting point. The pieces at their surfaces are in a melting condition, and are surrounded by a thin film of water just an infinitesimal degree above the freezing-point. The film has on the one side solid ice at the freezing-point, and on the other a warm atmosphere considerably above the freezing-point. The tendency of the ice is to lower the temperature of the film, while that of the air is to raise its temperature. When the two pieces are brought into contact the two films unite and form one film separating the two pieces of ice. This film is not like the former in contact with ice on the one side and warm air on the other. It is surrounded on both sides by solid ice. The tendency of the ice, of course, is to lower the film to the same temperature as the ice itself, and thus to produce solidification. It is evident that the film must either melt the ice or the ice must freeze the film, if the two are to assume the same temperature. But the power of the ice to produce solidification, owing to its greater mass, is enormously greater than the power of the film to produce fluidity, consequently regelation is the result.

VII.

LIST OF PAPERS WHICH HAVE APPEARED IN DR. A. PETERMANN'S *GEOGRAPHISCHE MITTHEILUNGEN* RELATING TO THE GULF-STREAM AND THERMAL CONDITION OF THE ARCTIC REGIONS.

The most important memoir which we have on the Gulf-stream and its influence on the climate of the arctic regions is the one by Dr. A. Petermann, entitled "Der Golfstrom und Standpunkt der thermometrischen Kenntniss des nord-atlantischen Oceans und Landgebiets im Jahre 1870." *Geographische Mittheilungen*, Band XVI. 1870.

Dr. Petermann has, in this memoir, by a different line of argument from that which I have pursued in this volume, shown in the most clear and convincing manner that the abnormally high temperature of the north-western shores of Europe and the seas around Spitzbergen is owing entirely to the Gulf-stream, and not to any general circulation such as that advocated by Dr. Carpenter. From a series of no fewer than 100,000 observations of temperature in the North Atlantic and in the arctic seas, he has been enabled to trace with accuracy on his charts the very footsteps of the heat in its passage from the Gulf of Mexico up to the shores of Spitzbergen.

The following is a list of the more important papers bearing on the subject which have recently appeared in Dr. Petermann's *Geogr. Mittheilungen*:—

An English translation of Dr. Petermann's Memoir, and of a few more in the subjoined list, has been published in a volume, with supplements, by the Hydrographic Department of the United States, under the superintendence of Commodore R. H. Wyman.

The papers whose titles are in English have appeared in the American volume. In that volume the principal English papers on the subject, in as far as they relate to the north-eastern extension of the Gulf-stream, have also been reprinted.

The System of Oceanic Currents in the Circumpolar Basin of the Northern Hemisphere. By Dr. A. Mühry. Vol. XIII., Part II. 1867.

The Scientific Results of the first German North Polar Expedition. By Dr. W. von Freeden. Vol. XV., Part VI. 1869.

The Gulf-stream, and the Knowledge of the Thermal Properties of the North Atlantic Ocean and its Continental Borders, up to 1870. By Dr. A. Petermann. *Geographische Mittheilungen*, Vol. XVI., Part VI. 1870.

The Temperature of the North Atlantic Ocean and the Gulf-stream. By Rear-Admiral C. Irminger. Vol. XVI., Part VI. 1870.

Meteorological Observations during a Winter Stay on Bear Island, 1865-1866. By Sievert Tobilson. Vol. XVI., Part VII. 1870.

Die Temperatur-verhältnisse in den arktischen Regionen. Von Dr. Petermann. Band XVI., Heft VII. 1870.

Preliminary Reports of the Second German North Polar Expedition, and of minor Expeditions, in 1870. Vol. XVII.

Preliminary Report of the Expedition for the Exploration of the Nova-Zembla Sea (the sea between Spitzbergen and Nova Zembla), by Lieutenants Weyprecht and Payer, June to September, 1871. By Dr. A. Petermann. Vol. XVII. 1871.

Der Golfstrom ostwärts vom Nordkap. Von A. Middendorff Band XVII., Heft I. 1871.

Kapitän E. H. Johannesen's Umfahrung von Nowaja Semlä im Sommer 1870, und norwegischer Finwalfang östlich vom Nordkap. Von Th. v. Heuglin. Band XVII., Heft I. 1871.

Die Nordpol-Expeditionen, das sagenhafte Gillis-land und der Golfstrom im Polarmeere. Von Dr. A. Petermann. 5 Nov. 1870.

Th. v. Heuglin's Aufnahmen in Ost-Spitzbergen. Begleitworte zur neuen Karte dieses Gebiets. Tafel 9. 1870. Band XVII., Heft V. 1871.

Die zweite deutsche Nordpolar-Expedition, 1869-70. Schlittenreise an der Küste Grönlands nach Norden, 8 März—27 April, 1870. Von Ober-Lieutenant Julius Payer. Band XVII., Heft V. 1871.

Die Entdeckung des Kaiser Franz Josef-Fjordes in Ost-Grönland, August, 1870. Von Ober-Lieutenant Julius Payer. Band XVII., Heft V. 1871.

Die Erschliessung eines Theiles des nördlichen Eismeeres durch die Fahrten und Beobachtungen der norwegischen Seefahrer Torkildsen, Ulve, Mack Qvale, und Nedrevaag im karischen Meere, 1870. Von Dr. A. Petermann. Band XVII., Heft III. 1871.

Die zweite deutsche Nordpolar-Expedition, 1869-70. Schlittenreise nach Ardencaple Inlet, 8—29 Mai, 1870. Von Ober-Lieutenant Julius Payer. Band XVII., Heft XI. 1871.

Ein Winter unter dem Polarkreise. Von Ober-Lieutenant Julius Payer. Band XVII., Heft XI. 1871.

Die Entdeckung eines offenen Polarmeeres durch Payer und Weyprecht im September, 1871. Von Dr. A. Petermann. Band XVII., Heft XI. 1871.

James Lamont's Nordfahrt, Mai—August, 1871. Die Entdeckungen von Weyprecht, Payer, Tobiesen, Mack, Carlsen, Ulve, und Smyth im Sommer, 1871.

Stand der Nordpolarfrage zu Ende des Jahres 1871. Von Dr. A. Petermann. Band XVII., Heft XII. 1871.

Das Innere von Grönland. Von Dr. Robert Brown. Band XVII., Heft X. 1871.

Captain T. Torkildsen's Cruise from Tromsö to Spitzbergen, July 26 to September 26, 1871. Vol. XVIII. 1872.

The Sea north of Spitzbergen, and the most northern Meteorological Observations. Vol. XVIII. 1872.

Results of the Observations of the Deep-sea Temperature in the Sea between Greenland, Northern Europe, and Spitzbergen. By Professor H. Möhn. Vol. XVIII. 1872.

The Norwegian Cruises to Nova Zembla and the Kara Sea in 1871. Vol. XVIII. 1872.

The Cruises in the Polar Sea in 1872. Vol. XVIII. 1872.

The Cruise of Smyth and Ulve, June 19 to September 27, 1871. Vol. XVIII. 1872.

Die fünfmonatliche Schiffbarkeit des sibirischen Eismeeres um Nowaja Semlja, erwiesen durch die norwegischen Seefahrer in 1869 und 1870, ganz besonders aber in 1871. Von Dr. A. Petermann. Band XVIII., Heft X. 1872.

Die neuen norwegischen Aufnahmen des nordöstlichen Theiles von Nowaja Semlja durch Mack, Dörma, Carlsen, u. A., 1871. Von Dr. Petermann. Band XVIII., Heft X. 1872.

Nachrichten über die sieben zurückgekehrten Expeditionen unter Graf Wiltschek, Altmann, Johnsen, Nilsen, Smith, Gray, Whymper; die drei Überwinterungs - Expeditionen; die Amerikanische, Schwedische, Österreichisch-Ungarische; und die zwei neuen: die norwegische Winter-Expedition und diejenige unter Kapitän Mack. Von Dr. A. Petermann. Band XVIII., Heft XII. 1872.

Konig Karl-Land im Osten von Spitzbergen und seine Erreichung und Aufnahme durch norwegische Schiffer im Sommer 1872. Von Professor H. Möhn. Band XIX., Heft IV. 1873.

Resultate der Beobachtungen angestellt auf der Fahrt des Dampfers "Albert" nach Spitzbergen im November und Dezember, 1872. Von Professor Möhn. Band XIX., Heft VII. 1873.

Die amerikanische Nordpolar-Expedition unter C. F. Hall, 1871-3. Von Dr. A. Petermann. Band XIX., Heft VIII. 1873.

Die Trift der Hall'schen Nordpolar-Expedition, 16 August bis 15 Oktober, 1872, und die Schollenfahrt der 20 bis zum 30 April, 1873. Von Dr. A. Petermann. Band XIX., Heft X. 1873.

Das offene Polarmeer bestätigt durch das Treibholz an der Nordwestküste von Grönland. Von Dr. A. Petermann. Band XX., Heft V. 1874.

Das arktische Festland und Polarmeer. Von Dr. Joseph Chavanne. Band XX., Heft VII. 1874.

Die Umkehr der Hall'schen Polar-Expedition nach den Aussagen der Offiziere. Von Dr. A. Petermann. Band XX., Heft VII. 1874.

Die zweite österreichisch-ungarische Nordpolar-Expedition unter Weyprecht und Payer, 1872-4. Von Dr. A. Petermann. Band XX., Heft X. 1874.

Beiträge zur Klimatologie und Meteorologie des Ost-polar-Meeres. Von Professor Möhn. Band XX., Heft V 1874.

Kapitän David Gray's Reise und Beobachtungen im ost-grönländischen Meere, 1874, und seine Ansichten über den besten Weg zum Nordpol. Original-Mittheilungen an A. Petermann, d.D., Peterhead, Dezember, 1874. Band XXI., Heft III. 1875.

VIII.

LIST OF PAPERS BY THE AUTHOR TO WHICH REFERENCE IS MADE IN THIS VOLUME.

On the Influence of the Tidal Wave on the Earth's Rotation and on the Acceleration of the Moon's Mean Motion.—*Phil. Mag.*, April, 1864.

On the Nature of Heat-vibrations.—*Phil. Mag.*, May, 1864.

On the Cause of the Cooling Effect produced on Solids by Tension.—*Phil. Mag.*, May, 1864.

On the Physical Cause of the Change of Climate during Geological Epochs.—*Phil. Mag.*, August, 1864.

On the Physical Cause of the Submergence of the Land during the Glacial Epoch.—The *Reader*, September 2nd and October 14th, 1865.

On Glacial Submergence.—The *Reader*, Deccember 2nd and 9th, 1865.

On the Eccentricity of the Earth's Orbit.—*Phil. Mag.*, January, 1866.

Glacial Submergence on the Supposition that the Interior of the Globe is in a Fluid Condition.—The *Reader*, January 13th, 1866.

On the Physical Cause of the Submergence and Emergence of the Land during the Glacial Epoch, with a Note by Professor Sir William Thomson.—*Phil. Mag.*, April, 1866.

On the Influence of the Tidal Wave on the Motion of the Moon.—*Phil. Mag.*, August and November, 1866.

On the Reason why the Change of Climate in Canada since the Glacial Epoch has been less complete than in Scotland.—*Trans. Geol. Soc. of Glasgow*, 1866.

On the Eccentricity of the Earth's Orbit, and its Physical Relations to the Glacial Epoch.—*Phil. Mag.*, February, 1867.

On the Reason why the Difference of Reading between a Thermometer exposed to direct Sunshine and one shaded diminishes as we ascend in the Atmosphere.—*Phil. Mag.*, March, 1867.

On the Change in the Obliquity of the Ecliptic; its Influence on the Climate of the Polar Regions and Level of the Sea.—*Trans. Geol. Soc. of Glasgow*, vol. ii., p. 177. *Phil. Mag.*, June, 1867.

Remarks on the Change in the Obliquity of the Ecliptic, and its Influence on Climate.—*Phil. Mag.*, August, 1867.

On certain Hypothetical Elements in the Theory of Gravitation and generally received Conceptions regarding the Constitution of Matter.—*Phil. Mag.*, December, 1867.

On Geological Time, and the probable Date of the Glacial and the Upper Miocene Period.—*Phil. Mag.*, May, August, and November, 1868.

On the Physical Cause of the Motions of Glaciers.—*Phil. Mag.*, March, 1869. *Scientific Opinion*, April 14th, 1869.

On the Influence of the Gulf-stream.—*Geol. Mag.*, April, 1869. *Scientific Opinion*, April 21st and 28th, 1869.

On Mr. Murphy's Theory of the Cause of the Glacial Climate.—*Geol. Mag.*, August, 1869. *Scientific Opinion*, September 1st, 1869.

On the Opinion that the Southern Hemisphere loses by Radiation more Heat than the Northern, and the supposed Influence that this has on Climate.—*Phil. Mag.*, September, 1869. *Scientific Opinion*, September 29th and October 6th, 1869.

On Two River Channels buried under Drift belonging to a Period when the Land stood several hundred feet higher than at present. —*Trans. Geol. Soc. of Edinburgh*, vol. i., p. 330.

On Ocean-currents: Ocean-currents in Relation to the Distribution of Heat over the Globe.—*Phil. Mag.*, February, 1870.

On Ocean-currents: Ocean-currents in Relation to the Physical Theory of Secular Changes of Climate.—*Phil. Mag.*, March, 1870.

The Boulder Clay of Caithness a Product of Land-ice.—*Geol. Mag.*, May and June, 1870.

On the Cause of the Motion of Glaciers.—*Phil. Mag.*, September, 1870.

On Ocean-currents: On the Physical Cause of Ocean-currents. Examination of Lieutenant Maury's Theory.—*Phil. Mag.*, October, 1870.

On the Transport of the Wastdale Granite Boulders.—*Geol. Mag.*, January, 1871.

On a Method of determining the Mean Thickness of the Sedimentary Rocks of the Globe.—*Geol. Mag.*, March, 1871.

Mean Thickness of the Sedimentary Rocks.—*Geol. Mag.*, June, 1871.

On the Age of the Earth as determined from Tidal Retardation. —*Nature*, August 24th, 1871.

Ocean-currents: On the Physical Cause of Ocean-currents. Examination of Dr. Carpenter's Theory.—*Phil. Mag.*, October, 1871.

Ocean-currents: Further Examination of the Gravitation Theory. —*Phil. Mag.*, February, 1874.

Ocean-currents: The Wind Theory of Oceanic Circulation.—*Phil. Mag.*, March, 1874.

Ocean-currents.—*Nature*, May 21st, 1874.

The Physical Cause of Ocean-currents.—*Phil. Mag.*, June, 1874. *American Journal of Science and Art*, September, 1874.

On the Physical Cause of the Submergence and Emergence of the Land during the Glacial Epoch.—*Geol. Mag.*, July and August, 1874.

INDEX.

THE END.

PRINTED BY VIRTUE AND CO., CITY ROAD, LONDON.

Printed in the United States
By Bookmasters